2nd International Conference on Design, Materials and Manufacturing (ICDMM 2017)

IOP Conference Series: Materials Science and Engineering
Volume 220

Beijing, China
23 – 25 June 2017

ISBN: 978-1-5108-4510-7
ISSN: 1757-8981

Printed from e-media with permission by:

Curran Associates, Inc.
57 Morehouse Lane
Red Hook, NY 12571

Some format issues inherent in the e-media version may also appear in this print version.

Printed by Curran Associates, Inc. (2017)

For permission requests, please contact the Institute of Physics at the address below.

Institute of Physics
Dirac House, Temple Back
Bristol BS1 6BE UK

Phone: 44 1 17 929 7481
Fax: 44 1 17 920 0979

techtracking@iop.org

Additional copies of this publication are available from:

Curran Associates, Inc.
57 Morehouse Lane
Red Hook, NY 12571 USA
Phone: 845-758-0400
Fax: 845-758-2633
Email: curran@proceedings.com
Web: www.proceedings.com

TABLE OF CONTENTS

STRUCTURAL DESIGN AND OPTIMIZATION OF FUNCTIONAL MATERIALS

PERFORMANCE MEASUREMENT OF NEW MATERIALS: MODELING, EXPERIMENT AND CALCULATION

ELECTRICAL MATERIALS AND DEVICE DESIGN

MECHANICAL ENGINEERING AND MANUFACTURING SYSTEM

IOP Conf. Series: Materials Science and Engineering **220** (2017) 011001 doi:10.1088/1757-899X/220/1/011001

PREFACE

It is our great pleasure to welcome you to 2017 2nd International Conference on Design, Materials and Manufacturing which has been held in Beijing, China during June 23-25, 2017. ICDMM 2017 is dedicated to addressing issues related to design, materials and manufacturing.

The major goal and feature of the conference is to foster communication among researchers and practitioners working in a wide variety of scientific areas with a common interest in improving Design, Materials and Manufacturing related techniques. Professors from Malaysia, Taiwan, USA and Singapore are invited to deliver keynote speeches regarding latest researches in their respective expertise areas. It is a great opportunity for the students, researchers and engineers to interact with the experts and specialists to get their advice or consultation on technical matters, teaching methods and strategies.

The proceedings presents a selection from papers submitted to the conference from universities, research institutes and industries. All of the papers were subjected to peer-review by conference committee members and international reviewers. The papers are selected depended on their quality and their relevancy to the conference. The volume tends to present to the readers the recent advances in the field of materials behavior, machining, biomedical manufacturing, modeling, analysis, and simulation of manufacturing processes, rapid manufacturing technologies, metrology and measurement, computer-aided design, manufacturing, and engineering etc.

We would like to thank all the authors who have contributed to this volume and also to the conference committee, reviewers, speakers, chairpersons, and all the conference participants for their support to ICDMM 2017.

ICDMM 2017
Organizing Committee
June 30, 2017

IOP Conf. Series: Materials Science and Engineering **220** (2017) 011002 doi:10.1088/1757-899X/220/1/011002

Conference Committee

Advisory Chair

Prof. Alias Mohd Noor, Universiti Teknologi Malaysia, Malaysia

Conference Chairs

Prof. Guanghsu Chang, Georgia Southern University, USA
Prof. Jieh-Shian Young, National Changhua University of Education, Taiwan
Prof. Haider Fa. Abdulamir, University Teknologi Mara Malaysia, Malaysia

Program Chairs

Prof. Chun-Lang Yeh, National Formosa University, Taiwan
Assoc. Prof. Yamuna Munusamy, UTAR, Malaysia

Technical Committee

Prof. Hirpa G. Lemu, University of Stavanger, Norway
Dr. Helen Wu, University of Western Sydney, Australia
Prof. Hassan A. Youness, Minia University, Egypt
Assoc. Prof. Khalilur Rhaman, BRAC University, Bangladesh
Assoc. Prof. Labia M. Daloub, University of Aljabel Algharbi, Lybia
Dr. Maksim Beresnev, Southern Federal University, Russia
Dr. Yu-Jen Liu, National Chung Cheng University, Taiwan
Prof. S. G. Ponnambalam, Monash University Sunway Campus, Malaysia
Prof. Kuldip Singh Sangwan, Birla Institute of Technology and Science Pilani, India
Prof. Sergey Borisovich Vasilyev, Petrozavodsk State University, Russia
Dr. Wirasak Smitthipong, Kasetsart University (KU), Thailand
Prof. Qumrul Ahsan, Universiti Teknikal Malaysia Melaka, Malaysia
Dr. Yuhazri M.Y., Universiti Teknikal Malaysia Melaka, Malaysia
Prof. Jaya Christiyan K G, M S Ramaiah Institute of Technology, India
Prof. Aurelia Meghea, Politehnica University of Bucharest, Romania
Asst. Prof. Praseetha P Nair, Government Engineering College, India
Dr. Yang Xi, Diodes Inc., USA
Assoc. Prof. Suresh Palanisamy, Swinburne University of Technology, Australia
Prof. Syed Masood, Swinburne University of Technology, Australia

IOP Conf. Series: Materials Science and Engineering **220** (2017) 011003 doi:10.1088/1757-899X/220/1/011003

Peer review statement

All papers published in this volume of *IOP Conference Series: Materials Science and Engineering* have been peer reviewed through processes administered by the proceedings Editors. Reviews were conducted by expert referees to the professional and scientific standards expected of a proceedings journal published by IOP Publishing.

IOP Conf. Series: Materials Science and Engineering **220** (2017) 012001 doi:10.1088/1757-899X/220/1/012001

Statistical analysis of porosity of 17-4PH alloy processed by selective laser melting

P Ponnusamy[1,2], S H Masood[1,3], D Ruan[1], S Palanisamy[1,2] and O A Mohamed[1]

[1]Faculty of Science, Engineering and Technology, Swinburne University of Technology, Victoria, Australia 3122
[2]Defence Materials Technology Centre, Victoria, Australia 3122

Abstract. Selective Laser Melting (SLM) is a powder-bed type Additive Manufacturing (AM) process, where parts are built layer-by-layer by laser melting of powder layers of metal. There are several SLM process parameters that affect the accuracy and quality of the metal parts produced by SLM. Therefore, it is essential to understand the effect of these parameters on the quality and properties of the parts built by this process. In this paper, using Taguchi design of experiments, the effect of four SLM process parameters namely laser power, defocus distance, layer thickness and build orientation are considered on the porosity of 17-4PH stainless steel parts built on ProX200 SLM direct metal printer. The porositywas found to be optimum at a defocus distance of -4mm and a laser power of 240 W with a layer thickness of 30 μm and using vertical build orientation.

1. Introduction

Selective Laser Melting (SLM) is an additive manufacturing process that produces metal parts by fusing metal powders through a high-powered laser beam. The fusing occurs through the melting and rapid solidification of metal powder by scanning a laser beam over metal powder spread on a platform via a 3D path created by processing software [1-3]. There is a growing need for SLM to create fully dense parts, where mechanical, thermal and other properties are comparable to those of wrought materials. However, the properties and quality of built parts depend a great deal on proper selection and optimisation of a combination of several key SLM process parameters for the specific metal powder used in the system.

SLM is a complex process with many process parameters, making it very challenging to evaluate each and every parameter [4, 5]. The SLM can produce near net shape complex geometries, which are not normally producible by conventional manufacturing methods. In SLM, it is possible to alter the final material properties by adjusting the process parameters. Some of the main parameters that are varied in SLM are laser power, scan speed, layer thickness, scanning space, laser diameter, focal distance, powder size, and powder temperature. These parameters can also vary according to materials used, machine types and build volumes[6]. The density and surface roughness of the parts are strongly affected by the variation of the parameters namely scanning speed, scan spacing and laser power.

Several researchers have investigated the influence of SLM process parameters on different mechanical properties of the SLM processed materials. Literature search shows that titanium and stainless steel are the most commonly used materials by SLM users. Spierings et al [7] have investigated the effects of several SLM process parameters such as laser power, scan speed, layer thickness and build orientation on E-modulus and ultimate strength of 17-4PH stainless steel parts made by SLM. Lu et al [8] have carried out a study on the effect of differing island scanning strategy of SLM process on the microstructure, mechanical property and residual stress of SLM Inconel-718 alloy. Childs et al [9] have investigated the relationship between energy and fabricated tracks in SLM using several types of stainless steel powders with different particle sizes. Di et al [10] have also carried out experimental study on the energy input and its influence on fabrication quality using 316 L stainless steel in SLM and obtained a high-density part with smooth surface.

In this paper, the focus is on the effect of four key process parameters of ProX200 SLM machine on the densification of 17-4PH stainless steel by eliminating porosity through the application of the Taguchi method with fractional factorial design approach. The four process parameters considered are the laser power, defocus distance, layer thickness and build orientation at three different levels. The 17-4PH is a martensitic precipitation hardened stainless steel material, which finds wide application in aerospace, chemical, petrochemical, food processing industries. The 17-4PH has outstanding combination of high strength, good corrosion resistance, and toughness. The chemical composition of 17-4PH has 17% Ni and 4% Cr and 4% Cu which provide high tensile strength, corrosive resistance and conductivity.

2. Material and methods

In this study, the 17-4PH stainless steel is used to investigate the effect of laser power, build orientation, layer thickness and defocus distance on the porosity of the parts produced by 3D Systems ProX200 SLM direct metal printer. A fibre laser with a maximum power of 300 W and a wavelength of 1070 nm is used in this machine. The porosity was evaluated by measuring the relative density of the samples. Cylindrical 17-4PH samples of size 25 mm height and15 mm diameter were produced on ProX200 as per ASTM E9 standard.

An experimental design was developed to study the interaction effects of the process parameters of the SLM process on the porosity of 17-4PH parts. The four parameters namely laser power (A), build orientation (B), layer thickness (C) and de-focus distance (D) were varied at three levels (-1, 0, +1) as shown in Table 1. Laser power was selected at 80 %, 85 % and 90% of the maximum laser power. The orientations of the built part were horizontal, inclined with respect to the X-axis in the X-Z plane, and vertical. Similarly, layer thicknesses were 30, 35 and 40 microns while the defocus distance selected was -6, -3 and -1 mm. In the ProX200 SLM machine, the defocus distance is the distance between the focal plane (which gives minimum laser beam diameter) and the build surface, where the powder is spread. The defocus distance can be positive or negative.

Table 1. The levels of SLM parameters in Experimental Design

Parameter	**Unit**	**Code**	**Levels**		
			-1	**0**	**+1**
Laser power	% of (300 W)	A	80	85	90
Orientation	degree	B	Horizontal	Inclined	Vertical
Layer Thickness	μm	C	30	35	40
Defocus distance	mm	D	-6	-3	-1

Instead of 4^3 levels, the experiment was narrowed down to the manufacturing and testing of 9 samples through the application of the orthogonal array in Taguchi method. The orthogonal array was based on the experimental design shown in Table 2. Minitab 16 software was used for the statistical analysis of the experimental data. The results of measured porosity from the experiments conducted according to the orthogonal array are also shown in Table 2. The objective was to minimise the porosity of the built part in the SLM process. Equation (1) was used to define Signal to Noise (SN) Ratio for each response, with the theme ‘smaller is better’.

$$\text{SN Ratio(Smaller is better)} = -10 \times Log10\left(\sum \frac{(Y^2)}{n}\right) \quad (1)$$

Porosity is measured using the relative density method, and is given by Equation (2).

$$\text{Porosity} = \left(1 - \frac{\rho 1}{\rho 2}\right) * 100 \quad (2)$$

where ρ1 is the measured density of the sample and ρ2 is the theoretical density of the material.

Table 2. Results of experiments for orthogonal array

Run	Coded factors				% Porosity
	A	B	C	D	
1	-1	-1	-1	-1	4.491
2	-1	-1	0	0	3.248
3	-1	-1	+1	+1	3.733
4	0	0	0	+1	4.142
5	0	0	+1	-1	4.698
6	0	0	-1	0	2.853
7	+1	+1	+1	0	3.250
8	+1	+1	-1	+1	3.606
9	+1	+1	0	-1	4.496

3. Results and discussion

The principal interest in SLM is the density of the part produced. The higher the density of the material, the lesser the porosity. From Table 2, the porosity was found to be at its lowest in experiment no. 6, where the sample was built using a combination of laser power at 85% of maximum power, inclined orientation, 30 μm layer thickness and -3 mm defocus distance, giving a dense part. Figure 1 and 2 show the main effects plots for percentage porosity and percentage SN ratios. As shown in figure 1, the increase in laser power from 80% to 85%, represented by code A, caused a considerable increase in the porosity. A similar trend can be observed for code B, where the inclination from a vertical to an inclined orientation led to an increase in porosity. Laser re-melting reduces porosity while improving density [11].

The plot of main effects suggests that the defocus distance -3 mm is the best distance for building the part; however, when the SN ratio is taken into account, a defocus distance of -6 mm is shown as optimum. In this study, the actual built part had -4 mm as the defocus distance for the fresh metal powder loaded into the SLM machine. This complies with the machine manufacturer's guidelines. The laser energy and heat transferred to the powder significantly affect the precision of fabrication if the defocus distance fluctuates [12].

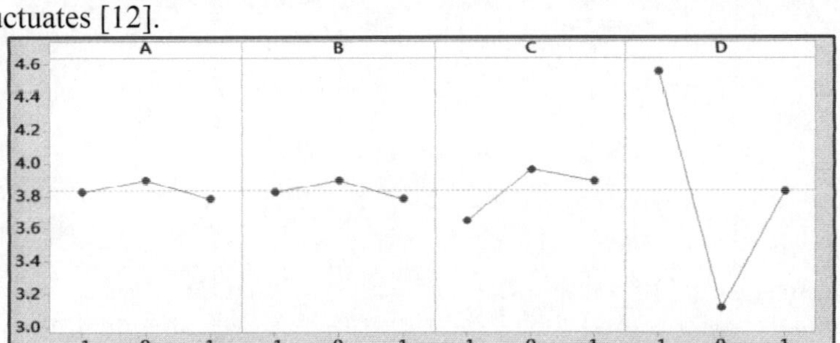

Figure 1. Main effects plot: Means of % porosity.

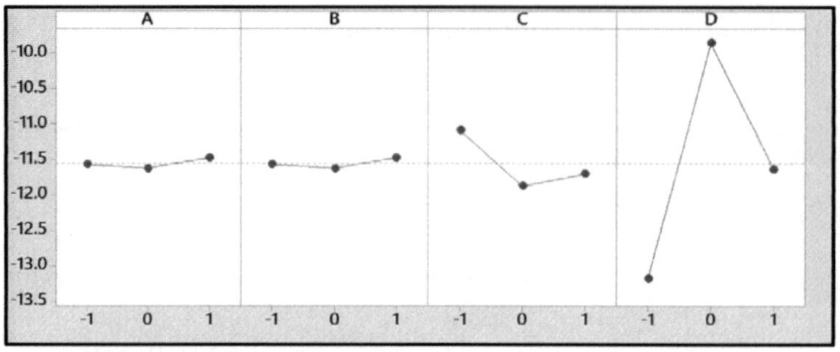

Figure 2. Main effects plot: SN ratios of % porosity.

In figure 2, the analysis of signal to noise ratio shows that at 85% of maximum laser power, the SN ratio is smaller. Similarly, the inclined orientation, layer thickness of 35 µm and defocus distance of -6 mm are observed to have the least SN ratio. Figure 2also shows that the vertical orientation has the main effect on porosity, and the SN ratio suggests that there will be maximum noise if vertical orientation is taken into consideration. However, a similar study has reported that vertically-built samples have an optimised combination of strength and ductility [13].

Figure 3 shows the interaction plots for porosity. In this plot, where the laser power (A) and build orientation (B) interact, the least porosity is observed at vertical orientation. Similar findings have been reported where struts built in a horizontal direction were found to have more pores than the vertically-built samples [14]. In the interactions between laser power and other parameters, a layer thickness of 30 µm, and a defocus distance of -3 mm were observed to cause less porosity. Similarly, with regard to the interaction between orientationand other parameters, a 30 µm layer thickness (the lowest-possible layer thickness) and a -3 mm defocus distance were observed to be better. It is essential that the layer thickness be the least possible, as the powder has to go through rapid heating and cooling cycles. Finally, the interaction of layer thickness and defocus distance also suggests that 30µm layer thickness and -3 mm defocus distance is the best combination, while the laser spot diameter remains the optimum for melting the metal powder.

In this study, a lower laser power of 240 W was sufficient to cause the melting. However, the literature review demonstrates that a lower laser power leads to the formation of a limited liquid phase. Therefore, the viscosity of the melt pool increases, which has a negative influence on the flowability of the liquid [15]. The consequences of these mechanisms are that caves and crevices can form inside the part, with many un-melted and half-melted particles. While high laser power with high scan speed leads to porosity, the low scan speed leads to the formation of crevices [16].

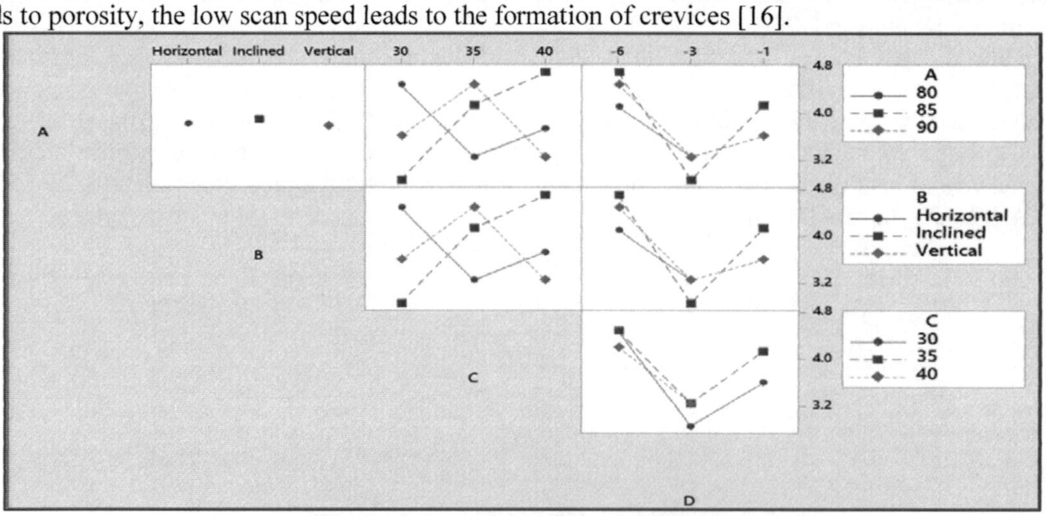

Figure 3. Interaction Plots for Porosity.

The least porosity among the built parts was found to be 2.8%, which is higher than the wrought sample porosity of 1.2%. The least noise or SN ratio has been noted at 35 μm layer thickness. When the layer thickness increases, porosity also increases, which would lead to crack formation associated with decrease in ductility and strength. Also, it is evident that density decreases with a larger layer thickness. The larger particles get entrapped in the thicker layer of powder, leading to porosity and causing gas bubbles to form around the pores. In addition, larger particles lead to shrinkage after solidification, and voids, cracks and layer separations can occur at grain boundaries. Therefore, it is advisable to use smaller particles, which would produce dense layers with better compaction. Larger particles could be excluded by the use of a recoater blade [17].

4. Conclusions

The effects of laser power, build orientation, layer thickness and defocus distance of fibre laser used in ProX200 SLM machine on the porosity of 17-4PH stainless steel parts have been investigated based on the orthogonal arrays arrived at using the Taguchi method. Using statistical methods, the following combination was found to be the optimum: the porosity and the surface roughness were measured at a defocus distance of -4 mm, laser power of 80% of 300 W, a layer thickness at 30 μm, and with a vertical build orientation. These process parameters have been validated by manufacturing samples using SLM, and the optimised results were obtained for the porosity.At the lowest level of porosity, the part built would have the highest relative density of all parts built. The experiment correspond tothe sample that had the highest density at 97.2%. The porosity of 2.8% could be due to the thermal history during manufacturing in the SLM. These pores can also affect the mechanical properties of parts built by SLM. In particular, the cracks can be initiated at these sites. Compared with a wrought sample of porosity 1.2%, the porosity of the wrought part was found to be less than for the SLM built part. As these parameters are not universally suitable for all SLM machines, it is advisable to perform more tests for the defocus distance. Other parameters could be deployed with the same values namely laser power, layer thickness and orientation as mentioned above for the 17-4PH material.

Acknowledgements

The authors gratefully acknowledge the support of Defence Materials Technology Centre (DMTC). The DMTC was established and is supported under the Australian Government's Defence Future Capability Technology Centre Program.

References

[1] Yap C Y, Chua C K, Dong Z L, Liu Z H, Zhang D Q, Loh L E and Sing S L 2015 Review of selective laser melting: materials and applications *Applied Physics Reviews* **2** 041101.

[2] Louvis E, Fox P and Sutcliffe C J 2011 Selective laser melting of aluminium components *J. of Materials Processing Technol.* **211(2)** 275-84.

[3] Kempen K, Thijs L, Van Humbeeck J and Kruth J P 2012 Mechanical properties of AlSi10Mg produced by selective laser melting *Physics Procedia* **39** 439-46.

[4] Olakanmi E 2013 Selective laser sintering/melting (SLS/SLM) of pure Al, Al–Mg, and Al–Si powders: Effect of processing conditions and powder properties *J. of Materials Processing Technol.* **213(8)** 1387-405.

[5] Kurzynowski T, Chlebus E, Kuźnicka B and Reiner J 2012 Parameters in selective laser melting for processing metallic powders *SPIE LASE. Int. Society for Optics and Photonics.*

[6] Pupo Y, Monroy K P and Ciurana J 2015 Influence of process parameters on surface quality of CoCrMo produced by selective laser melting *The Int. J. of Advanced Manufacturing Technol.* **80(5-8)** 985-95.

[7] Spierings A, Levy G and Wegener K 2014 Designing material properties locally with additive manufacturing technology SLM *ETH-Zürich.*

[8] Lu Y, Wu S, Gan Y, Huang T, Yang C and Junjie L 2015 Study on the microstructure,

mechanical property and residual stress of SLM Inconel-718 alloy manufactured by differing island scanning strategy *Optics & Laser Technol.* **75** 197-206.

[9] Childs T, Hauser C and Badrossamay M 2005 Selective laser sintering (melting) of stainless and tool steel powders: experiments and modelling *Proceedings of the Institution of Mechanical Engineers, Part B: J. of Engineering Manufacture* **219(4)** 339-57.

[10] Di W, Yongqiang Y, Xubin S, Yonghua C 2012 Study on energy input and its influences on single-track, multi-track, and multi-layer in SLM *The Int. J. of Advanced Manufacturing Technol.* **58(9-12)** 1189-99.

[11] Yasa E and Kruth J P 2011 Microstructural investigation of Selective Laser Melting 316L stainless steel parts exposed to laser re-melting *Procedia Engineering* **19** 389-95.

[12] Su X, Yang Y, Xiao D and Luo Z 2013 An investigation into direct fabrication of fine-structured components by selective laser melting *The Int. J. of Advanced Manufacturing Technol.* **64(9-12)** 1231-8.

[13] Guan K, Wang Z, Gao M, Li X and Zeng X 2013 Effects of processing parameters on tensile properties of selective laser melted 304 stainless steel *Materials & Design* **50** 581-6.

[14] Weißmann V, Bader R, Hansmann H and Laufer N 2016 Influence of the structural orientation on the mechanical properties of selective laser melted Ti6Al4V open-porous scaffolds *Materials & Design* **95** 188-97.

[15] Li R, Liu J, Shi Y, Wang L and Jiang W 2012 Balling behavior of stainless steel and nickel powder during selective laser melting process *The Int. J. of Advanced Manufacturing Technol* **59(9-12)** 1025-35.

[16] Gu H, Gong H, Pal D, Rafi K, Starr T and Stucker B 2013 Influences of energy density on porosity and microstructure of selective laser melted 17-4PH stainless steel *Proceedings of Solid Freeform Fabrication Symposium.*

[17] Abd-Elghany K and Bourell D 2012 Property evaluation of 304L stainless steel fabricated by selective laser melting *Rapid Prototyping Journal* **18(5)** 420-8.

IOP Conf. Series: Materials Science and Engineering **220** (2017) 012002 doi:10.1088/1757-899X/220/1/012002

Axial Bearing Capacity of Elliptical Concrete Filled Steel Tubular Stub Columns

Xu Youwu[1], Yao Jian[1,2*]

[1]Institute of Structural Engineering, Zhejiang University, Hangzhou, 310058, China
[2]College of Urban Construction, Zhejiang Shuren University, Hangzhou, 310058, China

*E-mail: yaojian5803@qq.com

Abstract. To study the axial bearing capacity of elliptical concrete filled steel tubular (ECFST) stub columns, the calculation theory and accuracy of existing formulas are analysed. The comparison of the calculation results of existing formulas and test results shows that the calculation precision is within the scope of engineering tolerance. Therefore, the existing formulas are proposed for practical project.

1. Introduction

As a new member of concrete-filled steel tubular (CFST) structural families, coupled with their aesthetical appeal and sound structural efficiency in bending, elliptical concrete-filled steel tubes (Fig. 1), referred to as ECFST, has been used as the main axial compression member in airport terminals and bridges. In recent years, the study on static behaviour of ECFST has been a current research focus upon the field of composite structure [1] The experimental study [2-7], finite element simulation [8-11], and the theoretical analysis [12-14] has been conducted by scholars at home and abroad, they enriched the research achievements one after another. And with the development of the steel processing technology, elliptical concrete-filled stainless-steel tubes [15-17], elliptical concrete-filled steel double-skin tubular columns [18] gradually attract more and more attention.

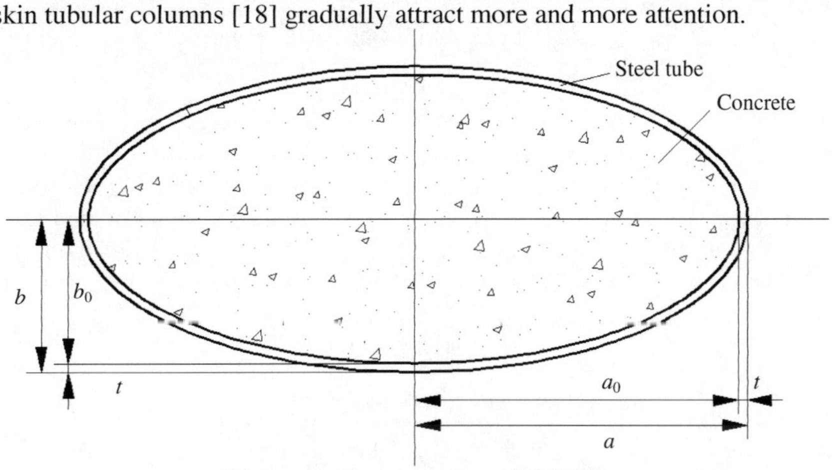

Figure 1. Cross section of ECFST.

Because of the limited researches and parameters on ECFST, no consistently recognitions on some properties of ECFST is reached, and the ECFST is temporarily not included in the design code of CFST structure in all countries. All these lack limit the application of ECFST in engineering practice.

Against this background, this paper presents the current formulas of the axial bearing capacity of ECFST stub columns. In the meantime, calculation for the latest axial loaded specimens is listed before the comparison of computation theory and accuracy of all the formulas. All these jobs can provide advice for structure design and make a little effort to fulfil the theory of ECFST.

2. Existing axial bearing capacity formulas of ECFST

Jamaluddin et al.[5] proposed a formula to calculate the axial bearing capacity N_u of ECFST stub columns based on experiment and the design rules in EC4 [19] for circular CFST,which is given by

$$N_u = \eta_s A_s f_y + \left[1 + \eta_c (t/D_e)(f_y / f_{co})\right] A_c f_{co} \tag{1}$$

In which, A_s, f_y and t are the areas, yield strength and thickness of the steel tube respectively, A_c and f_{co} are the areas and cylinder compressive strength of unconfined concrete. The diameter D_e of circular section is taken as the equivalent diameter for elliptical section, η_s is a factor to reduce the tube strength to account for hoop stress, whilst η_c indicates the effect of confinement in the concrete strength. They are given by

$$D_e = 2a^2 / b \tag{2}$$

$$\eta_s = 0.25(3 + 2\lambda) \tag{3}$$

$$\eta_c = 4.9 - 18.5\lambda + 17\lambda^2 \tag{4}$$

$$\lambda = \sqrt{\frac{A_s f_y + 0.85 A_c f_{co}}{(\pi^2 EI)/(\mu L)^2}} \tag{5}$$

$$EI = E_s I_s + 0.6 E_c I_c \tag{6}$$

In which, EI is the flexural rigidity of the elliptical cross section, λ is the slenderness ratio of ECFST, E_s and I_s are the elastic modulus and inertia moment of the steel tube, E_c and I_c are the elastic modulus and inertia moment of the core concrete, L is the length of ECFST, μ is the effective length factor.

Zhao and Packer [6] put the same formula to predict the axial capacity of ECFST stub columns as that of Jamaluddin et al, but they are different in calculating D_e and EI.

$$D_e = 2a\left\{1 + \left[1 - 2.3(t/2a)^{0.6}\right](a/b - 1)\right\} \tag{7}$$

$$EI = E_s I_s + E_c I_c \tag{8}$$

In Liu's paper [12]. The distribution rule of interaction force between steel tube and core concrete of ECFST stub columns under axial load was derived by the plastic equilibrium theory, and the lateral stress nephogram of core concrete is obtained using finite element simulation. Then the assumption of effectively confined zone distribution of core concrete is proposed on account of the two results above. Based on the idea of "unified theory" and existing unified formula of axial loading capacity for circular CFST stub columns, a practical unified calculation equation for ECFST stub columns is obtained as follows.

$$N_u = f_{sc}(A_s + A_c) \tag{9a}$$

$$f_{sc} = \frac{1 + 1.5(b/a)^{0.3}\xi}{1 + A_s/A_c} f_{ck} \tag{9b}$$

$$\xi = (f_y A_s)/(f_{ck} A_c) \tag{9c}$$

In which, f_{sc} is the "unified strength" of CFST, f_{ck} is the axial compressive strength of unconfined concrete, ξ is the factor of confinement effect.

Followed the example of Liu, on the basis of numerical simulation, data fitting and the idea of "unified theory", Shen [11] presented another formula to calculate the "unified strength".

$$f_{sc} = \left(A + B\xi + C\xi^2 + D\xi^3\right) f_{ck} \tag{10}$$

In which, A, B, C and D are the parameters obtained by statistical analyzing, A=1.3625, B=0.7080, C= 0.0624, D=0.0075.

Du [13] analyzed the axial limit stress of three dimensional stress state, with the introduction of the effective constraint factor of the size effect to consider the restraints of steel tube to concrete, also established the formula of ultimate bearing capacity of ECFST stub columns based on the "unified theory",which is given by

$$N_u = A_c f_{co} + \frac{\left(A_c k k_e + A_s \tau\right) f_y t_s}{r_i} \tag{11}$$

In which, k is the factor related to the internal friction angle (β) of concrete, k=3, r_i and t_s are inner radius and thickness of the steel tube respectively of the equivalent circular CFST which has the same area and steel ratio as the ECFST. k_e is the effective restraint coefficient, τ is the factor considering the impact of the intermediate principal stress and the corresponding normal stress.

Wu [14] put forward the design equation for ECFST stub columns by conducting section analysis and introducing an equivalent stress block according to the Chinese code for design of reinforced concrete structures,it's given as follows

$$N_u = \alpha_1 f_{cc} \frac{A_c}{\pi}\left(\theta - \frac{\sin 2\theta}{2}\right) + 0.89 f_y \frac{A_s}{\pi}(2\theta - 3) \tag{12a}$$

$$M_u = \begin{cases} \frac{2}{3}\alpha_1 f_{cc} A_c b_0 \frac{\sin^3\theta}{\pi} + 0.89 f_y A_s b\left[\sin\theta + \sin(3-\theta) + 0.24(a/b - 1)\right]/\pi \\ \text{bend around major axis} \\ \frac{2}{3}\alpha_1 f_{cc} A_c a_0 \frac{\sin^3\theta}{\pi} + 0.89 f_y A_s a\left[\sin\theta + \sin(3-\theta) - 0.24(a/b - 1)\right]/\pi \\ \text{bend around minor axis} \end{cases} \tag{12b}$$

In which, θ is the angle corresponding to the height of the equivalent rectangular stress block of the compressive zone. α_1 is used to describe the equivalent rectangular stress block of concrete, $\alpha_1 = 1.17 - 0.25a/b$. The compressive strength of core concrete f_{cc} is given by

$$f_{cc} = \gamma_U f_{co} + k_1 f_l \tag{13}$$

In which, γ_U is the reduction factor of size effect on strength of concrete, $\gamma_U = 1.67\sqrt{a_0 b_0}^{\,-0.112}$. k_1 is the restraint coefficient and is obtained by test results, $k_1 = 4.1(b/a)^{1.5}$. f_l is the confining stress of core concrete, obtained by introducing the ratio of the volume of steel tube ρ_s.

$$f_l = 0.095 \rho_s f_y \tag{14a}$$

$$\rho_s = \left[1.5(a_0 + b_0) - \sqrt{a_0 b_0}\right] t / (a_0 b_0) \tag{14b}$$

3. Comparison of the formulas

Table 1 and Fig.2 show the calculation results of existing specimens [3-6,12,14] by all the formulas, in which, N_{ut} is the experimental result, N_j is the calculation result of paper [5], while N_z for [6], N_s for [11],N_l for [12], N_d for [13] and N_w for [14].

As we can see, the formula of paper [5] gives the most accurate prediction, but the formulas of paper [13] and [14] are more conservative and secure for design.

Table 1. Statistical Analysis of Selected Specimens

	N_j/N_{ut}	N_z/N_{ut}	N_s/N_{ut}	N_l/N_{ut}	N_d/N_{ut}	N_w/N_{ut}
Mean	1.006	1.011	1.022	0.971	0.917	0.849
SD	0.069	0.070	0.096	0.078	0.076	0.071
COV	0.069	0.069	0.094	0.080	0.083	0.084

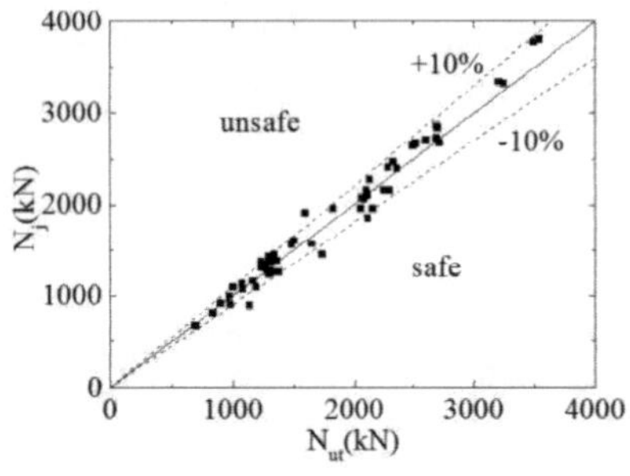

(a) comparison of N_j and N_{ut}

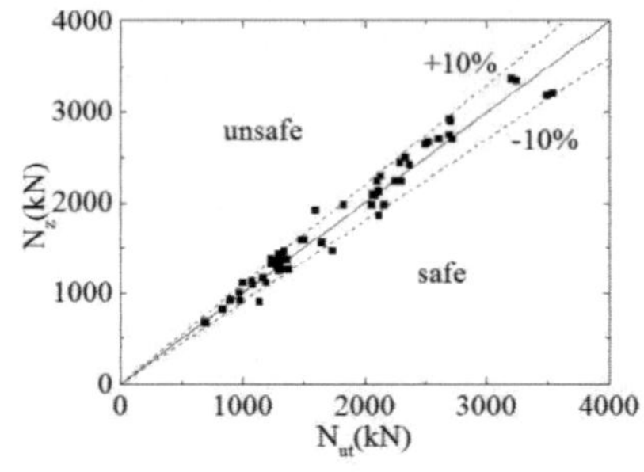

(b) comparison of N_z and N_{ut}

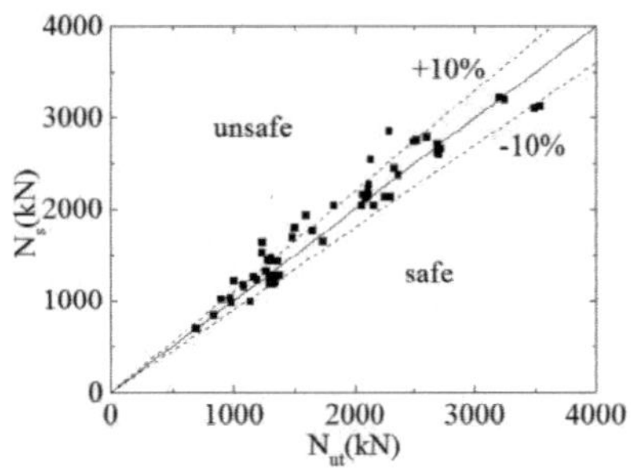

(c) comparison of N_s and N_{ut}

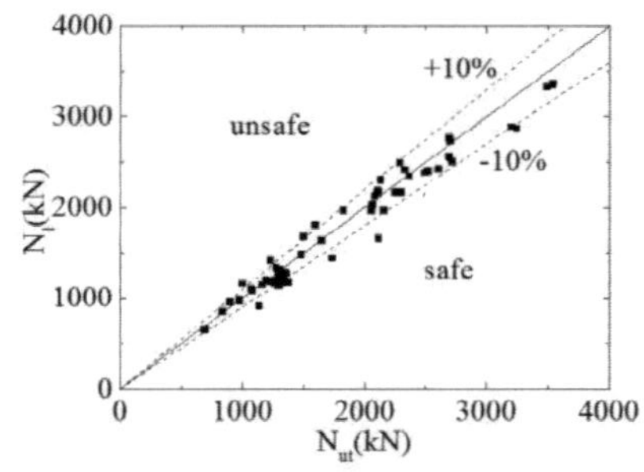

(d) comparison of N_l and N_{ut}

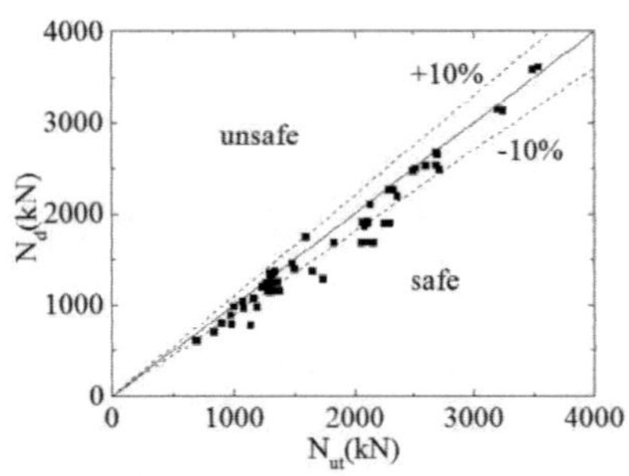

(e) comparison of N_d and N_{ut}

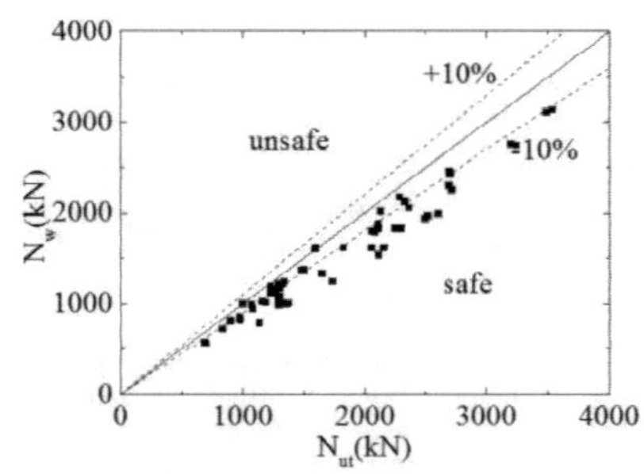

(f) comparison of N_w and N_{ut}

Figure 2. Comparisons between calculation and test results.

4. Summary
In a word, all the formulas above can meet the tolerable error of engineering, the formulas of paper [5] and [6] provide the most accurate predictions for the axial bearing capacity of ECFST stub columns. The formulas of paper [11] and [12] can also give reasonable predictions, and the calculation procedure are more simple. Though the formulas of paper [13] and [14] show the larger error, but the predictions are most conservative.

Acknowledgement
The authors are grateful for the financial support received from the Science and Technology Department of Zhejiang Province (Project No.: 2016C31021).

References
[1] Sun B, Tong L W 2013 Research development of compression properties of elliptical hollow section column *Journal of Architecture and Civil Engineering* **30** 76-81.
[2] Lam D, Testo N 2008 Structural design of concrete filled steel elliptical hollow sections *Proceedings of the Sixth International Conference on Composite Construction in Steel and Concrete* 252-262.
[3] Mccann F, Gardner L and Wei Q 2015 Concrete-filled elliptical section steel columns under concentric and eccentric loading *Proceedings of The Eighth International Conference on Advances in Steel Structures* 20150722.
[4] Yang H, Lam D and Gardner L 2008 Testing and analysis of concrete-filled elliptical hollow sections *Engineering Structures* **30** 3771-3781.
[5] Jamaluddin N, Lam D, Dai X H and Ye J 2013 An experimental study on elliptical concrete filled columns under axial compression *Journal of Constructional Steel Research* **87** 6-16.
[6] Zhao X L, Packer J A 2009 Tests and design of concrete-filled elliptical hollow section stub columns *Thin-Walled Structures* **47** 617-628.
[7] Liu F, Wang Y and Chan T M 2017 Behaviour of concrete-filled cold-formed elliptical hollow sections with varying aspect ratios *Thin-Walled Structures* **110** 47-61.
[8] Jamaluddin N, Lam D and Ye D 2009 Finite element analysis of elliptical hollow and concrete filled tube columns *International Journal of Integrated Engineering* **1** 95-101.
[9] Dai X, Lam D 2010 Numerical modeling of the axial compressive behavior of short concrete-filled elliptical steel columns *Journal of Constructional Steel Research* **66** 931-942.
[10] Ding F X, Li G, Yu Z W and Ou J P 2010 Nonlinear finite element analysis of axially loaded concrete-filled elliptical steel tubular stub columns *National Symposium on Basic Theory and Engineering Application of Concrete Structures* 243-246.
[11] Shen Q H, Wang J F, Wang W and Wang J 2015 Axial compressive behavior and bearing capacity calculation of ECFST columns based on numerical analysis *Progress in Steel Building Structures* **06** 68-78.
[12] Liu X C, Zha X X 2011 Study on behavior of elliptical concrete filled steel tube members I: stub and long columns under axial compression *Progress in Steel Building Structures* **01** 8-14.
[13] Du W C, Zhao J H, Zhang C G and Yin J 2016 Study on the axial ultimate bearing capacity of elliptical concrete filled steel tubular stub columns *Concrete* **04** 46-49.
[14] Wu J Y 2015 Structural behavior and design method of hollow and concrete-filled elliptical steel tubular columns *Zhejiang University*.
[15] Theofanous M, Chan T M and Gardner L 2009 Flexural behavior of stainless steel oval hollow sections *Thin-Walled Structures* **47** 776-787.
[16] Theofanous M, Chan T M and Gardner L 2009 Structural response of stainless steel oval hollow section compression members *Engineering Structures* **31** 922-934.
[17] Dai X, Lam D 2010 Axial compressive behavior of stub concrete filled columns with elliptical stainless steel hollow sections *Steel & Composite Structures* **10** 517-539.
[18] Han L H, Ren Q X and Li W 2011 Tests on stub stainless steel–concrete–carbon steel double-skin tubular (DST) columns *Steel Construction* **67** 437-452.
[19] CEN. EN 1994-1-1 Eurocode 4:2004 Design of composite steel and concrete structures-Part 1-1: General rules and rules for buildings *European Committee for Standardization Brussels.*

Electro-physical properties of thin films based on metal-containing polyacrylonitrile for application in low temperature gas sensors

Semenistaya T.V.[1*], Ivanenko A.V.[1]

[1]Engineering-technological Academy of the Southern Federal University, 44 Nekrasovskiy Lane, Taganrog 347928, Russia

*E-mail: semenistayatv@sfedu.ru

Abstract. The metal-containing (Cu, Co, Ag, Cr) polyacrylonitrile (PAN) thin films were fabricated using IR-pyrolysis under low vacuum conditions in different temperature and time modes. The thickness of the fabricated films was between 0.01÷0.68 μm. The metal-containing PAN films had the resistance values in the range from $2.9 \cdot 10^2$ to $5.1 \cdot 10^{11}$ Ohm. It has been investigated that the film thickness, resistance and gas sensitivity of the samples depends on the composition of the initial solution and on the process parameters of the film material's fabrication. It has been studied that the samples demonstrate gas-sensing properties to CO, NO_2, Cl_2 and gasoline vapours at room temperature.

1. Introduction

Polymeric materials have gained a wide theoretical interest and can be used for very different purposes and may offer unique possibilities [1-3]. Recent advances in polymer science and film fabrication have made polymeric material films useful, practical and economical in a wide range of sensor designs and applications [4, 5].

Conducting polymers have attracted considerable attention for room temperature gas sensing applications.

PAN is extremely popular and has attracted much attention due to its unique structure and ability to change it under heating, tolerance to most solvents and commercial availability [6]. PAN, which is a polymer with a chain of carbon atoms connected to one another, is turned into a semiconductor material under pyrolysis without changing its basic structure. PAN being subjected to carbonization under controlled conditions is known to be transformed into polymer with conjugated bonds and semiconductor properties. The method of fabrication of the electroconductive structure of PAN, hardware peculiarities and synthesis conditions ambiguously exert influence on the properties of the formed material. There are some methods of manufacturing the electroconductive structure of PAN such as: thermal pyrolysis and IR-radiation pyrolysis. Under the pyrolysis proceeding at temperatures higher than 300 °C, PAN is transformed into a polymer with a cyclic structure, containing the conjugated double bonds. Due to the π -orbital overlap of the neighbouring molecules of the conjugated structure, the π -electrons are delocalized along the entire chain, which provides semiconducting properties of PAN polymer. This method enables to manufacture films with a developed surface morphology and provides uniform distribution of modifying additives in a film. The presence of a transitional metal and its compounds completely excludes a stage of cyclization of PAN's nitrile groups. Being formed as a result of dehydrogenation of the main polymeric chain the sequence of conjugated bonds has greater length in comparison with the areas, fabricated in the absence of salts of transition metals, which increases the material conductivity [7, 8].

We are interested in PAN as the conjugated conducting polymer. PAN films can be considered as one of the most promising material in order to obtain chemoresistive sensors for environmental applications.

The composite films of metal-containing PAN, representing an organic matrix, which structure and properties change at influence of various temperatures, and the particles of a modifying additive dispersed in it, are a perspective material for microelectronics [9-14].

2. Experimental

The metal-containing PAN films were fabricated by pyrolysis method under the influence of incoherent IR-radiation under low vacuum conditions [14-17]. The following components were used: PAN (A ldrich 181315) as a conductive polymer matrix, copper chloride (II) $CuCl_2$, silver nitrate (I) $AgNO_3$, chrome chloride (III) $CrCl_3$ and cobalt chloride (II) $CoCl_2$ as a modifying additives for increasing the selectivity and adsorption activity of the films, dimethylformamide (DMF) as a solvent. The film-forming solutions were deposited on to polycrystalline aluminum oxide substrates using centrifugation method. Time-temperature modes of IR-annealing were selected experimentally, since the intensity and duration of infrared radiation provides an opportunity to control the properties of the film's material by changing the molecular structure of the PAN polymer. The radiation intensity at the first stage of IR-annealing corresponds to temperatures varying from 150 to 350 °C during a 5-60-minute time period, and the intensity of radiation at the second stage of the IR-annealing corresponds to temperatures of 350 – 800 °C during a 2-60-minute period.

The thickness of the films was measured by the interferential microscopy using interferometer MII-4. The electrical characteristics of the prepared samples were carried out in a setup equipped with heating element. Ag contacts were deposited on the film surface in order to investigate the electrical properties of the samples.

The steady-state gas distribution method was used for testing gas-sensing properties. A specially made setup equipped with a quartz chamber, sensor holder, gas and purge lines was used to maintain the desired level of detected gas concentration. Sensing characteristics were examined on basis of the measurement of the film resistance. The quartz chamber volume was around 700 cm^3 so that the changes in gas concentration was immediate. The gas sensitivity (S) was defined as the ratio $S = R_o - R_g/R_o$, when $R_o > R_g$, where R_o is the resistance in air, R_g is the resistance in the atmosphere of the detected gas. Operating temperature was chosen experimentally.

3. Results and discussion

The films growth was observed in accordance with the technological parameters and different weight concentration and type of a modifying additive in the initial solutions.

As observed from figure 1a, the preliminary drying stage (T_{dr} = 160 °C , t_{dr} = 30 minutes) leads to a considerable decrease in the Co-containing PAN film thickness values (20 - 40 %). The effect of the increase in temperature of the second stage of IR-annealing (350 °C, 450 °C, 500 °C) on the Co-containing PAN film thickness was showed on figure 1b. The PAN film thickness were between 0.01 ÷ 0.12 μm and the Co-containing PAN film thickness were in the range from 0.03 to 0.28 μm.

Figure 2 depicts the dependence of the Ag-containing PAN film thickness on different weight concentration of a modifying additive in the initial solutions that was between 0.22 ÷ 0.68 μm.

The Cu-containing PAN films were of the thicknesses in the range from 0.04 to 0.6 μm when fabricated in various temperature modes of the second stage of IR-annealing (figure 2b).

IOP Conf. Series: Materials Science and Engineering **220** (2017) 012003 doi:10.1088/1757-899X/220/1/012003

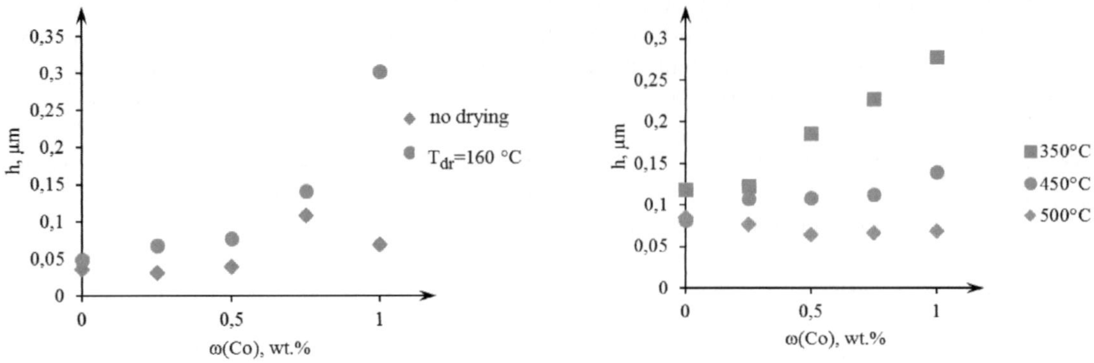

a) b)

Figure 1. The dependences of the PAN film thickness and the Co-containing PAN film thickness on the weight concentration of a modifying additive and drying stage (a): 1 – without drying, 2 – with drying; and on IR-pyrolysis temperature of the second stage of annealing (b): 1 – 350°C; 2 – 450°C; 3 – 500°C.

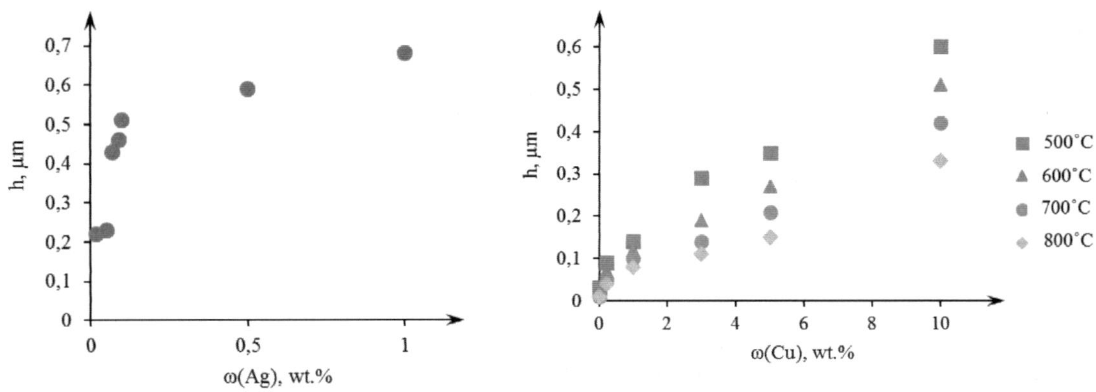

a) b)

Figure 2. The dependence of the Ag-containing PAN film thickness on the weight concentration of a modifying additive (a) and the dependences of the PAN film thickness and of the Cu-containing PAN film thickness on IR-pyrolysis temperature (b).

Prospects of the metal-containing organic polymer films' application as a gas-sensing layer of chemoresistive sensors is caused by PAN properties, which turned into a semiconductor material under pyrolysis [10]. In the work the influence of technological parameters of heat treatment on electrical properties of the fabricated films based on the metalorganic material were studied. The decrease in a modifying additive weight concentration in the initial solutions sharply worsened the conductivity of the Ag-containing PAN films (figure 3a), and it was not significantly reflected on the Cr-containing PAN film conductivity (figure 3b). The thin Cu-containing PAN films with the resistance values in the range from $4.0 \cdot 10^{2}$ to $2.7 \cdot 10^{11}$ Ohm were fabricated at various IR-annealing temperatures and using various weights of a modifying additive (figure 4). As observed from the figure 4 the films conductivity was improved significantly by the IR-annealing temperature growth and the increase of copper chloride (II) weight concentration in the initial solutions.

The investigations of the Co-containing PAN film resistance were carried out that revealed the specified dependence of the resistance values on the second stage IR-annealing temperature and on the weight concentration of a modifying additive. As observed from figure 5 the thin Co-containing PAN

IOP Conf. Series: Materials Science and Engineering **220** (2017) 012003 doi:10.1088/1757-899X/220/1/012003

films had the resistance values in the range from $2.9{\cdot}10^2$ to $4.7{\cdot}10^{10}$ Ohm. The resistance values of Cd-containing PAN films were between $10^6 \div 10^9$ Ohm. Thus, the studied composite material based on thermally treated PAN and Cu, Co, Ag, Cr, Cd compounds show a rich evolution of electric properties that inspires the idea that the metal-containing PAN films can be good materials for high performance chemoresistive sensors.

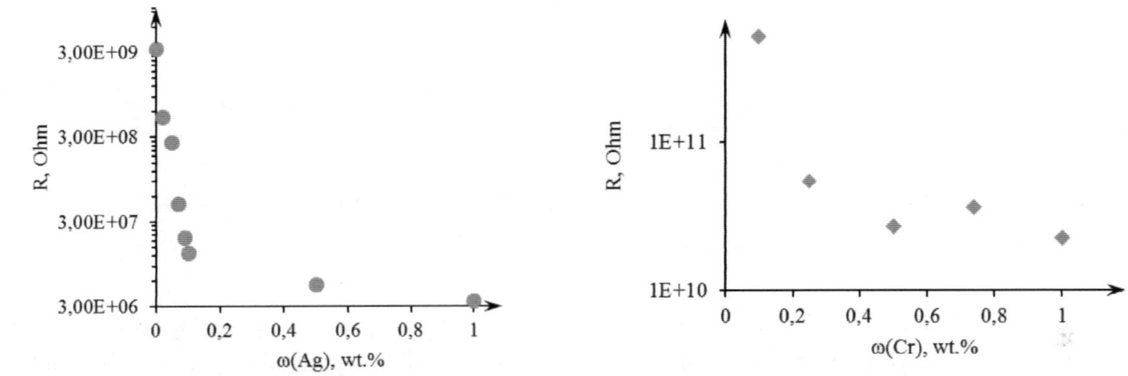

a) b)

Figure 3. The dependence of the Ag-containing PAN film resistance (a) and the Cr-containing PAN film resistance (b) on the weight concentration of a modifying additive.

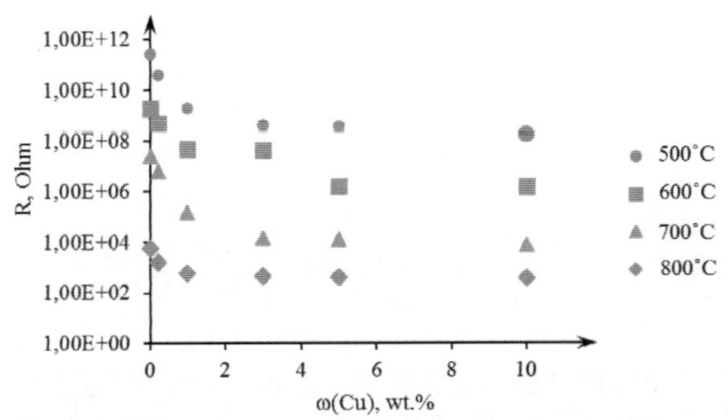

Figure 4. The dependence of the Cu-containing PAN film resistance on the weight concentration of a modifying additive and IR-pyrolysis temperature: 1 – 500 °C; 2 – 600 °C; 3 – 700 °C; 4 – 800 °C.

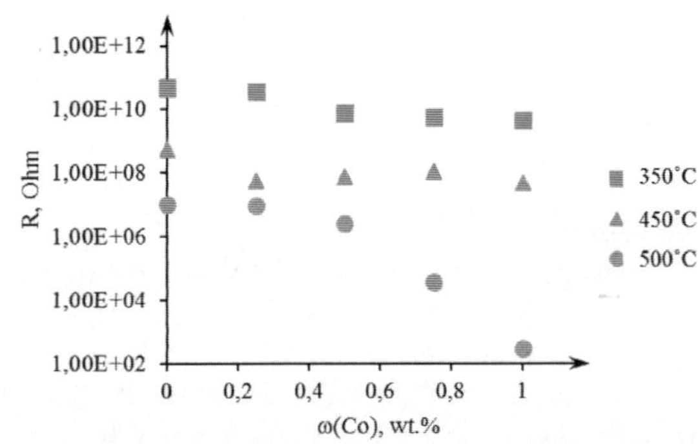

Figure 5. The dependence of the Co-containing PAN film resistance on the weight concentration of a modifying additive and IR-pyrolysis temperature of the second stage of annealing: 1 – 350 °C , 2 – 450 °C , 3 – 500 °C .

As a result of the carried-out researches it was established that Co-containing PAN films with rather high degree of resistance ($10^9 - 10^{11}$ Ohm) were sensitive to CO ($S = 0.44 \div 0.58$ r.u.) at working low temperatures ($18 \div 25$ °C) and the concentration of detected gas being in an interval of $250 \div 400$ ppm.

The results of the sensing tests of the Cr-containing PAN films revealed their sensitivity to 0.2% gasoline vapors in ambient air with $0.5 \div 0.9$ r.u. of the gas sensitivity value at room working temperature. The Ag-containing PAN films were found to be a highly sensitive chemoresistive sensor ($S = 0.68 \div 0.76$ r.u.) for chlorine Cl_2. The films gave a stable response for a sufficiently long time in the range of concentrations of detected gas between $0.14 \div 107$ ppm at working low temperatures ($18 \div 25$ °C). The fabricated Cu-containing PAN films used as a sensing layer for chemoresistive NO_2 sensor that operates at room temperature (figure 6). The sensor was found to detect nitrogen dioxide gas down to 36.5 ppm with a reversible and reproducible response ($S = 0.6 \div 0.8$ r.u.).

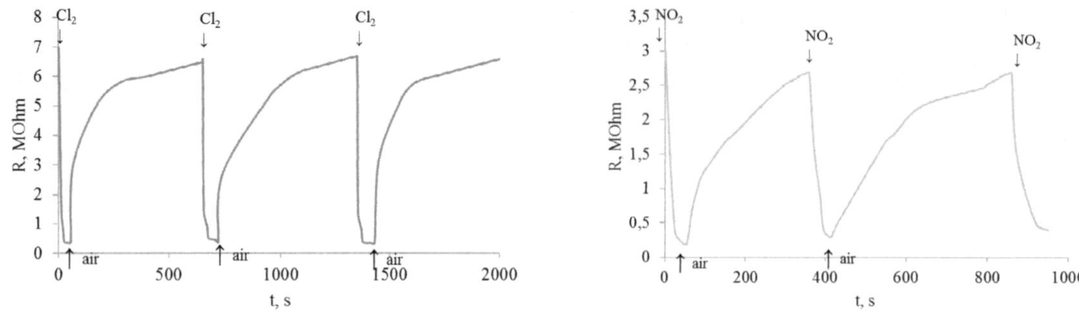

a) b)

Figure 6. Typical time-dependant curves of the Ag-containing PAN film resistance (a) and the Cu-containing PAN film resistance (b) upon periodic exposure to gas.

4. Conclusion

Thus, the gas sensitivity of the fabricated semiconductor thin films based on metal-containing PAN depends on the composition of the initial solution and on the process parameters of film material's forming. The fulfilled research inspires the idea that metal-containing PAN films can be good materials for high performance low temperature multisensory system.

Acknowledgments

This work was performed with the financial support of the research project No. VnGr-07 / 2017-21 within the framework of the internal grant of SFU. The equipment of the Centre of Collective Use of Equipment "Nanotechnologies" and Multi-Access Center of Microsystem Engineering and Integral Sensor Technology of Southern Federal University was used for this study.

References

[1] Lange U, Roznyatovskaya N V and Mirsky V M 2008 *Anal. Chim. Acta.* **614** 1.
[2] Li X, Wang Y, Yang X, Chen J, Fu H and Cheng T 2012 *Trends in Anal. Chem.* **39** 163.
[3] Iwan A and Sek D 2008 *Prog. Polym. Sci.* **33** 289.
[4] Javadpour S, Gharavi A, Feizpour A, Khanehzar A and Panahi F 2009 *Sens. and Act.* B **142** 152.
[5] Srivastava S, Kumar S and Vijay Y K 2012 *Int. J. Hydr. En.* **37** 3825.
[6] Surianarayanan M, Vijayaraghavan R and Raghavan K V 1998 *J. Polym. Sci.* **36** 2503.
[7] Chatterjee N, Basu S, Palit S K and Maiti M M 1995 *J. of polymer science: Part B: Polymer physics* **33** 1705.
[8] Jing M, Wang C and Wang Q, Bai Y and Zhu B 2007 *Polymer Degradation and Stability* **92** 1737.
[9] Korolev A N, Semenistaya T V, Al-Hadrami I S, Loginova T P and Bruns M 2010 *Persp. Mat.* **5** 52.
[10] Semenistaya T V, Petrov V V and Lu P 2013 *Adv. Mat. Res.* **804** 135.
[11] Bednaya T A, Konovalenko S P, Semenistaya T V and Korolev A N 2012 *Persp. Mat.* **5** 39.
[12] Lu P, Gorbatenko Yu V, Semenistaya T V, Vorobjev E V and Korolev A N 2011 *J. Nano- and microsystem technique* **9** 5.
[13] Petrov V V, Plugotarenko N K and Semenistaya T V 2013 *Chaotic Modeling and Simulation* **4** 609.
[14] Semenistaya T V, Petrov V V, Bednaya T A and Zaruba O A 2015 *Materials Today: Proceedings* **2(1)** 77.
[15] Semenistaya T V, Plugotarenko N K and Petrov V V 2015 *Applied Mechanics and Materials* **727-728** 145.
[16] Plugotarenko N, Semenistaya T and Petrov V 2015 *Modern Applied Science* **9(11)** 213.
[17] Semenistaya T V, Petrov V V, Kalazhokov Kh Kh, Kalazhokov Z Kh, Karamurzov B S, Kushkhov Kh V, Konovalenko S P 2015 *Surface Engineering and Applied Electrochemistry* **51(1)** 9.

IOP Conf. Series: Materials Science and Engineering **220** (2017) 012004 doi:10.1088/1757-899X/220/1/012004

Preparation of a Novel Nano or Submicron Tourmaline Ceramic and Its Ability to Inhibit the Activity of the Sulfate-Reducing Bacterium

Chun-ying Li[1*]

[1]School of Energy and Civil Engineering, Harbin University of Commerce, Harbin, 150028, China

*E-mail: heart.li@163.com

Abstract. In this study, novel nano- or submicron-scale tourmaline bacteriostatic ceramics in which nano or submicron tourmaline is one of the central materials, together with nano-zinc oxide, are prepared using ion exchange and solid-phase synthesis techniques. The material is then examined with IR, XRD and XPS and is tested for the ability to inhibit the activity of the sulfate-reducing bacterium (SRB). The calcination temperature of the ceramic is 600 °C, and the main components are present at 11.17% for Si2p, 3.12% for Mo3d, 3.47% for Zn2p3, 2.86% for Mn2p and 2.35% for Cu2p. Additionally, the density of this material is 1.4-3.5 g/cm^3, and its compressive strength exceeds 5.2 MPa, meeting the requirements of ceramic standards. Consequently, the bacteriostatic ceramic inhibits the activity of the sulfate-reducing bacterium effectively without inhibiting the removal of COD and NO_3^-. These results indicate that the sulfate-reducing bacterium maintains its functional metabolism, apart from its sulfate reduction potential, when using this bacteriostatic ceramic, thus achieving the goal of inhibiting the action of the sulfate-reducing bacterium.

1. Introduction

In China, the SRB[1] contained in the Daqing Oilfield wastewater has vastly increased its reproduction in the past decades, which increases the amount of sulfide and the suspended solids, causing serious accidents during production, such as the collapse of the electric field, and leading to losses every year due to the corrosion of approximately RMB 2.5 billion. Balancing the killing and the inhibition of the SRB has been a topic of interest for both researchers and workers [2]. New observations suggest that the SRB serves as the main supportive bacterium in the reservoir system, while it negatively affects the ground system of the oilfield. Thus, the focus lies in controlling the sulfate-reducing function appropriately to inhibit its sulfate-reducing activity while taking advantage of the useful aspects of this bacterium[3-5].

The goal of this research was to develop a novel nano or submicron tourmaline bacteriostatic ceramic; nano or submicron tourmaline was added to a ceramic to combine the characteristics of the tourmaline and nano materials, to enlarge the contact area and to strengthen the electrodes and the piezoelectricity of the tourmaline. Meanwhile, several metal salts and nano-zinc oxide with a high bactericidal capacity for the SRB were tested to reduce the cost, using mixed shale and zeolite as the carrier. The new bacteriostatic ceramic and its inhibition of the SRB are evaluated.

2. Materials and methods

2.1. Preparation of nano or submicron tourmaline bacteriostatic ceramic
Five grams each of nano or submicron tourmaline, zeolite and shale was added to 50 mL of deionized water at a 6:3:1 ratio and then stirred uniformly using a magnetic stirrer. A nitrate solution was prepared (0.03 mol/L) and adjusted to a specified pH value (4-8); molybdate (1%), silver chloride (0.5%), copper sulfate (0.5%) and nano zinc (2%) were added, and the solution was stirred in the dark with the dispersant. The mixture was then decompressed, pumped, filtered, separated and cleaned with deionized water; it was then dried in a thermostatic oven at 105°C. Then, pore-forming agents (CS) (5%), binder (SF) (1%) and enamel agent (15%) were added to the sludge in the dried sample and were ground for 10 min to form a ball. The samples dried naturally and were then calcinated at 600°C for 4 h to produce the finished product, which was then stored in a dry place[6-7].

2.2. SRB activity inhibition experiment
The experimental sludge taken from an anaerobic UASB reactor contained a large amount of SRB, with 135 mg/L, 135 mg/L and 500 mg/L of SO_4^{2-}, NO_3- and COD, respectively, and an alkalinity of 500 mg/L. The sludge was placed into a 500-mL bottle of serum and high-purity nitrogen was introduced to the culture, which was incubated on a shaker for 24 h. Then, 10 g of bacteriostatic ceramic was added to the culture. The sample was removed after 12 h to measure the activity of the sulfate reductase against SO_4^{2}-、NO_3-、pH、ORP and S^{2-}. The culture was incubated on a shaker (speed of 125 rpm and temperature of 36±1°C) and was sampled regularly.

2.3. Analysis and characteristics
Nano or submicron tourmaline was imported from the United States. The following techniques and equipment were used: Infrared Spectroscopy: Spectrum One B (FT-IR spectrometer); ICP: 5300DV (PE) and other plasma emission spectroscopy; X-ray photoelectron spectroscopy (PH15700), X-ray diffraction (D-MAX/rB rotating anode X-ray diffraction instrument), sulfate concentration: the application of ion determination of Shimadzu, Japan; the measurement of sulfate reductase activity; the Ostrowski method; and pH and oxidation-reduction potential (ORP): measurement with WTW-in lab.

3. Results and discussion

3.1. Characteristics of copper-carrying nano or submicron tourmaline

3.1.1. Ceramic characteristics. Figure 1 shows that the diameter of the nano or submicron tourmaline bacteriostatic ceramic is between 5~8 mm. The scanning electron micrograph of the surface indicates a honeycomb structure with many voids and an even distribution.

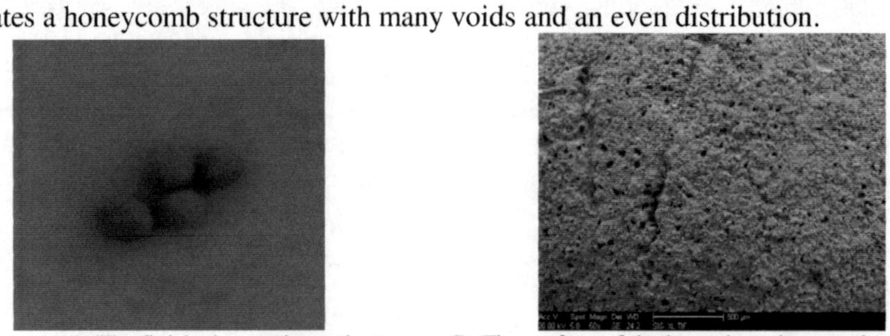

A. The finished ceramic product B. The surface of the bacteriostatic ceramic after calcination

Figure 1. The characterization of the tourmaline bacteriostatic ceramic.

3.1.2. Infrared Spectroscopy (IR). The complicated bacteriostatic ceramic consists of nano or submicron tourmaline, shale, zeolite, zinc oxide, copper chloride, manganese chloride and molybdate, among others, producing an extremely complex infrared spectrum (IR).

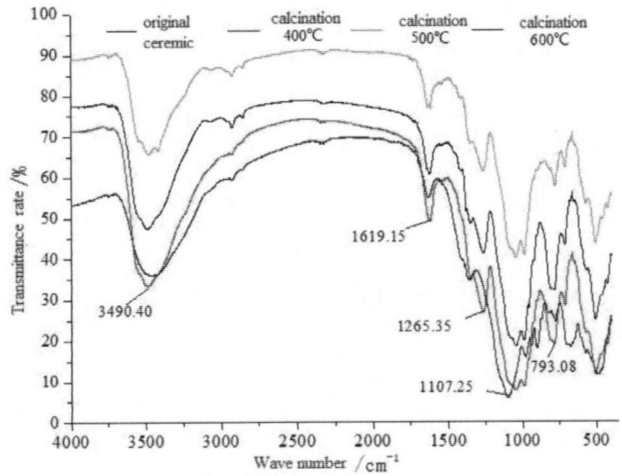

Figure 2. The IR map of the ceramic product at different temperatures.

Figure 2 shows that the OH stretching vibration absorption emerges at 3556.84 cm^{-1} for the tourmaline during calcination. The adsorption peak of the bacteriostatic ceramic is found between 3495 and 3470 cm^{-1} as the calcination temperature increases. The peak of the OH stretching vibration shifts, indicating that the ceramic experiences the association of hydroxy compounds during the preparation.

3.1.3. X-ray diffraction (XRD) analysis. The changes in the XRD spectrum were analyzed for the different calcination temperatures because of the complex elements of the bacteriostatic ceramic (Figure 3). The crystal form changes prominently and \ more intensely during the entire process. Based on the hardness of the ceramic and the nature of the tourmaline, 600°C was identified as the calcination temperature.

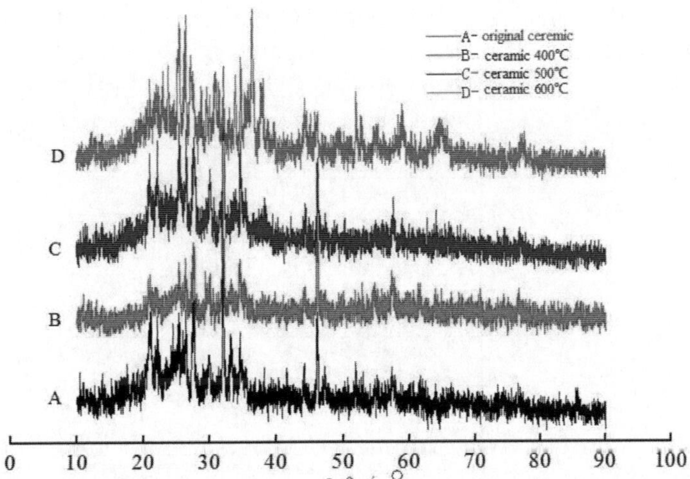

Figure 3. The XRD map of the ceramic under different calcination temperatures.

3.1.4. X-ray photoelectron spectroscopy (XPS). The whole ceramic is dominated by the nano or submicron tourmaline and is supplemented by the relevant bacteriostatic elements—nano-zinc oxide, copper chloride, manganese chloride, molybdate and the synthesis glazes and sludge generated during the formation of the ceramic with complexing factors. Figure 4 shows the main bacteriostatic elements in the ceramic by spectroscopy, which mainly includes Mo3d、Zn2p3、Mn2p and Cu2p (11.17% for

Si2p, 3.12% for Mo3d, 3.47% for Zn2p3, 2.86% for Mn2p and 2.35% for Cu2p). The strong inhibition ability of the individual elements of the ceramic indicates the stronger power of the nano or submicron tourmaline.

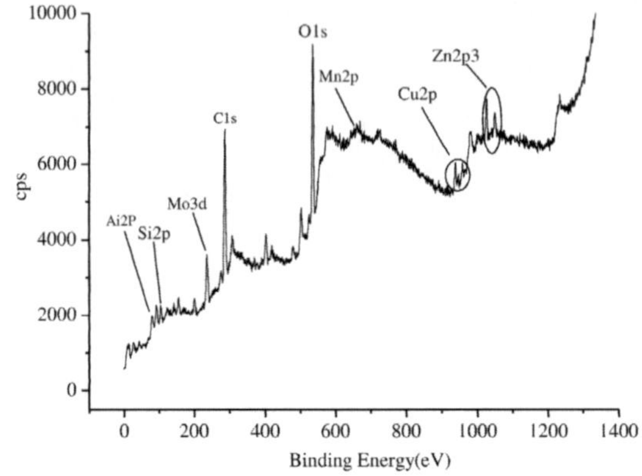

Figure 4. The XPS graph of the bacteriostatic ceramic.

3.2. Effectiveness of the ceramic at inhibiting SRB activity

3.2.1. Inhibition of SRB activity by the ceramic. The activity of the sludge gradually increases without the bacteriostatic ceramic, reaching 1.05 U/mL after 48 h. When the sulfate is almost entirely consumed, the activity begins to decrease. In comparison, the sludge treated with the bacteriostatic ceramic shows inhibited activity of the sulfate sludge: no hydrogen sulfide gas is generated or sulfate reduction occurs, achieving the desired effect.

3.2.2. The ORP of the bacteriostatic ceramic on the sulfate-reducing sludge and its influence on pH. Figure 5 indicates that the ORP increases immediately after adding the bacteriostatic agent, reaching -26 mV when the activity of the sulfate reductase is inhibited completely by the bacteriostatic ceramic. After 60 h, the ORP remains stable at 6.1 mV, and the ORP of the sulfate reductase was unable to grow. The bacteriostatic ceramic could improve the ORP significantly and could inhibit the growth of the SRB to stop the sulfate reduction. There is no major change in the pH compared with the original sample.

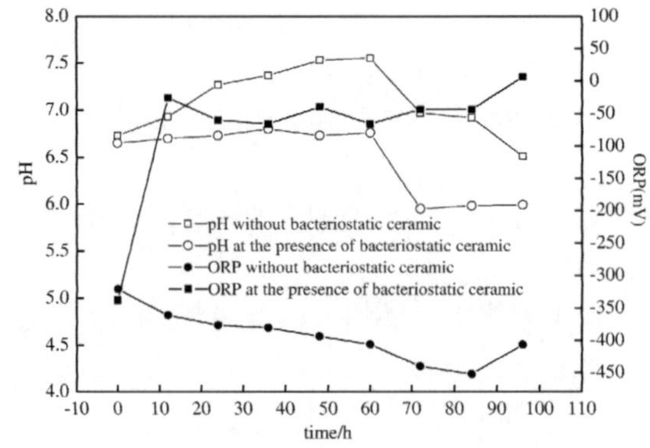

Figure 5. The influence of the bacteriostatic ceramic on the ORP and pH of the sulfate-reducing sludge.

3.2.3. Influence of the bacteriostatic ceramic on SO_4^{2-} and sulfide. Figure 6 shows that the concentration of SO_4^{2-} for the sample decreases to 7.6 mg/L from 134.2 mg/L and that the removal rate is 94.33% after 96 h; conversely, the concentration of S^{2-} rises from 5.2 mg/L to 30.7 mg/L. However, the concentration of SO_4^{2-} for the culture medium with the ceramic decreases to 116 mg/L from 134 mg/L without significant fluctuations, with a removal rate of 13.4%.

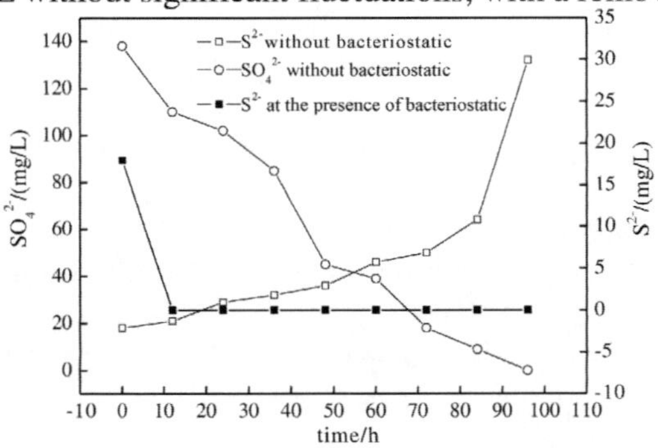

Figure 6. The influence of the bacteriostatic ceramic on the levels of SO_4^{2} and sulfides.

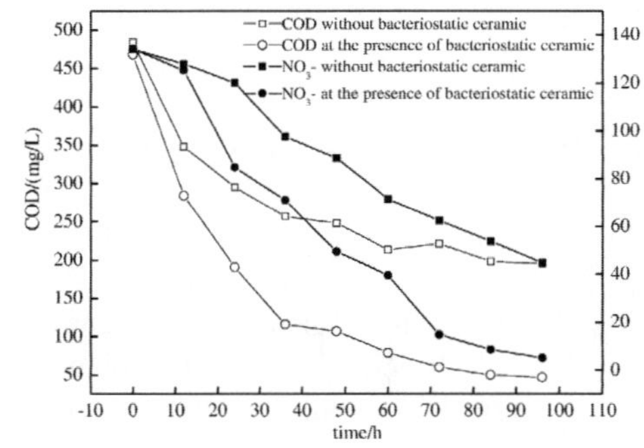

Figure 7. The influence of the bacteriostatic ceramic on the levels of COD and NO_3^-.

3.2.4. Influence of the ceramic on COD and NO_3^-. Figure 7 shows that the concentration of COD in the sample is reduced to 195 mg/L from 484 mg/L, with a removal rate of 59.71%. In comparison, the COD of the sample with ceramic is reduced to 47 mg/L from 468 mg/L at a rate of 89.95%. The concentration in both samples is reduced, but the ceramic sample has a faster removal rate; for comparison, the concentration of NO_3 in the untreated sample drops to 44.8 mg/L from 134 mg/L for a removal rate of 66.56%, while the NO in the ceramic-treated sample is reduced to 5.2 mg/L from 134.2 mg/L, an effective rate of 96.13%. The values of both the COD and NO_3^- are lower, indicating that only the sulfide is inhibited.

4. Conclusions

The novel nano or submicron tourmaline bacteriostatic ceramic, which uses nano or submicron tourmaline as one of the central materials, together with nano-zinc oxide, was prepared using ion exchange and solid-phase synthesis techniques.

1) The calcination temperature of the ceramic is 600 °C, and the content of the main components are 11.17% for Si2p, 3.12% for Mo3d, 3.47% for Zn2p3, 2.86% for Mn2p and 2.35% for Cu2p. Additionally, the density of this ceramic is 1.4-3.5 g/cm³, and the compressive strength exceeds 5.2 MPa, which meets the requirements of ceramic standards.
2) The bacteriostatic ceramic inhibits the activity of the sulfate-reducing bacterium effectively without generating hydrogen sulfide gas. The efficient removal of COD and NO_3^- indicates that the sulfate-reducing bacterium maintains its functional metabolism, apart from the inhibition of its sulfate reduction ability, creating a targeted inhibition of SRB activity.

Acknowledgements

We are grateful for the financial supports received from the state key laboratories namely, the Urban Water Resource and Environment Laboratory (Harbin Institute of Technology) (No.2017TS06), Natural Science Foundation of Heilongjiang Province of China (B201415, QC2015010), and Research Specific Fund Project for Talents with Technological Innovation (Type of Youth Reserve Talents'Planning) (No.2015RQQXJ064) and 2015 Projects of Artistic and Scientific Planning (Grant No.2015A03).

References

[1] Starosvetsky D, Starosvetsky J, Armon R, Ein-Eli Y 2010 A peculiar cathodic process during iron and steel corrosion in sulfate reducing bacteria (SRB) media *Corrosion Science* **52** 1536-1540.

IOP Conf. Series: Materials Science and Engineering **220** (2017) 012004 doi:10.1088/1757-899X/220/1/012004

[2] Wei L, Ma F, Zhao G 2010 Composition and dynamics of sulfate reducing bacteria during the waterflooding process in the oil field application *Bioresource Technology* **101** 2643-2650.
[3] Chidthaisong A, Conrad R 2000 Turnover of glucose and acetate coupled to reduction of nitrate, ferric iron and sulfate and to methanogenesis in anoxic rice field soil *FEMS Microbiology Ecology* **31** 73-86.
[4] Zumft W G 1993 The biological role of nitric oxide in bacteria *Arch. Microbiol.* **160** 253-264.
[5] Ricardo B J, Hamdy H E 1992 Biochemical control of sulfide production in waste water collection systems *National Conference on Environmental Engineering* 327-333.
[6] Taylor B F, Oremland R S 1979 Depletion of adenosine triphosphate in Desulfovibrio by oxyanions of group VI elements *Curr. Microbiol.* **3** 101–103.
[7] Lo K V, Liao P H 1990 Anaerobic treatment of baker's yeast wastewater. I. Start-up and sodium molybdate addition *Biomass* **21** 207-218.

IOP Conf. Series: Materials Science and Engineering **220** (2017) 012005 doi:10.1088/1757-899X/220/1/012005

The research on delayed fracture behavior of high-strength bolts in steel structure

Li Guo dong[1], Li Nan[1]

[1]Equipment Management Department, Suzhou Nuclear Power Research Institute, Shenzhen, 518000, China

Abstract. High-strength bolts have been widely used in power plants. However, the high-strength bolts which being employed in pumping station, steel structure and pipeline anti-whip structure have been found delayed fracture for many times in a power plant, this will affect the reliability of steel fracture and bring blow risk caused by falling objects. The high-strength bolt with delayed fracture was carried out fracture analysis, metallurgical analysis, chemical analysis, mechanical analysis, as well as bolts installation analysis, it can be comprehensively confirmed that the direct cause of high-strength bolts delayed fracture is the stress corrosion, and the root cause of high-strength bolts delayed fracture should be the improper installation at the initial and the imperfect routine anti-corrosion maintenance.

1. Introduction

High strength bolts are widely used in main steel structure and secondary steel structure of power plants. Compared with the welding connection, high strength bolts have many advantages, such as simple structure, reliable connection, quick construction etc.

A number of 10.9 high-strength bolts have been found broken in steel structure of power plant, in addition, broken number of high-strength bolts have more than 10 in pumping station plant steel structure and pipe anti-whip structure. Preliminary inspection found that all of the bolts were delayed failure, the shortest failure time was 3 years, the longest time was 20 years. The original design of steel structure demand that the use life of high strength bolts should be equal to the life of steel structure, if large-scale high strength bolts fracture, will affect the reliability of steel structure, high strength bolts fall from the steel structure, there is a blow risk, so it is necessary to study the reason of delayed fracture of high strength bolts.

In this paper, The high strength bolts being analyzed were manufactured by Shen Guang Shanghai high strength bolt factory, the bolts has been used more than 15 years, bolt material is 20MnTiB, bolts surface has phosphating treatment, bolts size are M20, The failure bolt break at the screw near the nut, as shown in Figure 1.

Before leave factory, manufacturer had sampled each batch of bolts, the sampling results meet the requirements of the national standard of GB/T 1228~1231 91. The appearance of the bolts, accuracy of thread, form and location tolerance or other dimensions conform to the requirements of GB5779.1~3 - 86, GB/T1228~1231 - 91, GB/T3631~3633 - 95. The chemical composition, mechanical properties test (tensile strength, yield strength, elongation, shrinkage, impact toughness), wedge load test, friction coefficient also meet the relevant standards.

IOP Conf. Series: Materials Science and Engineering **220** (2017) 012005 doi:10.1088/1757-899X/220/1/012005

Figure 1. High-strength bolts fracture location.

In this paper, the delayed fracture behavior of high strength bolts is studied by means of fracture analysis, metallographic examination, chemical composition analysis and hardness test. And the high strength bolts are randomly selected from the steel structure, and will be carried out comparative analysis to find out the reasons for the delayed fracture of bolt.

2. Experimental methods and results

2.1. Experimental object

A broken high strength bolt, as well as the same type, no broken high-strength bolts.

2.2. Fracture examination

Base on the analysis of macroscopic examination and microscopic examination, the failure characteristics and failure type of bolt are determined.

The failure bolt broke at the position of the screw and nut, the fracture is no plastic deformation, vertical with the screw. The fracture surface has been found trace of corrosion, the divergent lines starting at the corrosion location. It is shown that the corrosion location is source of crack, and the macroscopic examination is shown in Figure 2.

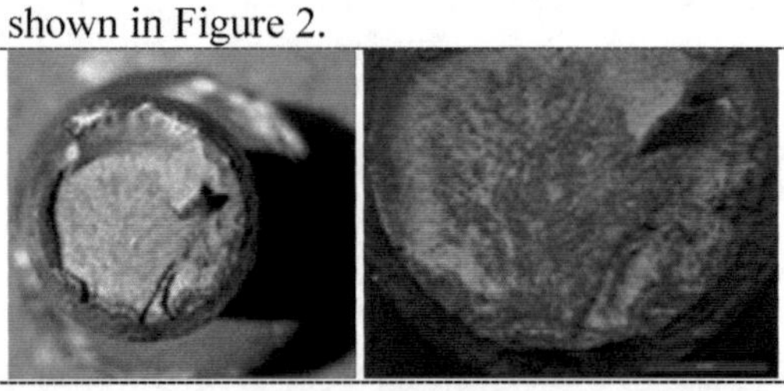

Figure 2. Macroscopic examination of failure bolt.

Cleaning fracture of failure bolt, the microscopic characteristics of fracture surface is observed. The source area of the crack is located at the bottom of screw thread, as shown in figure 3-a, the phosphating coating on the bottom surface of thread has been destroyed, corrosion products exist in the source region, the fracture shows intergranular fracture characteristics. The middle part of fracture is crack propagation zone, as shown in figure 3-b, there are intergranular feature and a few dimples in the fracture, the fracture extension area existe tear morphology and hydrogen embrittlement behavior. The opposite side of the crack source is the fracture zone, as shown in figure 3-c, the final fracture zone is dimple morphology, which is the ductile tearing area of failure bolt.

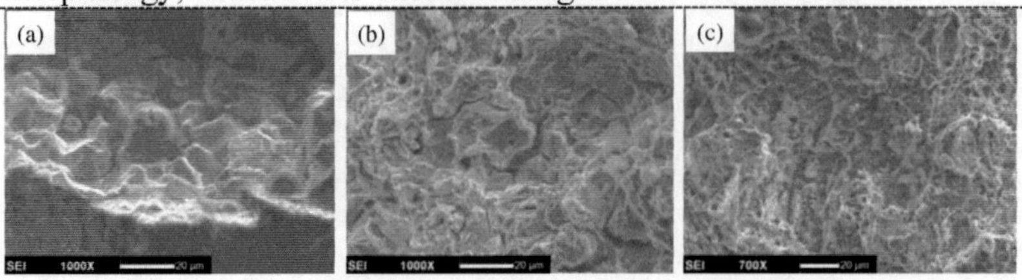

Figure 3. Microscopic examination of failure bolt: (a) The source region of crack, (b) The growth Area of Crack, (c) The final fracture area.

2.3. Metallographic analysis

The metallographic analysis shows that there are many intergranular cracks near the crack source, as shown in figure 4-a. There are corrosion pits on the adjacent surface of the section(the bottom of the thread adjacent to the fracture location), see figure 4-b, it can be seen that the intergranular crack originated from the corrosion pits, the crack length is about 0.14mm, and has not penetrated through the center of the bolt. According to the above crystal crack, the main crack has the same pattern, that is, the main crack comes from the corrosion pit; In addition, there are many secondary cracks near the main crack, when the main crack expands, bifurcation phenomenon occurs, therefore, comprehensive judgment the crack is caused by stress corrosion.

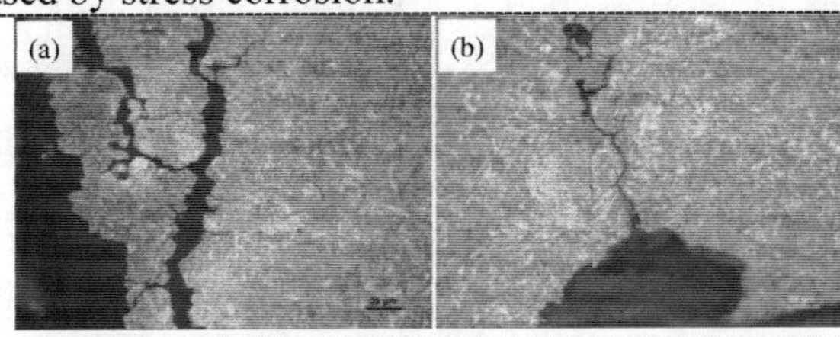

Figure 4. The intergranular cracks of failure bolt: (a) Fracture surface, (b) neighboring regions.

The metallographic structure of fracture bolt and sampling bolt is tempered sorbite, The metallographic structure is normal, as shown in figure 5-a, there is a strip distributed along the axial direction in each bolt, See figure 5-b, the size and quantity of inclusions meet the national standards.

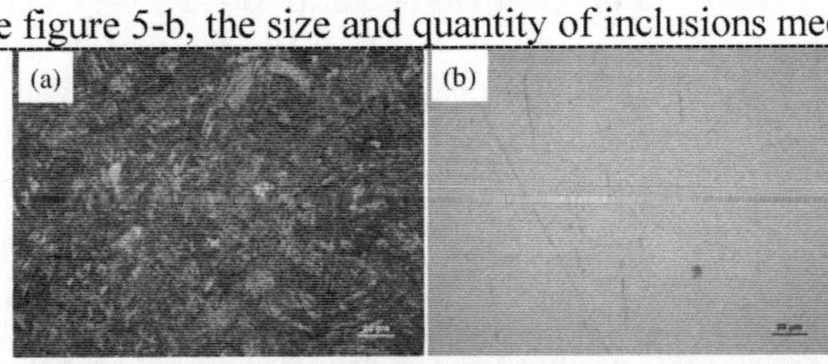

Figure 5. The Metallurgical of failure bolt: a microstructure b inclusions.

2.4. Chemical composition analysis

The results of chemical composition analysis of the failed bolt and the broken bolt are shown in table 1, each bolt material is 20MnTiB, chemical composition meets the requirements of GB/T 3077-1999 standard [1], Sulfur content less than 0.035, Phosphorus content less than 0.035.

2.5. Hydrogen content analysis

Samples of hydrogen content were obtained from the surface and center of the failed bolt and sampling bolt, measurement results are shown in table 2. The analysis shows that the hydrogen content of broken bolt is higher than sampling bolt, the hydrogen content in the middle part of the bolt is higher than the surface. In the expansion area of bolt fracture, there are traces of hydrogen embrittlement characteristics, this is consistent with the hydrogen content test.

Hydrogen embrittlement can be judged for the generalized stress corrosion in the core of failure bolt, the reason is follow. Hydrogen embrittlement characteristic is not found in the source of crack initiation site, and the hydrogen content of crack source is low, the crack propagation region has hydrogen embrittlement characteristic and high content of hydrogen, therefore, the hydrogen embrittlement caused by the application environment, which should be caused by stress corrosion cathodic process.

The hydrogen embrittlement mechanism in crack propagation: the cathodic process of stress corrosion is releasing hydrogen process, the hydrogen is diffused to the core of bolt, and the hydrogen

IOP Conf. Series: Materials Science and Engineering **220** (2017) 012005 doi:10.1088/1757-899X/220/1/012005

content in the heart of bolt is higher than the surface, the crack propagation process has characteristics of hydrogen embrittlement, therefore, the hydrogen embrittlement can also be defined as generalized stress corrosion, but on the whole, the main reason of failure should be stress corrosion.

Table 1. Bolts chemical composition (wt/%)

No.	C	Mn	Si	B	Ti	S	P
Failure bolt	0.21	1.46	0.31	0.0026	0.07	0.023	＜0.02
sample1	0.20	1.38	0.26	0.0023	0.06	0.020	＜0.02
sample2	0.21	1.46	0.31	0.0026	0.07	0.024	＜0.02
sample3	0.20	1.52	0.32	0.0026	0.08	0.026	＜0.02
sample4	0.21	1.50	0.28	0.0031	0.05	0.008	＜0.02
sample5	0.24	1.47	0.30	0.0026	0.07	0.024	＜0.02
GB/T 3077-1999	0.17~ 0.24	1.30~ 1.60	0.17~0. 37	0.0005~ 0.0035	0.04~ 0.10	＜ 0.035	＜0.035

Table 2. bolts hydrogen content (mg/kg)

No.	position	testing 1	testing 2	testing 3
Failure bolt	Near surface	0.9	0.8	/
	core	3	2	/
sample1	Near surface	0.2	0.3	0.3
	core	0.4	0.3	/
sample2	Near surface	0.5	0.4	0.6
	core	0.3	0.5	/
sample3	Near surface	0.2	0.2	0.5
	core	0.3	0.4	/
sample4	Near surface	0.5	0.3	0.4
	core	0.5	/	/

2.6. Mechanical properties analysis

The hardness, tensile properties and impact properties of high strength bolts are measured by hardness tester, mechanical testing machine and impact testing machine.

The results are as follows:

The hardness of the broken bolt and sampling bolt are 33~39HRC; the tensile strength of the bolt (σb) sampling are 1085Mpa ~ 1152Mpa; The yield strength (σ0.2) are 1036Mpa ~ 1083Mpa; Elongation (δ5) are 14.7% ~ 15.9%; Section shrinkage (ψ) are 66.3% ~ 70.3%; The impact energy of the sampling bolt (Ak) are more than 47J, and the impact toughness value (ak) are more than 58.75J/cm2. The mechanical properties of the broken bolt and non-broken bolts are all in accordance with the requirements of GB/T 1231-2006 standards [2].

3. Analysis and discussion

The above analyses show that the fracture of high strength bolt in steel structure is caused by stress corrosion. The material and mechanical properties of broken bolt meet the relevant national standards.

There is corrosion product in the source region of broken bolt, and the fracture is intergranular, there are many small cracks along the grain boundaries in the source region. The corrosion pits are found at the bottom of the thread adjacent to the fracture surface, and the intergranular cracks are initiated from the corrosion pits, therefore, comprehensive judgment stress corrosion is the direct cause of bolt broken.

The broken bolt is subject to stress corrosion, according to forming mechanism and influencing factor of stress corrosion, as well as the installation and environment of broken bolt, it is speculated that bolt failure is related to the corrosion and abnormal tensile stress.

3.1. Bolt corrosion

The steel structure of the power plant is installed outside the reactor building, the operating environment is marine atmospheric environment, there are Cl, S and other corrosive media in the air, so there is corrosion medium in the environment in which the bolt is located.

After the Broken bolt is tightened, sealed with putty in nuts, nut and two with mounting surface clearance, and painted anti-corrosion treatment to prevent wet air and rainwater into the bolt hole. Prolonged exposure to air and rain, paint and putty layer aging and cracking, the bolt is not tight seal. Once the corrosion medium reached the high stress position of the bolt, in the long term interaction between corrosion medium and stress, the phosphating layer of the high stress position of bolt is consumed gradually, the bolt matrix is directly exposed to the corrosive medium, and the bolt matrix was corroded.

Poor maintenance of bolt's paint leads to corrosion medium contact with bolt. The initial design of the steel structure didn't consider the subsequent anticorrosion work, maintenance personnel cannot reach some steel structure position, therefore, the power plant maintenance department cannot maintain the position which cannot reach in the steel structure.

It is observed that there are different degrees of corrosion in most sampling bolts, the phosphating coating of sampling bolts have been destroyed, and the protective effect of the paint layer has disappeared, bolts are easily corroded by atmosphere and rain.

3.2. Too much force on the bolt

According to bolt construction record, steel structure high strength bolt serious over screw, at least over 12%, the pre tension and the torque coefficient of high strength bolts are not satisfied with the national standard. According to the domestic standard, the pre tension of the 10.9 class M20 friction type high-strength bolt is 170KN[3], the steel structure high-strength bolt pretension is 190.4KN, the reason for the pre tension higher is that the reference installation file is not appropriate. The torque coefficient of the installation of high strength bolt is the factory test value (0.146), not checked again in installation environment, the torque coefficient of the bolt is not adjusted by experience, which leads to the fact that the construction torque coefficient is larger than the actual torque coefficient.

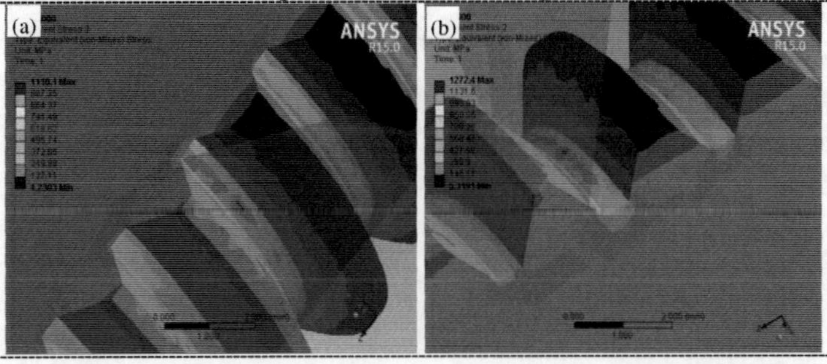

Figure 6. Finite element analysis: (a) normal tightening torque, (b) tightening torque exceeds.

The finite element analysis shows that the normal average equivalent stress is 727Mpa, the average equivalent stress after super twist reaches 884MPa, and the stress value increases by 21.6%, as shown in figure 6. The equivalent stress (884 MPa) and the yield strength (940 MPa) of the bolt after the super twist is about 0.94, according to the experience of the bridge industry, the status of the bolts must be replaced. After super twist, bolt force will increase, so that bolts are in a state of insecurity.

When the bolt is subjected to stress corrosion, the super twist will promote the development of corrosion, therefore, super twist is an important factor of bolt stress corrosion damage.

4. Analysis conclusions and solutions

In this paper, the manufacture, installation and operation of fracture bolt are studied, the reason for the delayed fracture of the bolt is found out.

According to the above analysis, the direct reason for the delayed fracture of steel bolt is the stress corrosion, this isn't consistent with some of the conclusions given in the literature[4-7]. The root cause of delayed fracture is poor maintenance and improper installation of bolt, this result in the stress corrosion of bolt.

Preventive measures which based on the root cause of delayed fracture can prevent the high strength bolt from breaking, such as:

- Replace all too tightened bolts, construction process needs to strictly comply with national standards. Before the construction of high strength bolts, check on the torque coefficient, and select the correct construction pre tension, to ensure that the construction torque is suitable. High precision torque wrench should be used in the construction of high strength bolts, after the completion of construction of high strength bolts, spot checks should be performed, Timely tightening of the less screwed bolts, super twist should be replaced.
- Improve the anti-corrosion measures of high-strength bolts. Daily inspection found high strength bolt screw thread damage or rust, it should be immediately replaced
- Selection better corrosion resistance bolts. The replacement of high strength bolts should be based on the structural characteristics of steel, in severely corrosive and hard to reach places, should use the better corrosion resistance of Dacromet high strength bolt, increase the corrosion resistance of high strength bolts.

References

[1] National quality and technical supervision GB/T 3077-1999. 2000 Alloy structural steel Beijing: China Standard Press.

[2] China National Standardization Management Committee GB/T 1231-2006. 2006 Specifications of high strength bolts with large hexagon head, large hexagon nuts, plain washers for steel structures Beijing: China Standard Press.

[3] Ministry of Construction of the People's Republic of China JGJ82-91. 1992 Specification for design, construction and acceptance of high strength bolt connection of steel structure Beijing: Publisher unknown.

[4] Zhang J 2011 Fracture Analysis of 20MnTiB High Strength Bolts *Heat Treatment Technology and Equipment* **5** 90-92.

[5] Wu J, Zhao K, Shen C 2015 Analysis on Fracture Reason of High Strength Bolt Made by 40ACR Steel *Physical Testing and Chemical Analysis (Part A:Physical Testing)* **3** 13-17.

[6] Song Z, Zheng H 2012 Research Status of Brittle Fracture of High Strength Steel Bolt *Research & Application of Building Materials* **1** 15-19.

[7] Niu X, Zhang R, Xu X 2012 Analysis on Fracture Reasons for 40Cr High-strength Bolt *Hot Working Technology* **10** 25-30.

IOP Conf. Series: Materials Science and Engineering **220** (2017) 012006 doi:10.1088/1757-899X/220/1/012006

Research on wire rope deformation distribution of WR-CVT

Wu Zhang[1*], Wei Guo[1], Chuanwei Zhang[1], Zhengxiong Lu[1], Xiaobin Xu[1]

[1]School of Mechanical Engineering, Xi'an University of Science and Technology, Xi'an, Shaanxi, 710054, China

*E-mail: zhangwu0828@163.com

Abstract. A wire rope continuously variable transmissions (WR-CVT) has been introduced in the paper, in view of its less research, this paper mainly studied the deformation distribution of 6×7+IWS bending wire rope. The results shown that in the same section, half of the side strands are in a stretched state and half are in a compressed state. When the transmission ratio i=2.35, the maximum deformation and the minimum deformation are decrease when section U1 to U2, U3 transition. Wire deformation distribution when the transmission ratio i=0.42 is similar to that of i=0.2.35. Wire deformation amount and the deformation difference decrease as the transmission ratio decreases, this shows that the increase in the bending radius of the wire will make the wire deformation more uniform, and the reduction of the deformation difference will also reduce the wear. This study provides a basis for the study of fatigue and wears failure of WR-CVT components.

1. Introduction

CVT(Continuously Variable Transmission) as one of the trend of auto technological development, which has the advantages of saving fuel, easier operation, comfortable drive and producing less pollutions[1]. In a comparison of the fuel consumptions of vehicles with different transmission types on the US market for the US Environmental Protection Agency city cycle[2], it was seen that, generally, vehicles with CVT achieve a higher fuel economy than vehicles with stepped AT. Páczelt, I [3] developed a finite element program that takes into account factors such as nonlinear contact, Poisson effect, friction and wear, which can perform small displacement and deformation analysis of wire ropes in a variety of loads including stretching, torsion and bending. Ma Jun's [4] research results show that the stress and strain of the wire are higher than that of the core wire. The stress of the steel wire in the different position is spirally distributed in the longitudinal direction. Wu juan's [5] research results show that the right by the same twist wire rope steel wire stress, deformation and wire The magnitude of the deformation difference is greater than the right cross twist wire rope.Wire rope CVT (WR-CVT) has been designed in this paper, WR-CVT as shown in Figure 1 and focus on the WR-CVT closed wire rope deformation distribution.

2. Closed wire rope geometry model

The wire rope model has been designed in this paper. 6×7+IWS wire rope contains 1 core strand and 6 side strands, one strand contains a center wire and six side wires. It contains four different geometric forms of wire, they are core strand center wire, core strand side wire, outer strand center wire and outer strand side wire. Core strand center wire is the axis of closed wire rope, core strand side wire and outer strand center wire are single helix, and outer strand side wire is second helix. According to the right-hand rules to establish a Cartesian coordinate system O-xyz as a closed wire rope mathematical model of the global coordinate system[6,7], the centerline equation for each wire in a closed wire rope is given[8-10]. The 3D solid model of 6×7+IWS bending wire rope was established by Pro/E. 6×7+IWS wire rope section and 3D solid model is shown in Figure. 2.

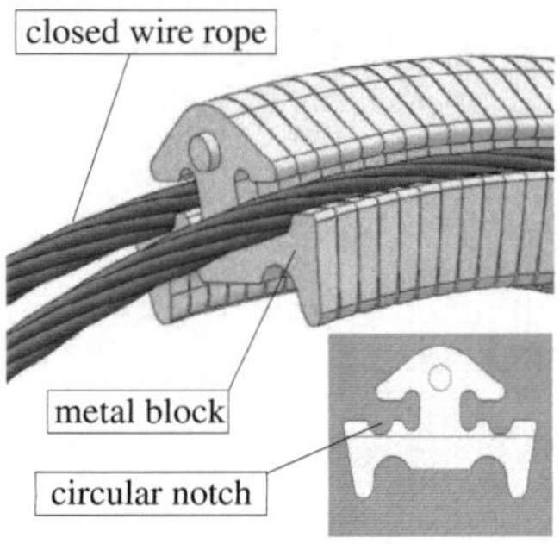

Figure 1. WR-CVT.

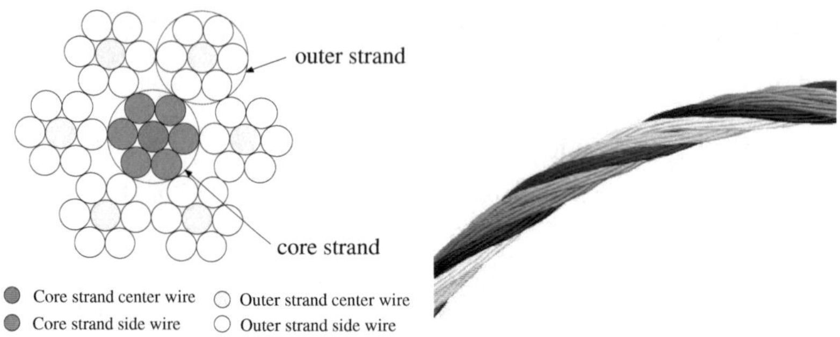

Figure 2. 6×7+IWS wire rope section and 3D solid model.

3. Wire deformation analysis

In order to better observe the change of the wire deformation in the wire along the direction of the core, the local coordinate system along the axis of the cord is established in the bending part of the rope ring. And intercept three sections U_1, U_2, U_3 and U_1', U_2', U_3', Section *O-O* and *O-O'* are the reference plane, as shown in Figure.3. In the finite element model of the wire rope, one end applies the surface constraint, restrains all degrees of freedom in three directions, *x*, *y*, *z*, and applies the surface load at the other end. Wire number and path schematic as shown in Figure.4.

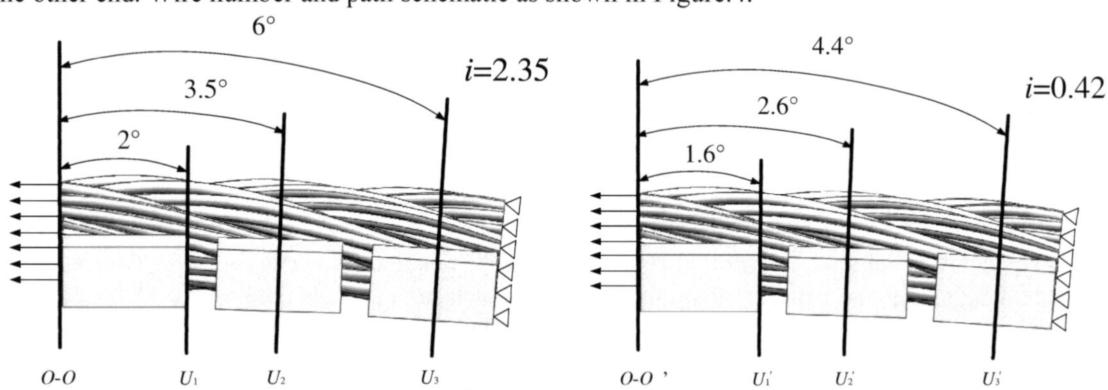

Figure. 3. Section position

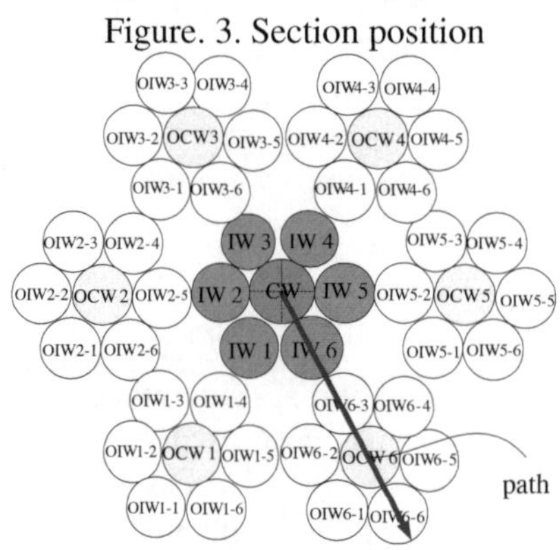

Figure 4. Wire number and path schematic.

When the transmission ratio i=2.35, the deformation of the wire is shown in Figure 5. It can be seen that the cross section of the datum plane is the loading end of the wire rope, and the deformation of the wire closer to the loading end is larger.

In the same section, half of the side strands are in a stretched state and half are in a compressed state. And the amount of deformation of the core strand center wire is substantially the same for the strand 1, 6 and the strand 3, 4. However, the strand 3, 4 deformation are larger than strand 1, 6 which the strand 3, 4 are far greater than the strand 1, 6 of the circular notch.

In section U_1, OIW4-4 wire deformation is the maximum, its value is 0.135mm, OIW1-2 wire deformation is the minimum, its value is 0.04mm, and deformation difference is 0.095mm. In section U_1, OIW3-3 wire deformation is the maximum, its value is 0.07mm, OIW6-5 wire deformation is the minimum, its value is 0.018mm, and deformation difference is 0.052mm. In section U_3, OIW3-4 wire deformation is the maximum, its value is 0.028mm, OIW6-3 wire deformation is the minimum, its value is 0.011mm, and deformation difference is 0.017mm. Deformation difference is the relative misalignment of the two wires, which will cause the micro-wear of the inner wire, in the load conditions unchanged, the greater the amount of wrong, the more serious wear and tear.

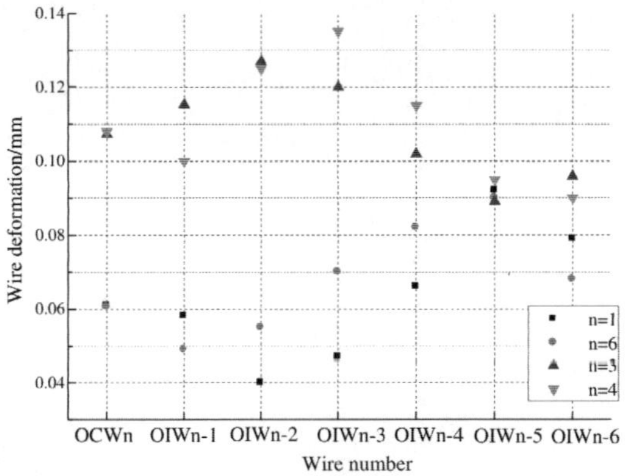

(a) Section U_1

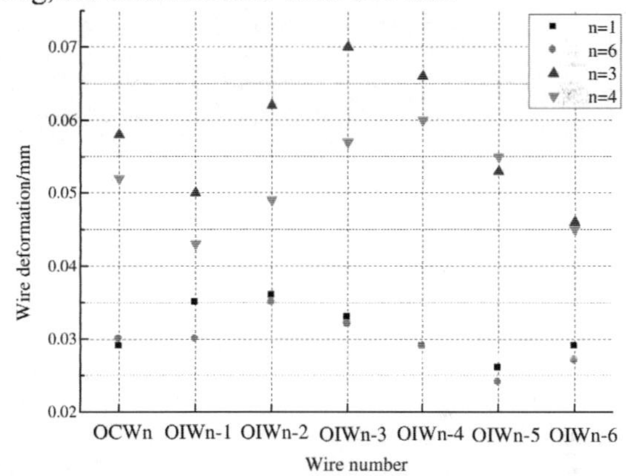

(b) Section U_2

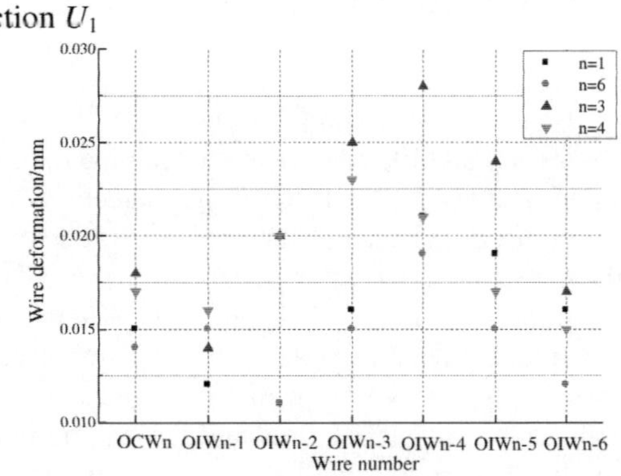

(c) Section U_3

Figure 5. Wire section deformation (i=2.35).

When the transmission ratio i=0.42, the deformation of the wire is shown in Figure 6. It can be seen that on the whole distribution of wire deformation is similar to that of i=0.2.35, the amount of wire deformation decreases as the transmission ratio decreases, and the difference between the maximum deformation amount and the minimum deformation amount of the wire is also reduced. This shows that the increase in the bending radius of the wire will make the wire deformation more uniform, and the reduction of the deformation difference will also reduce the wear.

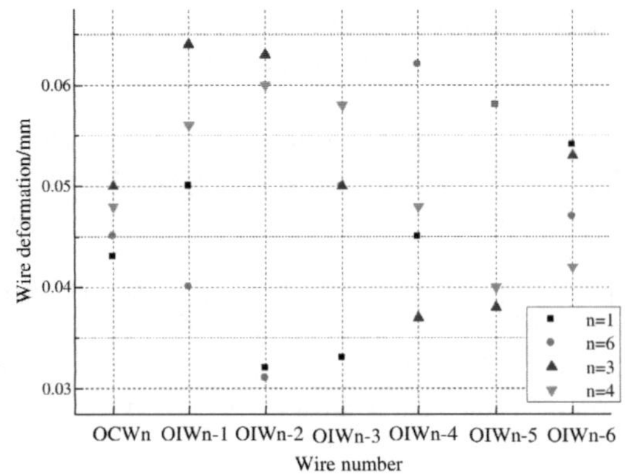

(*a*) Section U_1'

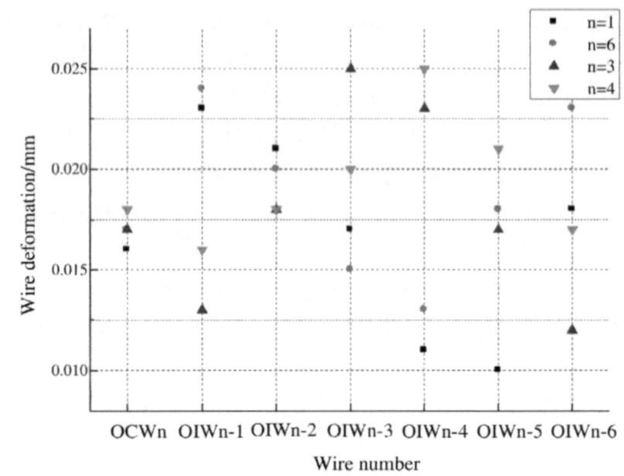

(*b*) Section U_2'

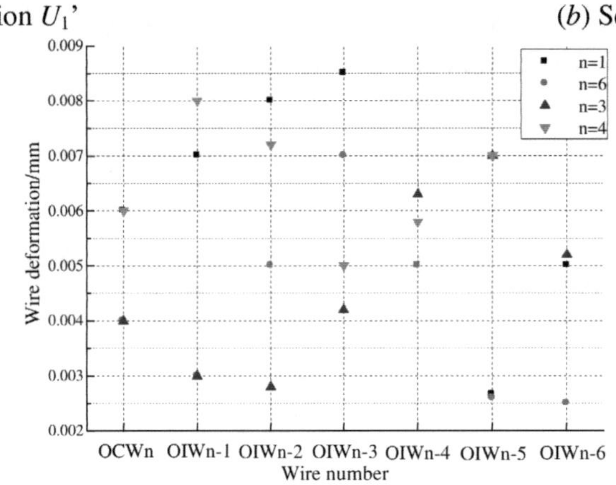

(*c*) Section U_3'

Figure 6. Wire section deformation (i=0.42).

4. Conclusion

The wire rope deformation distribution of WR-CVT has been studied in this paper. The 6×7+IWS wire rope geometrical model has been established. The deformation distribution of the wire rope on the three sections is analyzed. The results shown that in the same section, half of the side strands are in a stretched state and half are in a compressed state. When the transmission ratio i=2.35, in section U_1, OIW4-4 wire deformation is the maximum, its value is 0.135mm, OIW1-2 wire deformation is the minimum, its value is 0.04mm, and deformation difference is 0.095mm. The deformation difference between the maximum and the minimum deformation decrease when section U_1 to U_2, U_3 transition. When the transmission ratio i=0.42, wire deformation distribution is similar to that of i=0.2.35, wire deformation amount and the deformation difference decreases as the transmission ratio decreases, this shows that the increase in the bending radius of the wire will make the wire deformation more uniform, and the reduction of the deformation difference will also reduce the wear.

Acknowledgement

The research presented here was supported by the National Natural Science Foundation of China (Grant No.51505373). The authors are grateful for the support.

Nomenclature

CVT	continuously variable transmission
WR-CVT	wire rope CVT

i	transmission ratio
DOF	Degrees-of-freedom
$(O: x, y, z)$	the Cartesian coordinate system
O-O, U_1, U_2, U_3 O-O', U_1', U_2', U_3'	section

References

[1] Srivastava N, Haque I 2008 Clearance and friction-induced dynamics of chain CVT drives *Multibody System Dynamics* **19(3)** 255-280.

[2] Sluis F V D, Velde A V D, Dongen T V, Spijk G J V, Heeswijk A V 2007 Efficiency optimization of the push belt CVT SAE paper 2007-01-1457.

[3] Páczelt I, Beleznai R 2011 Nonlinear contact-theory for analysis of wire rope strand using high-order approximation in the FEM *Computers and Structures* **89(11-12)** 1004-1025.

[4] Ma J, Ge S R, Zhang D K 2012 Research on Three-dimensional Contact Model and Stress Distribution between Wires of Steel Strand *China Mechanical Engineering* **23(7)** 864-868.

[5] Wu J, Kou Z, Liu Y H, Wu G X 2014 Distribution of equivalent stresses and deformations for 6 strand Warrington seale rope with an independent wire rope core *Journal of China Coal Society* **39(11)** 2340-2347.

[6] Lee W K 1991 An insight into wire rope geometry *International Journal of Solids & Structures* **28(4)** 471-490.

[7] Erdönmez C 2013 n-tuple complex helical geometry modeling using parametric equations *Engineering with Computers* **30(4)** 715-726.

[8] Stanova E, Fedorko G, Fabian M, Kmet S 2011 Computer modelling of wire strands and ropes Part I: Theory and computer implementation *Advances in Engineering Software* **42(6)** 305-315.

[9] Stanova E, Fedorko G, Fabian M, Kmet S 2011 Computer modelling of wire strands and ropes Part II: Finite element based applications *Advances in Engineering Software* **42(6)** 322-331.

[10] Guo W, Lu Z X, Zhang W 2015 Geometric modeling theory of bent wire rope based on Pro/E *China Mechanical Engineering* **26(17)** 2363-2368.

IOP Conf. Series: Materials Science and Engineering **220** (2017) 012007 doi:10.1088/1757-899X/220/1/012007

Synthesis and Characterization of High Energy Sheet Materials Based on HMX / RDX and Hydroxyl Terminated Polybutadiene

Karim Elsharkawy[1]*, Lin Guo[1], Elhussein Taha[2] and Hany Fouda[1]

[1]Beijing University of Aeronautics and Astronautics, Beijing, China
[2]Military Technical College, Cairo, Egypt

*E-mail: karimelsharkawy@buaa.edu.cn

Abstract. In this paper three types of thin sheets of highly energetic materials were prepared and characterized. The first based on 1,3,5,7-tetranitro-1,3,5,7-tetrazocane (HMX). The second type based on 1,3,5-trinitro-1,3,5-triazinane (RDX). Both types contain polyurethane (PU), formulated by hydroxyl terminated polybutadiene (HTPB) and Isophorondiisocyanate (IPDI). The third type based on (RDX) and polyisoprene (PI) as high elastomeric material. The first and second types of thin sheets were prepared by applying the casting technique while the third type was prepared by slurry technique then followed by rolling of the prepared beads of the RDX coated by PI. These high energy sheet materials were cured in oven at 60°C. The measured explosive properties of the prepared sheets were discussed and showed that the sensitivity to impact and friction of the prepared sheets explosives materials were markedly decreased when compared to pure HMX or pure RDX, but the sensitivity to heat was close to that of pure RDX. In spite of the markedly decrease in the sensitivity of these sheets, the explosive characteristics were nearly not affected the sheets have very good stress-strain values.

1. Introduction

Sheet explosive materials are plastic bonded explosives (PBXs) in sheet form which comprised of flexible materials including of explosives/energetic materials which are distributed in a polymer matrix based on polyurethane or highly flexible elastomeric binder as polyisoprene [1-4]. Thin sheets of high energetic materials (HEMs) afford the high energy to achieve the goals of the system [5].

The flexibility of these formulations is utilized in wide important fields (civil and military applications) as in explosive reactive armors (ERA). ERA consisting of sandwiched sheet explosive between metal casings offers effective protection to the armored tanks by defeating kinetic energy projectiles and the shaped charge warheads [6-9]. Many researches were focused on the synthesis, characterization and applications of PBXs including sheet explosive materials to accomplish high mechanical, physical properties and lower sensitivity without losing explosive properties of these formulations [10-16].

In the present investigation RDX and HMX were selected to be the explosive filler. Polyurethane was selected as the first binder type due to its flexibility wide range which depends mainly on the type and content of DOZ, IPDI, filler and HTPB. Polyisoprene PI was selected as the second high flexible binder because it, either natural or synthetic, forms after mastication process and sulfur vulcanization, the model for the ideal elastomeric properties including rapid extensibility to great elongation, high stiffness and strength when stretched and also rapid and complete retraction on release of external stress [17].

2. Experimental work

2.1. Materials
In this research the entire chemical used are of high purity materials as; filler (HMX, RDX), binder (PI, PU based on HTPB), crosslinking agent (IPDI), plasticizer (DOZ), Bonding agent (MAPO), solvent (dichloroethane (DCE)) and solid lubricant (Teflon (PTFE)).

2.2. Preparation of PBX sheet based on HMX or RDX and PU (HTPB)
The calculated amounts of explosive, HTPB, IPDI, RDX, DOZ and MAPO were accurately weighed. These PBXs sheets based on PU were prepared by using the casting technique [18]. HTPB was then placed in a mixing kettle with two third of DOZ weight and mixed together. HMX or RDX was then added in small portions during mixing for 30 minutes .Thoroughly mixing was conducted for 10 minutes after complete addition of HMX or RDX to ensure complete coating. The residual amount of IPDI, DOZ and MAPO calculated, on basis of NCO/OH= 1.1, were added to the mixture and mixing continued for about 15 minutes. The prepared matrix (paste) was then pressed into a mold of Teflon and cured for 10 days at about 55-60°C. The obtained sheet was 6 mm thickness and its final composition by weight percentage was 88% HMX or RDX, 9.2% PU and 2.8% DOZ.

2.3. Preparation of PBX sheet based on RDX and PI
That PBX sheet based on PDX and PI was prepared by using the slurry technique followed by rolling and curing processes [19]. The first step is to dissolve the PI completely in DCE and then RDX was slowly added to the laquer. After 30 minutes about 500(ml), more than the volume of DCE, of distilled water was poured drop-wise and continue stirring for 30 minutes. Then increase the temperature to evaporate DCE, after that the formed beads were filtered off, washed with water and dried at 70°C. Teflon as a solid lubricant, DOZ as plasticizer and small amount of distilled water as wetting agent were added and thoroughly mixed with the RDX beads coated with PI to be ready for the rolling process. Finally using two heavy duties calendars and put the explosive beads between them to produce a sheet of thickness 4 mm and the rolling process was repeated 10 times at 25°C. We used Sulfur as vulcanizing agent in the first run which was added with some minor additives as activator and accelerator. Finally the produced sheet was cured at 55-60°C for about 48 hours.

2.4. Evaluation of PBXs sheets formulations

2.4.1. Sensitivity tests. The sensitivity to heat was obtained by measuring the ignition temperature for the prepared PBXs sheets using Chilworth deflagration test apparatus [20]. To determine the ignition temperature, the temperature was uniformly increased [5 °C/min] until the explosion conversion occurred.Sensitivity to impact was determined by using IKaMaschinenbau apparatus, using 5 Kg falling weight [21]. Sensitivity to friction was determined using BAM test apparatus. The test was conducted by the percentage of initiation by changing the loading of the pistil [20].

2.4.2. Detonation velocity. The Exploment -Fo-2000- Multi Channel test apparatus were used to determine the detonation velocity. For the sheet explosives formulations based on PU binder; the detonation velocities were obtained by measuring the time interval for a detonation wave to travel a known distance between the two fiber optic probes in microseconds which was displayed with the calculated detonation velocity in (m/s). The prepared PBX samples were pressed in a PVC mould tubes of 190 (mm) length and 22 (mm) inside diameter at a density 1.6 (g/cm^3).For the sheet explosive based on PI binder, we used compressed pure PDX as a booster and three sheets of the prepared PBXs sheets formulations to form a sheet of 12 (mm) thickness.

IOP Conf. Series: Materials Science and Engineering **220** (2017) 012007 doi:10.1088/1757-899X/220/1/012007

2.4.3. Stress-strain characteristics. The universal testing machine WDW-100 which supplied by Fanguri technology company, China was used to measure the stress-strain characteristics of the PBXs sheet samples before and after curing.

3. Results and discussion

The results of sensitivity tests and detonation velocity test of all the prepared PBXs sheets formulations were compared to pure HMX and pure RDX as shown in table 1.

Table 1. Results of different explosive characteristics for the prepared sheet explosives based on PU and PI compared with pure HMX and pure RDX

Explosive Characteristics	Pure HMX	Pure RDX	Sheet Explosive Based on HMX / PU	Sheet Explosive Based on RDX / PU	Sheet Explosive Based on RDX / PI
Sensitivity to Impact (J)	7.4	7.5	14.1	14.5	45
Sensitivity to friction (N)	120	120	>360	>360	324
Sensitivity to heat (Ignition temp. ^{o}C)	287	222	252	209	235
Detonation velocity (m/sec)	9100	8750	7689	7180	8100

It is clear from table 1 that the sensitivity to impact was markedly decreased when the RDX crystals were coated by the polyisoprene thin layers. The sensitivity to impact of the sheet explosive based on RDX/PI was decreased by 600% when compared with that of pure RDX. For the sheet explosives formulations based on HMX and RDX was show also much less sensitivity than that of pure HMX and RDX.

The sensitivity to friction of the prepared sheets of explosives based on RDX/ PU and HMX/PU showed no initiation even when applying the maximum friction force (360 N), while for the sheet explosive based on PI was significantly less than pure HMX and pure RDX.

The sensitivity to heat expressed by the values of the ignition temperature of the prepared sheets explosives when compared with that of pure RDX are found to be close to each other.

From the obtained sensitivity results of the prepared sheets of explosives we can conclude that the prepared sheets of explosives satisfy low vulnerability explosives which are highly safe during storage, rolling and applications.

The obtained results of the detonation velocity for the prepared sheets of explosives which represent the explosive performance were slightly decreased than the value obtained for pure HMX and pure RDX by a slight decrease. The slight decrease in the value of the detonation velocity of the prepared sheets of explosives means that in spite of the markedly decrease in the sensitivity of these sheets, the explosive characteristics were nearly not affected.

The stress-strain characteristics results are shown in table 2. The results emphasize the high mechanical properties, ductility and flexibility after curing. It is clear from the table the values of the stress and strain were greatly improved for the sheet explosive depends on PI after curing and the obtained sheet has excellent flexibility and ductility properties. The stress-strain values of the sheets explosives depend on HMX or RDX with PU binder are also accepted.

Table 2. Stress Strain Characteristics of the Prepared Sheets Explosives

Sheet explosive	Characteristics	Before Curing	After Curing
Sheet Explosive Based on HMX / PU	Maximum Stress (N/cm^2)	--	5.8
	Maximum Strain (%)	--	8.8
Sheet Explosive Based on RDX / PU	Maximum Stress (N/cm^2)	--	5.5
	Maximum Strain (%)	--	8.3
Sheet Explosive Based on RDX / PI	Maximum Stress (N/cm^2)	5.87	8.58
	Maximum Strain (%)	43	144

4. Conclusion

In the present study, three thin sheets of explosives of highly energetic materials based on RDX / polyisoprene or polyurethane as polymeric binders and HMX or RDX as high energetic explosive were successfully prepared and showed high ductility and excellent flexibility properties after curing. The stress-strain values of the sheets explosives depend on HMX or RDX with PU binder are also accepted. In the same time these sheets have high energetic properties which are close to that of pure RDX, detonation velocity of 8100(m/s) for the sheet explosive based on PI, also the sheet explosive based on HMX and PU showed high detonation velocity of 7689 (m/s)which is greater than that based on RDX and PU by 7.1%. The sensitivity to heat was also decrease but with a less extend than that of pure RDX and are found to be close to each other, while the sensitivity to impact and friction of the prepared sheets were markedly decreased when compared with that of pure HMX and pure RDX which decrease the vulnerability so these sheets are highly safe during storage, rolling and applications.

References

[1] Nouguez B, Mahe B and Vignaud P 2009 Cast PBX related technologies for IM shells and warheads *Sci. & Tech. Energetic. Materials* **70(5)** 135-139.

[2] Suresh K, Mahadev B and Mrityunjay K 2016 Experimental studies on advanced sheet explosive formulations based on 2,4,6,8,10,12-Hexanitro- 2,4,6,8,10,12-hexaazaisowurtzitane (CL-20) and Hydroxyl Terminated Polybutadiene (HTPB), and comparison with a RDX-based system *Central European Journal of Energetic Materials* **13(1)** 135-147.

[3] Ekhlas A 2002 Preparation and characterization of explosive sheets used in reactive armours M.Sc. Thesis MTC, Cairo, Egypt.

[4] Brousseau P, Thiboutot S and Diaz E 2001 New melt-cast explosives based on energetic thermoplastic elastomers *32nd International Annual Conference of ICT* Karlsruhe, Germany.

[5] Suresh K, Mrityunjay K and Vasant J 2016 Experimental studies on a high energy sheet explosive based on RDX and Bis (2,2-dinitropropyl) Formal/Acetal (BDNPF/A) *Central European Journal of Energetic Materials* **13(3)** 557-566.
[6] Yadav H, Bohra B M, Joshi G D, Sundaram S G, Kamat P V 1995 Study on basic mechanism of reactive armour *Def. Sci. Journal* **45(3)** 207-212.
[7] Held M 2001 Disturbance of shaped charge jets by bulging armour *Propellants Explos. Pyrotech.* **26** 191-195.
[8] Warford J M 1988 Reactive armour-new life for soviet tanks *Armor* **97(1)** 6.
[9] Yadav H 1988 Flyer plate motion by thin sheet of explosive *Propellants Explos. Pyrotech.* **13** 17-20.
[10] Radwan M, Mostafa H and Elsharkawy K 2006 Preparation and characterization of thin sheets of energetic materials based on polyisoprene *3rd International Conference on Chemical & Environmental Engineering* Cairo, Egypt 449-455.
[11] Radwan M 2001 Sensitivity and brisance of PBXs based on HMX and fluorel *32nd International Annual Conference of ICT* Karlsruhe, Germany.
[12] Sjoberg P, Hytti H, Strandberg R, Kariniemi A and Muilu M 2002 PBX with insensitive RDX, a new EIDS. substance for under water use *33rd International Annual Conference of ICT* Karlsruhe,Germany.
[13] Kaye S 1978 Encyclopedia of explosives and related items Part 2700 **8**.
[14] Radwan M 2001 Explosive characteristics of aluminized PBXs based on HMX and polyurethane binder *32nd International Annual Conference of ICT* Karlsruhe, Germany.
[15] Gerber P, Hubner C and Menke K 2000 Characterization of the Matrix filler properties of the GAP/RDX propellant system *31st International Annual Conference of ICT* Karlsruhe, Germany.
[16] Bunyan P, Cunnliffe B, Torry A and Bult H 1997 Stability studies on end modifications poly GLYN *Insensitive Munitions and Energetic Materials Technology Symposium* Orlando.
[17] Brandrup J and Immergut E H 1989 *Polymer Handbook* Third edition, John Wiley & Sons, New York.
[18] Talawar M B, Jangid S K, Nath T, Sinha R K, Asthana S N 2015 New directions in the science and technology of advanced sheet explosive formulations and the key energetic materials used in the processing of sheet explosives: Emerging trends *Central European Journal of Energetic Materials* **300** 509-528.
[19] Kasprzyk D J, Bell D A, Flesner R L and Larson S A 1999 Characterization of a slurry process used to make a plastic bonded explosive *Propellants, Explosives, Pyrotechnics* **24** 333-338.
[20] Suceska M 1995 *Test methods for explosives* Springer, ISBN 0-387-94555-5.
[21] United Nations 1999 *Recommendations on the transportation of dangerous goods, manual of tests and criteria* Third edition, United Nations, New York and Geneva.

IOP Conf. Series: Materials Science and Engineering **220** (2017) 012008 doi:10.1088/1757-899X/220/1/012008

Influence of Annealing Temperature on the Structure and Performance of YBCO Thin Film on MgO Substrate

L Guo[1*], L Wang[1], and S Cao[1]

[1]Science and Technology on Vacuum Technology and Physics Laboratory, Lanzhou Institute of Physics, P. O. Box 94, Lanzhou, Gansu, China

*E-mail: gl_510@163.com

Abstract. Annealing treatment is adopted for studying the changes of structure and low temperature superconducting performance of YBCO thin film on MgO Substrate. X-ray diffraction (XRD) analysis results show that c axis orientation of YBCO thin film is optimized with the rise of annealing temperature. It reaches the optimum value at 720 ℃. The c axis orientation of YBCO thin film shows recession when the annealing temperature is higher than 720 ℃. The change trend of increasing firstly and decreasing subsequently is also discovered with annealing temperature, the optimal critical current density value is obtained at about 730 ℃. Research results show that the YBCO thin film structure and low temperature superconducting performance can be effectively improved through optimizing annealing temperature.

1. Introduction

YBCO high temperature superconducting thin film, as a superconducting material, has been a hot topic of research in the field of electronic materials. Its application has pushed the emergence of high temperature superconductor research upsurge within the world scope. Though the superconducting mechanism of YBCO high temperature superconducting thin film is not completely mastered by people from the perspective of material science, its research progress hereof in practical engineering application is not affected, and high temperature superconducting technology is rapidly developed into a technology and industry[1-3].

Because YBCO high temperature superconducting thin film has extremely low microwave surface resistance, the insulation substrate used for thin film growth belongs to monocrystal (100), which is consistent with the selection of materials during production of microwave passive components. Application of microwave passive components (such as filters, resonators, delay line, etc.) made of high temperature superconducting thin films in high sensitivity receiver front end and satellite communication ground receiving station, etc. is one of main application directions of current high temperature superconducting technology[4-6]. However, development of high-quality YBCO epitaxial thin films on low loss substrates (such as sapphire, $LaAlO_3$ and MgO, etc.) is the key of the devices[7-8]. Therefore, the selection of the substrate is very important aiming at preparation of high-temperature superconducting thin films. MgO has a cubic crystal structure. Its lattice constant is a=4.21Å. Despite it has large lattice mismatch with YBCO, dielectric constants of MgO are isotropic, it has very low dielectric loss, which is very suitable for application of microwave devices. YBCO thin film prepared by MgO as the substrate can be used for preparing microwave passive devices with excellent performance. MgO is adopted as the substrate for settling YBCO thin films by pulsed laser deposition, magnetron sputtering, thermal evaporation and E-beam evaporation, which has been reported. However, The YBCO thin film structure and performance with MgO thin film as substrate are lower than YBCO thin films prepared with other materials (such as $LaAlO_3$, etc.) as substrate due to lattice mismatch. The research shows that annealing is a fairly effective method to optimize YBCO structure and performance. In the paper, annealing method is adopted to study the influence of annealing temperature on structure and performance of YBCO thin films with MgO as the substrate.

2. Experiment

Self-developed high vacuum tube annealing furnace is adopted for carrying out annealing researching on YBCO thin films at the MgO substrate. The vacuum degree of the annealing furnace is up to 5×10^{-4} Pa. YBCO thin films are deposited by pulse laser deposition equipment. The applied substrate belongs to a (001) crystal orientation monocrystal MgO substrate. The dimension is 10mm×10mm. The thickness is 0.5mm. The selected target purity is higher than 99.95%. When PLD is used for plating YBCO thin films, MgO substrate temperature is 600 ℃, oxygen pressure is stabilized at ~2.5mTorr. The deposited thin film undergoes high temperature annealing treatment. The annealing condition is divided into oxygen atmosphere annealing. The oxygen pressure is 1Torr, the annealing temperature is 650 ℃ ~ 800 ℃, and the annealing time is 3h.

Philip X'PERT MPDX is adopted for XRD scanning. Thin film phase composition and thin film crystallographic orientation are analyzed. J-sacn Leipzig system is adopted for testing critical current density and the test condition is 77K.

3. Results and discussion

3.1. Influence of annealing temperature on YBCO thin film structure

We fix annealing pressure and annealing time process conditions, YBCO thin films are annealed under different temperature, and concrete process conditions are shown in table 1 in order to study the influence of annealing temperature on YBCO thin film structure at the MgO substrate, and explore appropriate annealing temperature parameters.

Table 1. Annealing process parameters

Sample number	Annealing temperature	Annealing atmosphere	Annealing time
S1	650℃	1Torr	3h
S2	700℃		
S3	750℃		
S4	800℃		

Figure 1 shows θ-2θ scanning spectra of growing YBCO thin films under different temperatures. When the annealing temperature is reduced to 650 ℃, YBCO thin film grows at a-axis orientation. Micro small peak (005) (figure 1a) is available at 38 °. YBCO thin films are gradually changed from a-axis to c-axis orientation with the increase of temperature. The a-axis diffraction peak intensity is weaker and weaker, c-axis diffraction peak intensity is higher and higher, and c-axis preferred orientation is the best at 750 ℃. The substrate temperature is increased continuously. When the temperature is increased to 800 ℃, other orientation diffraction peak of YBCO thin film, non-YBCO diffraction peak and other impurity peaks appear in the thin film. The thin film may suffer from decomposition (figure1d).

The c-axis orientation degree in the thin film is reflected by the size of FWHM. FWHM is smaller, c-axis orientation degree is higher, and the volume content occupied by grains in c-axis tilt direction. We conducted XRD ω scanning analysis on YBCO thin films (005) with higher c-axis preferred orientation. The scanning spectra is shown in figure 2. The FWHM of sample S1(650℃) and sample

S3(750℃) is similar, which is about 0.9 °(figure 2a,c). However, it is much larger than the FWHM of the sample S2(700℃)(figure 2b). It is obvious that the annealing temperature has greater influence on YBCO thin film surface external orientation. When the substrate temperature is 750 ℃, although c-axis preferred orientation degree is high, a-axis grain is less, the volume content occupied by grains of the c-axis tilt orientation is still larger, and the FWHM is larger. FWHM is increased after 750 ℃. It is obvious that more and more grains grow away from c-axis. The FWHM of the sample S4(800℃) reaches 1.29°(figure 2d). It further illustrates that too high temperature is very unbeneficial for thin film surface external orientation. The orientation degree of the thin film is the best at 700℃, and the FWHM is only 0.67 °.

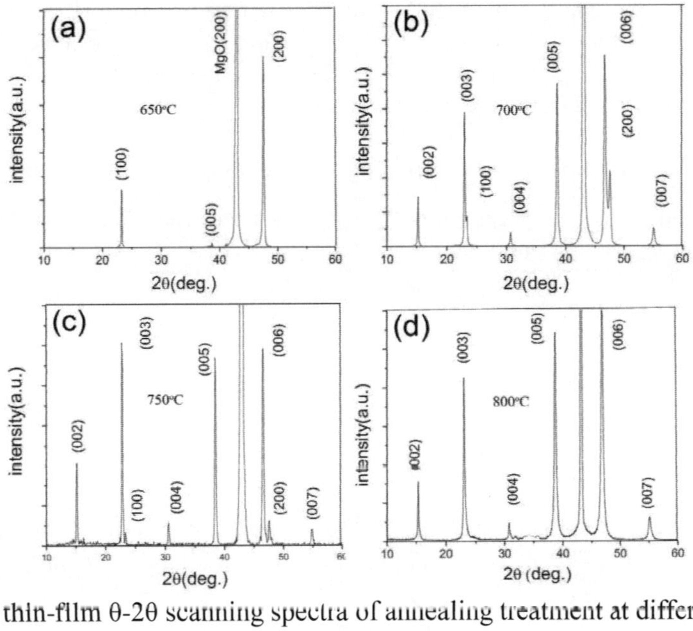

Figure 1. YBCO thin-film θ-2θ scanning spectra of annealing treatment at different temperatures.

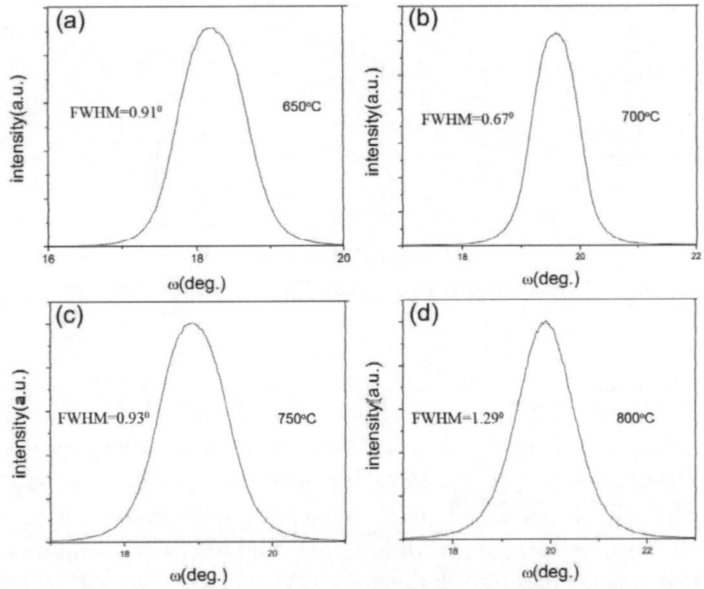

Figure 2. YBCO thin-film ω scanning spectra of annealing treatment at different temperatures.

3.2. The influence of annealing temperature on YBCO thin film superconducting performance

S1 to S4 samples respectively undergo critical current density Jc tests. The influence of annealing on YBCO thin film performance is discussed. Main results are shown in table 2. When annealing temperature is 650 ℃, Jc is 1.55 mA/cm^2. Jc is gradually increased with the increase of the annealing temperature. When the annealing temperature is 700 ℃, Jc is 2.63 mA/cm^2. When the annealing temperature is increased to 750 ℃, Jc reaches 2.93 mA/cm^2. The annealing temperature is continuous increased, Jc shows a trend of decrease, Jc is only 1.43mA/cm^2 at 800 ℃. The change curve is shown in figure 3. The optimum annealing temperature of YBCO thin film on the MgO substrate is about 730℃ by fitting curve. Jc size of YBCO thin films is changed with annealing temperature mainly because annealing leads to changes of YBCO thin film lattice orientation. Jc size change with the annealing temperature is basically consistent with the change of c-axis orientation grain proportion with the annealing temperature. It further proves that grain c axis orientation is the key for YBCO thin film to get excellent low temperature superconducting performance.

Table 2. YBCO thin film critical current density under different annealing temperatures

Sample number	Annealing temperature	Critical current density （Jc, mA/cm^2）
S1	650℃	1.55
S2	700℃	2.63
S3	750℃	2.92
S4	800℃	1.43

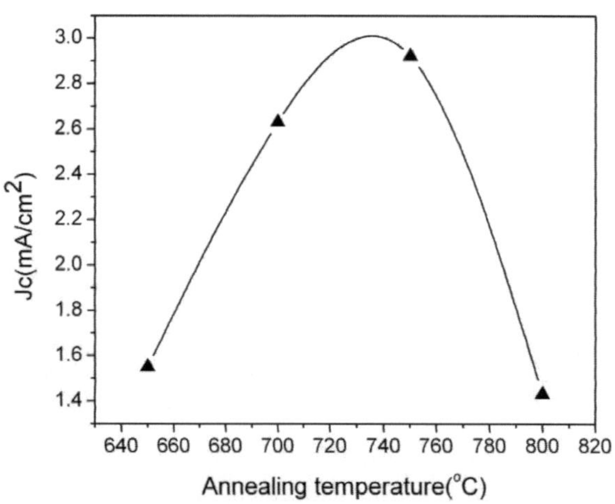

Figure 3. Change curve of YBCO thin film critical current density with annealing temperature.

4. Conclusion

Annealing treatment is adopted for studying the changes of structure and low temperature superconducting performance of YBCO thin film at MgO substrate under different annealing temperatures. X-ray diffraction (XRD) analysis results show that c axis orientation of YBCO thin film is optimized with the rise of annealing temperature. It reaches the optimum value at 720 ℃. The c-axis orientation of YBCO thin film shows recession when the annealing temperature is higher than 720 ℃. Similar change laws are displayed in the test of low temperature critical current density. The change trend of increasing firstly and decreasing subsequently is also discovered with annealing temperature, the optimal critical current density value is obtained at about 730 ℃. Research results show that the annealing temperature is one more sensitive physical quantity. The YBCO thin film structure and low temperature superconducting performance can be effectively improved through optimizing annealing temperature.

References

[1] Avenhaus B, Porch A, Laneaster M J, et al. 1995 *IEEE Transactions on Applied Superconductivity* **2** 1737.
[2] Chen M, Donzel L, Lakner M, et al. 2004 *Journal of the European ceramic society* **24** 1815.
[3] Bagarinao K D, Yamsaki H, Nie J C, et al. 2005 *IEEE Transactions on Applied Supereonductivity* **15** 2962.
[4] Bindi M, Botarelli A, Gauzzi A, et al. 2004 *Superconductor Science and Technology* **17** 512.
[5] Kim C J, Jun B H, Kim H J, et al. 2005 *Physica C：Superconductivity and Its Applications* **426** 915.
[6] Ohshima S, Shirakawa M, Kitamura K, et al. 2004 *Chinese Journal of Physics* **42** 425.
[7] Wang Y S, Guan X J, Zhang H Y, et al. 2010 *Science China, Technological Sciences* **53** 2239.
[8] Wang L H, Shu Y H, Fan J 2012 *Science China, Technological Sciences* **55** 1.

IOP Conf. Series: Materials Science and Engineering **220** (2017) 012009 doi:10.1088/1757-899X/220/1/012009

A new modification method of a Cetyl Trimethyl Ammonium Bromide/Nano-ZnO and Multi-walled Carbon Nanotubes Electrode for Determination of Anti Doping in Urine

Zhiwei ZHAO[1], Shuai MU[1], Jie ZHENG[1], Lingyan GU[1], Guijun SHEN[1] and Yuan SHEN[2*]

[1]College of Environmental and Chemical Engineering, Dalian University, Dalian 116622, China
[2]Talent Development Office, Dalian University, Dalian 116622, China

*E-mail: sgj0501@163.com

Abstract. The preparation and application of Cetyl Trimethyl Ammonium Bromide/Nano-ZnO and Multi-walled Carbon Nanotubes (CTAB/ ZnO-MWNTs) Modified Electrodes was studied, establishing a new electrochemical method for determination of carteolol hydrochloride in urine. After its pre-enrichment by adsorption and extraction on modified electrodes, electrochemical behaviors of carteolol hydrochloride on the modified electrodes were studied by CV and DPV. The response is linear at the range of $1\times10^{-3} \sim 2\times10^{-1}$ g/L, with a detection limit of 2×10^{-4} g/L. Under appropriate conditions, the content of carteolol hydrochloride in urine can be determined directly by the method, which had strong anti-interference ability and the recovery is 96.5% - 110.5%. In addition, extraction and adsorption behaviors of the modified electrodes for carteolol hydrochloride were studied by chronocoulumetry, and the results showed that extraction during the enrichment process played a major role at low concentrations, and contribution of surface adsorption became greater with the increase of concentrations.

1. Introduction

Carteolol hydrochloride (CH) is a beta blocker which has intrinsic sympathomimetic activity. It is used extensively in clinical practice for treatments with hypertension, arrhythmia, myocardial infarction and cardiac neurosis[1]. CH has been listed as forbidden drugs for athletes by WADA in some sport events. At present, there are several analytical techniques for determination of CH, such as liquid chromatograph-electrospray ionization mass spectrometry analysis[2,3], HPLC[4], NMR spectroscopy[5] and fluorescence method[6]. However, these methods are usually complex, time-consuming, solvent intensive, or require expensive devices. Because of these reasons, development of a simple, highly sensitive, selective, easily miniaturized, low cost, and rapid analytical method for the determination of CH is of great significance. To our best knowledge, electrochemical techniques can offer all the advantages.

Carbon nanotubes has received considerable attention by its particular mechanical, electrical, chemical properties and special tubal molecular structures in recent years[7,8]. Nano-ZnO has been used extensively in electroanalytical chemistry research because of its special properties, such as electrochemical activity, catalytic activity and biocompatibility. Chemical sensors and biosensors based on this material with high sensitivity/selectivity have obtained extensive applications[9,10], and

Cationic surfactant Cetyl Trimethyl Ammonium Bromide (CTAB) has a function of hydrophobic adsorption[11]. In this study, a kind of CTAB/ZnO-MWNTs modified electrodes was prepared. The ability of the modified electrodes for direct detection of pH in urine samples was evaluated. The studies showed that the modified electrodes had better catalytic oxidation performance for CH comparing with bare glassy carbon electrode.

2. Materials and methods

2.1 Reagents

The CATB was purchased from Shanghai Aibi Chemical Co, Ltd. The MWNTs (diameter 8-15 nm, length 50 μm, purity > 95%) was purchased from Chengdu Organic Chemicals Co, CAS. Potassium ferricyanide was purchased from Beijing Chemical Plant. Carteolol Hydrochloride(CH), Terbutaline Sulfate(TES), Clenbuterol(CLE) and salbutamol(SAL) were purchased from Beijing Shanglifang Association Chemical Industry.Co, CAS. Phosphate buffer solution (PBS) was home-made. All the chemicals in this study were of analytical grade. All the solutions were prepared with twice distilled deionized water.

2.2 Apparatuses

Electrochemical measurements were carried out on a CHI 600C Electrochemical Analyzer (CHI Co., Shanghai, China). A standard three-electrode cell was employed. The working electrodes were either bare glassy carbon electrode (GCE) or modified electrode. A platinum wire and Ag/AgCl were used as counter and reference electrodes, respectively. All of the electrodes were from CHI Co. And all the potentials referred to the Ag/AgCl reference electrode. The pH values were measured with a PHB-1 precision acidity meter(Instrument Co., Ltd Shanghai Yu Long). All experiments were carried out at room temperature.

2.3 Preparation of modified electrodes

A 3-mm (in diameter) GCE was sanded with metallographic abrasive papers, then sequentially polished by 0.1, 0.3 and 0.05-μm alumina pastes on suede pad, and finally fully cleaned in an ultrasonic cleaner with double distilled water, diluted nitric acid (volume ratio 1:1), alcohol, and double distilled water, respectively. The freshly cleaned GCE was electrochemically activated in 0.5 M H_2SO_4 at a scanning rate of 50 mV s^{-1} and in the potential range from -0.5 to +1.4 V for 20 cycles[12].

37 mg of $Zn(CH_3COO)_2 \cdot 2H_2O$ and 100 mg of polyethylene glycol were added into 15 mL twice distilled deionized water. 8 mL of 0.04 mmol·L^{-1} $Li(OH)_2 \cdot H_2O$ solution was slowly added in the above mixture at 50 ℃, then treated by ultrasound at a constant temperature for 10 minutes. After centrifugal separation, the precipitation was washed for several times with twice distilled deionized water and dried at 70 ℃ under vacuum. The final product nano-ZnO was placed in porcelain crucible for 2 hours at 300 ℃. The product was cooled down at room temperature before use.

MWNTs were ultrasonicated in mixed acids (98% H_2SO_4/HNO_3 3:1v/v) for 12 h, and then washed with distilled water and 0.01 M NaOH solution until pH reached 7.0[13]. The washed MWNTs were centrifuged and dried under vacuum at room temperature.1 mg of nano-ZnO and 2 mg of prepared MWNTs were added in 10 mL DMF and prepared by ultrasonic dispersion for 30 minutes. The stable dispersion liquid was obtained.

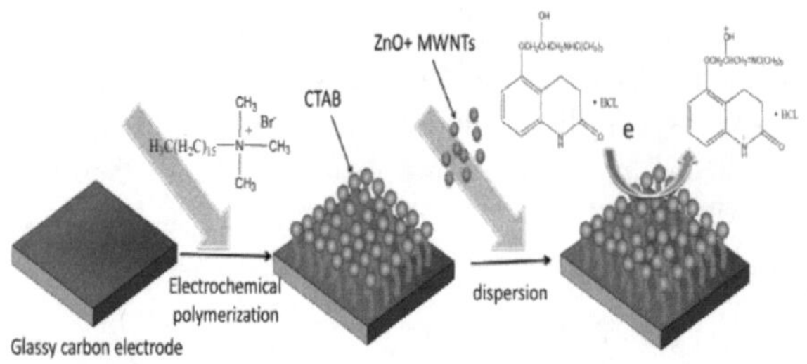

Figure 1. Preparation of CTAB/ ZnO-MWNTs modified electrodes.

As was shown in the Fig.1, GCE was chosen as the basal electrode, and the pretreated electrode was put in PBS (pH = 5.6) containing 5 mmol/L CTAB. The electrodes were treated by cyclic voltammetry in the potential range from -1.6 ~ +1.8 V for 15 cycles until stable voltammetric responses were obtained. After the electropolymerization, the modified electrodes were rinsed fully with distilled water and dried at room temperature. Then nano-ZnO and MWNTs dispersion liquid was dropped on CTAB/GCE by micropipette. After dried under infrared lamp, the CTAB/ZnO-MWNTs/GCE was obtained. The electrodes have good electrical catalytic ability for CH.

2.4 Experimental methods

The electrodes were placed in buffer solution before testing, in the potential range from -0.5 ~ 1.2V for several cycles until stable voltammetric responses were obtained. Then the electrodes were dipped into sample solution to enrich for a while, using differential pulse and cyclic voltammetry to study the electrochemical behaviors of CH on the electrodes. Potential sweep range was 1.0 ~ 1.5 V, with scanning speed 100 mV/s, pulse height 50 mV, and pulse width 50ms. The potential step range of chronocoulometry was -1.0 ~ 1.0V, enriched for 20 min with pulse width 20 s.

3. Results and discussions

3.1 The polymerization and characterization of modified membrane

The cyclic voltammetry curve of GCE electrochemical polymerization in CTAB is shown in Fig.2, indicating that oxidation peak b current decreased and disappeared sharply around potential 1.5 V, due to the oxidation of Br- to Br_2. Based on the reducing reactivity of the active groups in the reduction processes, peak c indicates the reduction peak of Br_2, and the reduction product was Br-. Because the concentration of Br_2 was low in solution, the reduction peak of Br_2 was low. The concentration of cetyl trimethyl ammonium bromide kept almost unchanged, so peak a and peak d were slightly lower. With the number of scanning cycle increased, peak current decreased constantly, indicating polymer film formed gradually on the GCE surface.

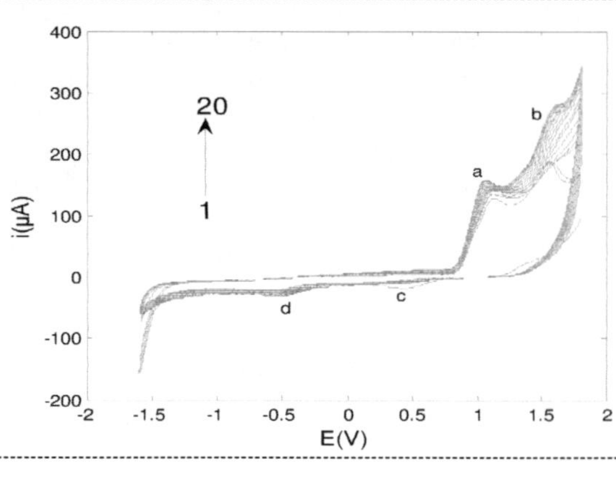

Figure. 2 Cyclic voltammograms for the polymerization of CTAB on bare GCE in 0.2M PBS (pH 5.6) containing 5×10^{-3} M CTAB.

In the potential range from -0.2 ～ +0.7 V at a scanning rate of 100 mV s^{-1} , GCE, CTAB/GCE, and CTAB/ ZnO-MWNTs/GCE were scanned with cyclic voltammetry in $K_3Fe(CN)_6$ solution, respectively. In Fig. 3, the peak current of curve c is lower than that of curve b. It was caused by CTAB polymer film. The peak current of curve a obviously was the highest, indicating CTAB/ ZnO-MWNTs/GCE electrodes can transfer electrons effectively.

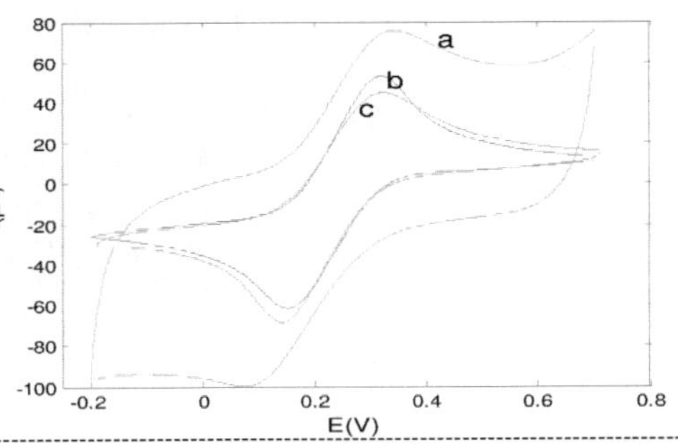

Figure. 3 Cyclic voltammograms at CTAB /ZnO-MWNTs/GCE(a), bare GCE (b), CTAB / GCE(c) in the 5×10^{-3} M potassium ferricyanide; scan rate: 100 mV/s; supporting electrolyte: 0.2 M PBS pH 7.0.

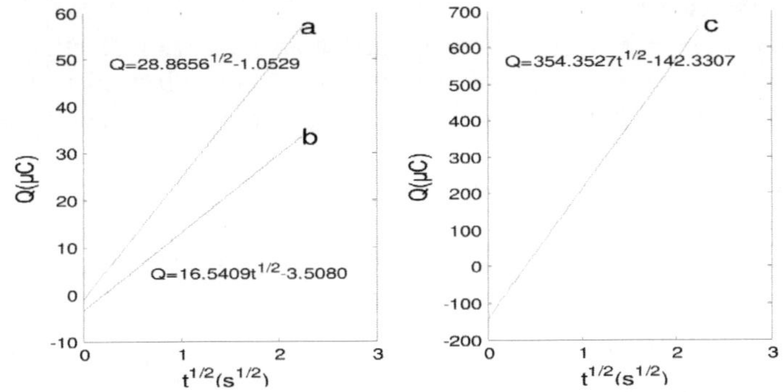

Figure 4. Q-$t^{1/2}$ curves at bare GCE (a), CTAB / GCE (b) and CTAB /ZnO-MWNTs/GCE(c) in the 5x10-3 M potassium ferricyanide; scan rate: 100 mV/s.

3.2 The comparison of different electrodes

Curves of GCE (a) , CTAB/GCE (b), or CTAB/ZnO-MWNTs/GCE (c) drawn in 5.0×10^{-3} mol/L of K3[Fe(CN)6] solution (Fig. 4), according to Anson formula[14]:

$$Q=2nFCA\pi^{-1/2}D^{1/2}t^{1/2}+Q_{ads}+Q_{dl} \quad (1)$$

A is the effective surface area of working electrodes, C is concentration of electrolyte, n is electronic transmission number, D is diffusion coefficient of electrolyte. N and D of K3[Fe(CN)6] is 1 and 7.6×10^{-6} $cm^2\cdot s^{-1}$ [15], respectively. Based on the slope of the linear relationship of Q-$t^{1/2}$, the effective surface areas of GCE, CTAB/GCE, and CTAB/ZnO-MWNTs/GCE could be calculated, which were 0.0192 、 0.0111 and 2.361 cm^2, respectively. These results demonstrated that ZnO-MWNTs had enlarged the surface areas of GCE.

Fig. 5 shows the CV curves when GCE 、 CTAB/GCE 、 ZnO-MWNTs/GCE and CTAB/ZnO-MWNTs/GCE measured the CH standard solution of the same concentration. The peak current of CTAB/GCE (curve 2) was higher than that of GCE (curve 1), indicating that CTAB had better electrochemical response to CH. Although the peak current of ZnO-MWNTs/GCE (3) increased, the electrochemical response to CTAB/ZnO-MWNTs/GCE (curve 4) was worse, indicating that CTAB and ZnO-MWNTs had sensitizing effect on CH. Fig. 5 also shows that there were three steps that in the oxidation and reduction reactions of CH on the CTAB/ZnO-MWNTs/GCE, which were quasi-reversible processes.

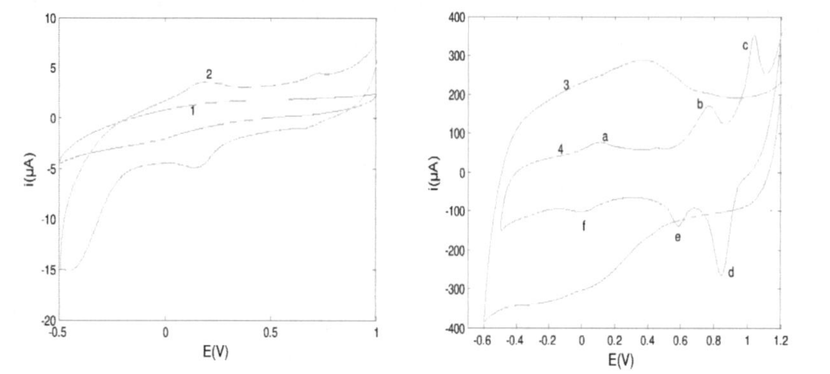

Figure 5. Cyclic voltammograms at bare GCE (1), CTAB / GCE (2), ZnO-MWNTs/GCE (3) and CTAB /ZnO-MWNTs/GCE (4), in the CH of the same concentration; can rate: 100 mV/s; supporting electrolyte: 0.2 M PBS pH 6.6.

3.3 Electrochemical behaviors of CH on CTAB /ZnO-MWNTs/GCE

At a scanning rate of 50 ～ 150 mV/s, the electrochemical mechanism of 0.2 g/L CH was studied on modified electrode surface using cyclic voltammetry (Fig. 6). The oxidation and reduction reactions of CH were under similar conditions. So the reactions of CH on modified electrodes were quasi-reversible processes. Tab. 1 shows the relationship and electrode processes between current peaks Ip of a, b, c and scanning speed.

The electron transfer number of CH electrochemical reaction on CTAB/ZnO-MWNTs/GCE can be measured by cyclic voltammetry.

For peak a, $E_{pa}(a)=0.1107V$, $E_{pa/2}(a)=0.053V$, for b, $E_{pa}(b)=0.766V$, $E_{pa/2}(b)=0.815V$, for c, $E_{pa}(c)=1.051V$, $E_{pa/2}(c)=1.081V$.

But for reversible process: $|Epa—Epa/2|=0.0565/\alpha n(\alpha=1)$ (2)

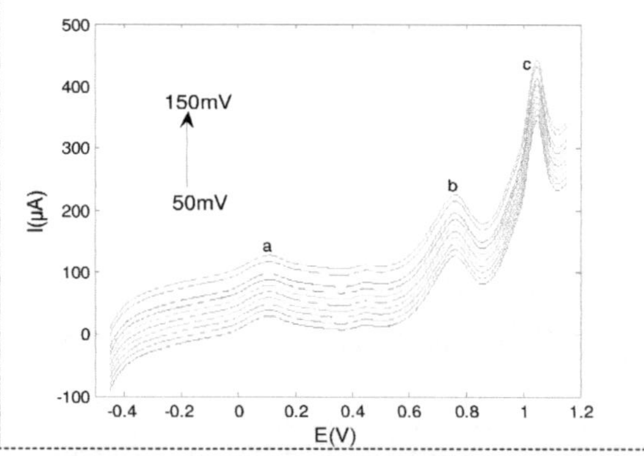

Figure 6. Cyclic voltammograms of 0.2 g/L CH in 0.2 M PBS of pH 6.6 at CTAB /ZnO-MWNTs/GCE with different scanning rates (V): 50, 60, 70, 80, 90,100, 110, 120, 130, 140 and 150 mV/s.

E_{pa} is peak potential, $E_{pa/2}$ is half peak potential, α is electron transfer coefficient, and n is number of electron transfer. The number of electron transfer of peak a, b and c were 1, 1and 2, respectively.

The pH of buffer solution has important effect on electrochemical response of CH. Therefore, voltammetry behaviors of CH in PBS buffer solutions with different pH values. The peak profile is the best at pH 6.6. With pH changed. The peak potentials of a, b and c changed accordingly, with the changes of –0.0524 V·pH-1, –0.0524 V·pH-1, and -0.0563 V·pH-1, respectively. The changes were close to the theoretical value –0.059 V·pH-1, indicating that the number of electron transfer and proton transfer was 1:1.

3.4 The adsorption and extraction behaviors of CH on CTAB/ZnO-MWNTs/GCE

The electric quantity-time curve of CTAB/ZnO-MWNTs/GCE was determined by chronocoulometry after pre-enrichment. Using Anson formula, total electric quantity Q could be calculated, which was generated after potential step. When the time of enrichment is certain, Anson formula[16] can be simplified as :

$$Q = Q_{ext} + Q_{ads} + Q_{dl} \quad (3)$$

Q_{ext} is electric quantity of extraction, and its value is proportional to extraction quantities. Q_{ads} is electric quantity generated by surface adsorption. Q_{dl} is the electric quantity of electric double-layer charges. After 20 min of enrichment, electric quantities were determined in several different concentrations of CH standard solutions. The Q_{ext} and the Q_{ads} +Q_{dl} of CTAB/ZnO-MWNTs/GCE were calculated according to Q-t1/2 curves of different CH concentrations. Tab. 1 shows that at low concentrations, the enrichment of electrode to CH mainly came from extraction reaction. With the increase of CH concentrations, the relative contribution of surface absorption to all enrichment process gradually increased.

Table 1. Extraction and adsorption capacity of the different concentrations of CH

Concentration of CH（g/L）	Q_{ext} (C)	$Q_{ads}+Q_{dl}$ (C)	$Q_{ext}/(Q_{ads}+Q_{dl})$(C)
0.2	11419.71	242.39	47.11
0.1	11676.35	240.43	48.56
0.05	12965.11	114.36	113.38
0.02	13644.49	81.97	166.46
0.01	13797.71	77.26	178.60
0.005	19066.93	47.98	397.42
0.001	20668.95	14.52	1423.07

3.5 Selectivity experiment

0.2 g/L CH, 0.1 g/L TES, 0.1 g/L CLE, 0.1 g/L SAL and 0.1 g/L mixed solution (CH+TES+CLE+SAL) were prepared using PBS buffer solution at pH 6.6. After pre-enrichment, the solutions were scanned with cyclic voltammetry. In Fig. 7, for peak c, CTAB/ZnO-MWNTs/GCE almost had no response to SAL (curve 5), and very small to CLE (curve 3). The current peaks of the mixed solution (curve 1) and CH (curve 2) differed little, indicating that the electrodes had good selectivity to CH.

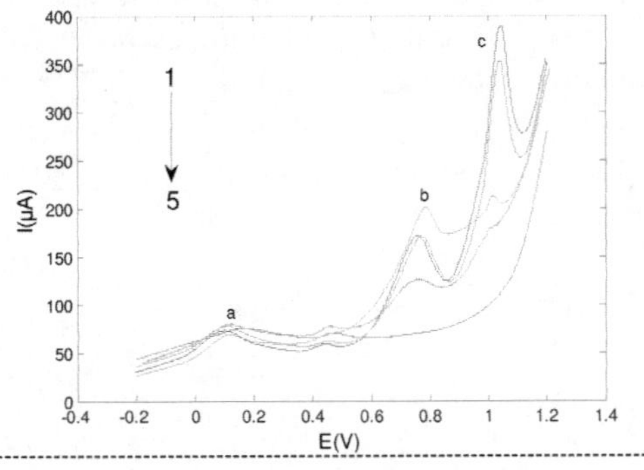

Figure 7. The selective electrode of CTAB/ZnO-MWNTs/GCE for CH+TES+CLE+SAL(1), CH(2), CLE(3), TES(4), SAL(5).

3.6 Working curve and stability of electrode

Under the best experimental conditions, CH standard solutions of different concentrations were tested using CV and DPV. Fig.8 and Fig. 9 show that peak c presented good linear relationship. Tab. 2 shows its linear range and linear regression equation. When the same electrode was used to determine the same 0.2 g/L CH standard solution ten times within a day, the relative standard deviation (RSD) of the current peak was 2.9%, indicating that the determination of electrode had good reproducibility. Stability tests were also carried out at room temperature. The current response decreased slightly after a period of 17 days of exposing the CTAB/ ZnO-MWNTs/GCE in air, which indicated that the CTAB/ ZnO-MWNTs/GCE had a good long-time stability.

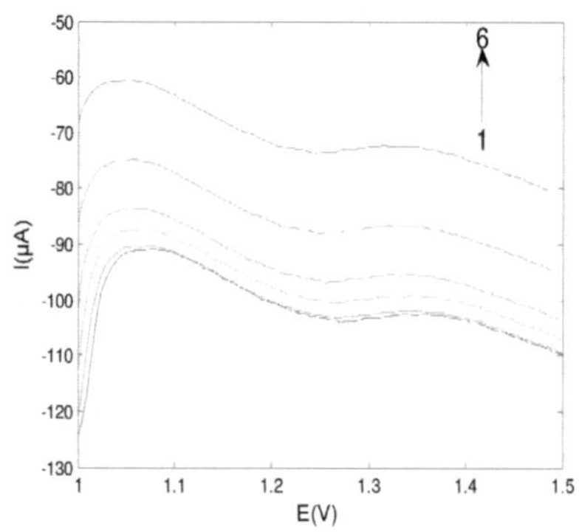

Figure 8. Differential pulse voltammetry in 0.2 M PBS of pH 6.6 on CTAB /ZnO-MWNTs/GCE with different concentrations of CH at 0.002, 0.01, 0.02, 0.05, 0.1, 0.2 g/L (1-6).

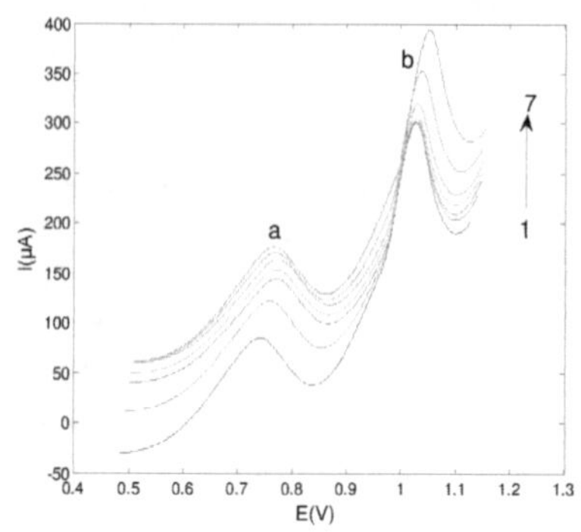

Figure 9. Cyclic voltammograms in 0.2 M PBS of pH 6.6 on CTAB /ZnO-MWNTs/GCE with different concentrations of CH at 0.002, 0.01, 0.02, 0.05, 0.1, 0.2 g/L (1-7).

Table 2. Linear ranges, equations and correlation coefficients for determination of CH

Method	Linear range(g/L)	Regression equation	Correlation coefficient
CV	$1\times10^{-3}\sim2\times10^{-1}$	I_p=477.52C_{CH}(g/L)+299.93	0.9937
DPV	$2\times10^{-3}\sim2\times10^{-1}$	I_p=-154.18C_{CH}(g/L)-30.064	0.9976

3.7 Sample analysis and recovery rate of standard addition experiment

The empty urine was picked form the volunteer who did not took carteolol hydrochloride .5 ml of urine was placed at room temperature for 5 min. 2 mL methanol was added into the urine and centrifuged for 10 min. The supernatant was added into PBS solution, and the pH was adjusted to 6.6. Then recovery rate of standard addition experiment was determined by CV methods to test the sample in parallel for five times. The results in Tab.3 show that recovery rate of standard addition for CTAB/ZnO-MWNTs/GCE was 98.4% ~ 108.1%, and the relative standard deviation was 1.97% ~ 4.29%, indicating that the electrodes have potential application value.

Table 3. Detection results of recovery rate in sample

Sample	Added (g/L)	Found(g/L)	Recovery(%)	RSD (%， n=5)
1	2×10^{-3}	0.00216	108.1	2.38
2	5×10^{-3}	0.00522	104.5	1.97
3	1×10^{-2}	0.00984	98.4	4.29

4. Conclusion

Using CTAB/ZnO-MWNTs to modify electrode, CTAB/ZnO-MWNTs/GCE were obtained. It can be used to determine carteolol hydrochloride in urine. The results of determination would not be influenced under the conditions when other doping reagents such as clenbuterol, terbutaline sulphate and salbutamol are present. The results demonstrated that the modified electrodes haav high sensitivity, good selectivity, rapid response, good stability as well as reproducibility. Furthermore, the proposed method provides advantages such as time-saving and simplicity of the reagents and apparetus. The modified electrode has hopeful application prospect.

References

[1] Netland P A, Weiss H S, Stewart W C, Cohen J S and Nussbaum L L 1997 Cardiovascular effects of topical carteolol hydrochloride and timolol maleate in patients with ocular hypertension and primary open-angle glaucoma *American journal of ophthalmology* **123** 465.

[2] Thevis M, Opfermann G and Schanzer W 2001 High speed determination of beta-receptor blocking agents in human urine by liquid chromatography/tandem mass spectrometry *Biomedical Chromatography* **15** 393.

[3] Zhang Y, Yang Y, Zhu R J, Li X R, Chu M and Qin Y 2012 Upl-ms/MS measures 30 chemicals in the lower pressure class and health care products *Chinese Pharmaceutical Journal* **47** 141.

[4] Chu S Y 1978 HPLC determination of 8 - hydroxycarteolol in plasma and urine using electrochemical detection *Journal of pharmaceutical sciences* **67** 1623.

[5] Zhang A J, Tao R H, Bi H S, Wang Y X and Zhao C J 2015 The absolute content of nine types of lol drugs, such as katilol hydrochloride, was measured by NMR *Journal of Shenyang Pharmaceutical University* **32** 40.

[6] Shrivastav P S, Buha S M and Sanyal M 2010 Detection and quantitation of β-blockers in plasma and urine *Bioanalysis* **2** 263.

[7] Shahrokhian S, Kamalzadeh Z, Bezaatpour A and Boghaei D M 2008 Differential pulse voltammetric determination of N-acetylcysteine by the electrocatalytic oxidation at the surface of carbon nanotube-paste electrode modified with cobalt salophen complexes *Sensors and Actuators B: Chemical* **133** 599.

[8] Abbar J C, Malode S J, Nandibewoor S T 2012 Electrochemical determination of a hemorheologic drug, pentoxifylline at a multi-walled carbon nanotube paste electrode *Bioelectrochemistry* **83** 1.

[9] Sun W, Zhai Z, Wang D, Liu S, and Jiao K 2009 Electrochemistry of hemoglobin entrapped in a Nafion/nano-ZnO film on carbon ionic liquid electrode *Bioelectrochemistry* **74** 295.

[10] Zhang C, Wang G, Liu M, Feng Y, Zhang Z and Fang B 2010 A hydroxylamine electrochemical sensor based on electrodeposition of porous ZnO nanofilms onto carbon nanotubes films modified electrode *Electrochimica Acta* **55** 2835.

[11] Zhang X L, Peng Y F and Hu C G 2008 Hexadecyl trimethyl ammonium bromide modified carbon paste electrode linear scanning voltammetry to determine uretic acid in urine *Journal of Analytical Science* **24** 641.

[12] Zhao Z Z, Shen G J, Mu S, and Gong X J 2015 preparation and characteristics of a molecularly imprinted electrode for Dioxopromethazine Hydrochloride *Chinese J Analysis Laboratory* **34** 178.

[13] Kang W Z, Shen G J and Wang R H 2015 Electrochemical behavior of metformin hydrochloride at a β-CD and MWNTs modified glassy carbon electrode *Chinese J Analysis Laboratory* **34** 99.

[14] Anson F C 1964 Application of Potentiostatic Current Integration to the Study of the Adsorption of Cobalt (III)- Ethylenedinitrilo (tetraacetate) on Mercury Electrodes *Analytical Chemistry* **36** 932.

[15] Adams R N 1969 *Electrochemistry at solid electrodes. Monographs in Electroanalytical Chemistry and Electrochemistry* Marcel Dekker, Inc, In New York, Aufl XIII, p 102.

[16] Wang S F and Peng T Z 2000 Adsorption / extraction behavior and electrochemical determination of Mela Thorning for carbon paste electrode *Chinese Journal of Analytical Chemistry* **28** 1350.

IOP Conf. Series: Materials Science and Engineering **220** (2017) 012010 doi:10.1088/1757-899X/220/1/012010

The crystal morphology effect of Iridium tris-acetylacetonate on MOCVD iridium coatings

Jing Shi[1], Yupeng Hao[1], Xiaodong Yu[1,2] and Chengwen Tan[1,2]

[1]School of Material Science and Engineering, Beijing Institute of Technology, Beijing 86100081, China
[2]China Astronaut Research and Training Center, Beijing 86100081, China

Abstract. Iridium tris-acetylacetonate is the most commonly used precursor for the metal organic chemical vapour deposition (MOCVD) of iridium coating. In this paper, the crystal morphology effect of iridium tris-acetylacetonate on iridium coatings prepared by MOCVD was studied. Two kinds of Ir(acac)$_3$ crystalline powder were prepared. A precursor sublimation experiment in a fixed bed reactor and an iridium deposition experiment in a cold-wall atmospheric CVD reactor were designed. It is found that the volatility of the hexagonal columnar crystals is better than that of the tetragonal flake crystals under the experimental conditions. It's due to the hexagonal columnar crystals exposed more crystal faces than the tetragonal flake crystals, increasing its contact area with the transport gas. An adequate supply of iridium tris-acetylacetonate during the pre-deposition period contributed to obtain an iridium coating with a smooth and uniform continuity surface.

1. Introduction

The refractory metal iridium (Ir) is well-known for its superior high-temperature strength and remarkable resistance against different types of corrosions, especially oxidation [1]. This combination of properties makes Ir a potential material for use as protective coatings in high-temperature anaerobic environments [2-5]. Several methods are available in fabricating Ir, such as physical vapour deposition (PVD), plasma spraying (PS) and metal organic chemical vapour deposition (MOCVD). Among these methods, MOCVD is an effective method to prepare high-purity Ir of particular shapes with high density [6].

Several Ir organic molecular compounds have been investigated as precursors for Ir MOCVD process [7-9], among which, iridium tris-acetylacetonate is the most commonly used in Ir deposition studies and preparations. Iridium tris-acetylacetonate (Ir(acac)$_3$) is an air-stable crystalline powder with a melting point of 265℃ and a decomposition temperature of 290℃. The decomposition products of Ir(acac)$_3$ are CO_2, H_2O and other non-toxic gas, which will not corrode metal substrate and operating equipment. These characteristics of this Ir complexes effectively lower the demands for apparatus and increase the ease of operation during the MOCVD process.

Although there are some seemingly advantages of Ir(acac)$_3$ as being the precursor for Ir MOCVD process, yet only several studies have been reported, much less on the effect of crystal morphology of Ir(acac)$_3$ crystals on iridium coatings prepared by MOCVD. In this work, two kinds of Ir(acac)$_3$ crystalline powder were prepared. A precursor sublimation experiment in a fixed bed reactor and an iridium deposition experiment in a cold-wall atmospheric CVD reactor were designed to find out the relationship between the deposition and precursor.

IOP Conf. Series: Materials Science and Engineering **220** (2017) 012010 doi:10.1088/1757-899X/220/1/012010

2. Experimental

$Ir(acac)_3$ crystalline powder in different shapes were prepared in anhydrous ethanol and methylene chloride, respectively. The anhydrous ethanol solution with the concentration of 2g/100mL was prepared at the boiling point of anhydrous ethanol, and crystallized at 0℃, 10℃, 20℃, 30℃ for 30min, adding magneton with a speed of 200r/min, 400r/min, 600r/min, 800r/min or ultrasound with a frequency of 40Hz, 60Hz, 80Hz, 100Hz. The methylene chloride solution with the concentration of 1g/8mL was prepared at 20℃ and crystallized at 20℃ by volatilization method or -60℃ by low temperature cooling method for 30min, respectively. The $Ir(acac)_3$ crystalline powder was collected from the solution by filtering, and dried in a vacuum oven at 80℃ for 4 hours. The morphology of the $Ir(acac)_3$ crystalline powder was observed via scanning electron microscopy (SEM) (Hitachi S-4800), accelerating voltage of 20 keV.

The evaporation of $Ir(acac)_3$ crystalline powder in different shapes was tested prior to deposition in a fixed bed reactor. Defined amount of $Ir(acac)_3$ crystalline powder and SiC sand was weighed via electronic balance (Hengzheng, FA2004N) and mixed uniformly, put into the fixed bed reactor, which was heated at 255℃. Stopped heating every 10min to weigh the mass of the mixture, draw the curve of the loss of the $Ir(acac)_3$ and time.

Iridium coatings were deposited onto the Φ6×25 polycrystalline molybdenum substrates, in a cold-wall CVD reactor under atmospheric pressure. The set-up is illustrated in Figure 1. $Ir(acac)_3$ was evenly heated and controlled using a high-temperature quartz valve. The substrate was vertically suspended in the reactor and heated via a high-frequency induction heating (Xi'anTianfeng, IGPS-30KW-200KHz). The temperature of the substrate was measured using an infrared thermometer (Xi'an Innocent, ATS). The flow rate of Ar (99.999 wt.%) was determined using an electronic mass-flow meter (Beijing Sevenstar, D07). Iridium coatings were deposited using $Ir(acac)_3$ as the precursor at 480°C to 520°C under 1.96×103 Pa $Ir(acac)_3$ partial pressure. The morphology of the coatings was observed via scanning electron microscopy (SEM) mentioned above.

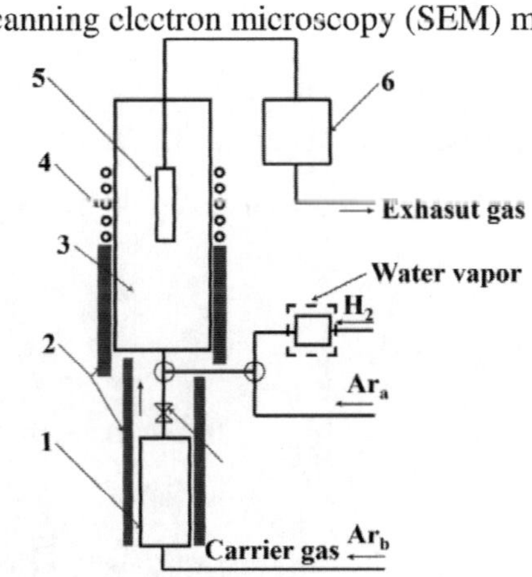

Figure 1. A schematic diagram of deposition device:
(1) Evaporation chamber
(2) Heating belt
(3) Deposition chamber
(4) Induction heating
(5) Substrate
(6) Collection chamber.

3. Results and discussion

3.1. Shape of the $Ir(acac)_3$ crystalline powder

Figure 2 shows SEM image of morphology of $Ir(acac)_3$ crystalline powder at different temperatures in anhydrous ethanol system. The average size of particles in this experimental group was 300μm-500μm. The particles crystallized at 0℃, 10℃, 20℃ were coarse and broken, with poor crystal development integrity and obvious layered growth morphology, hardly to see the complete crystalline face, while the particles crystallized at 30℃ had better crystal development integrity and clear crystalline face.

According to previous study [10], the disperse coarse and broken $Ir(acac)_3$ particles can not be evenly heated in the fixed bed reactor, which has great effect on the evaporation of the precursor. A lack supply of the precursor resulted in coating cracks, shedding and other defects in Ir coatings. The experiment results demonstrated that the temperature of 30℃ is more favorable for the crystallization of the Ir(acac)$_3$ particles in this experimental group.

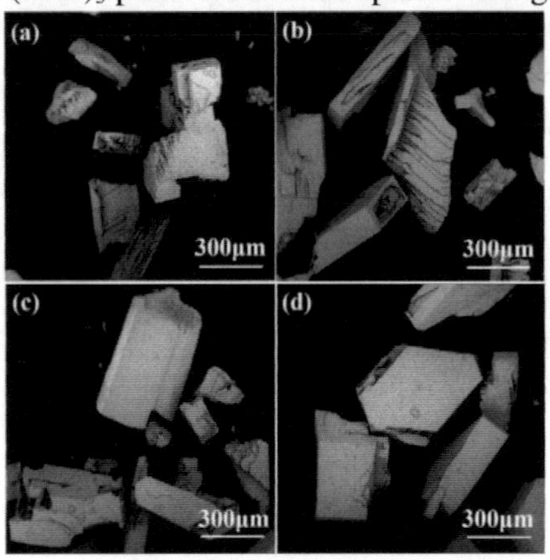

Figure 2. SEM image of morphology of Ir(acac)$_3$ crystalline powder at different temperatures in anhydrous ethanol system: (a) 0℃, (b) 10℃, (c) 20℃, (d) 30℃.

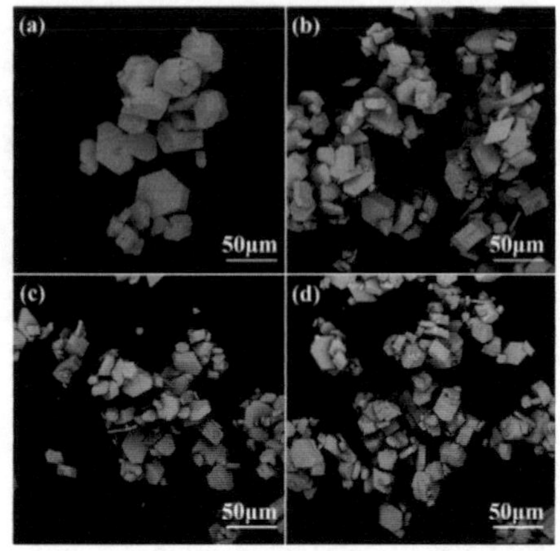

Figure 3. SEM image of morphology of Ir(acac)$_3$ crystalline powder at different magneton speeds in anhydrous ethanol system: (a) 200r/min, (b) 400r/min, (c) 600r/min, (d) 800r/min.

Figure 3 shows SEM image of morphology of Ir(acac)$_3$ crystalline powder at different magneton speeds in anhydrous ethanol system. The average size of particles in this experimental group was 30μm-40μm. The particle size decreases as the magneton speed increases. Most of the particles crystallized at 200r/min were hexagonal columnar crystals, which had good crystal development integrity and complete crystalline face. Some crystals in irregular shape such as rhombus were crystallized at 400r/min, 600r/min and 800r/min. For the consistency of the crystal development integrity and morphology of particles, only when the magnetic speed was 200r/min can obtain most hexagonal columnar crystals.

Figure 4. SEM image of morphology of Ir(acac)$_3$ crystalline powder at different ultrasonic frequencies in anhydrous ethanol system: (a) 40Hz, (b) 60Hz, (c) 80Hz, (d) 100Hz.

Figure 5. SEM image of morphology of Ir(acac)$_3$ crystalline powder at different temperatures in methylene chloride system: (a) 20℃, (b) -60℃.

Figure 4 shows SEM image of morphology of Ir(acac)$_3$ crystalline powder at different ultrasonic frequencies in anhydrous ethanol system. The average size of particles in this experimental group was 20μm-30μm. The particles in this experimental group were in good crystal development integrity, except the particles crystallized at 100Hz, which had some unformed small particles adhering on the

IOP Conf. Series: Materials Science and Engineering **220** (2017) 012010 doi:10.1088/1757-899X/220/1/012010

crystals. The addition of ultrasound enhanced the collision of $Ir(acac)_3$ molecules, which accelerated the nucleation of the $Ir(acac)_3$ crystals, but made it difficult for the particles to grow in a stable environment, causing the crystalline powder finally obtained was very fine. This fine crystalline powder was easy to adhere to the wall of the MOCVD equipment, which was not conducive to the evaporation of precursor.

Above all, the anhydrous ethanol solution with the concentration of 2g/100mL prepared at the boiling point of anhydrous ethanol and crystallized at 30℃ for 30min at 200r/min, was chosen to obtain the hexagonal columnar crystals with good crystal development integrity and regular shape.

Figure 5 shows SEM image of morphology of $Ir(acac)_3$ crystalline powder at different temperatures in methylene chloride system. The particles crystallized in this experimental group were all tetragonal flake crystals, but the particles crystallized at 20℃ by volatilization method was much bigger than those crystallized at -60℃ by low temperature cooling method. The size of the particles crystallized at 20℃ were not in the same order of magnitude with that of the particles obtained in anhydrous ethanol system, so we chose the particles crystallized at -60℃ as precursor in Ir MOCVD process.

3.2. Volatility of the Ir(acac)₃ crystalline powder

Figure 6 shows a curve of volatility of the $Ir(acac)_3$ crystalline powder in different shapes. The loss of $Ir(acac)_3$ and time were in a good linear relationship for both of the hexagonal columnar crystals and the tetragonal flake crystals. the sublimation rate of hexagonal columnar crystals was 0.0342g/min, the sublimation rate of tetragonal flake crystals was 0.0265g/min. It is found that the volatility of the hexagonal columnar crystals is better than that of the tetragonal flake crystals under the experimental conditions. The volatility of $Ir(acac)_3$ is related to its properties, sublimation temperature and mass transfer area. In this experimental condition, the properties and sublimation temperature was consistent for both samples. It's due to the hexagonal columnar crystals exposed more crystal faces than the tetragonal flake crystals, increasing its contact area with the transport gas.

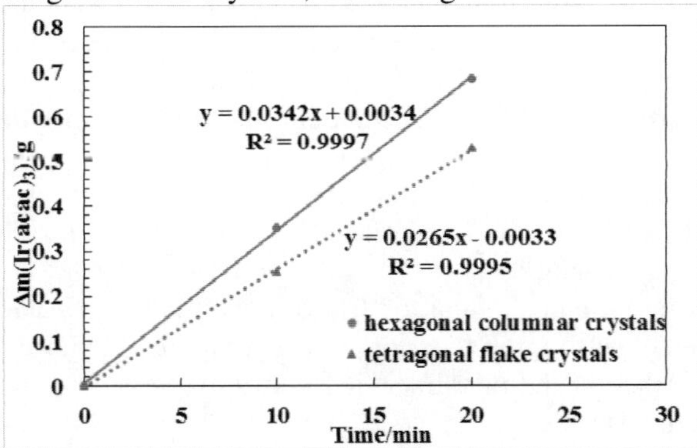

Figure 6. A curve of volatility of the $Ir(acac)_3$ crystalline powder in different shapes: (a) hexagonal columnar crystals, (b) tetragonal flake crystals.

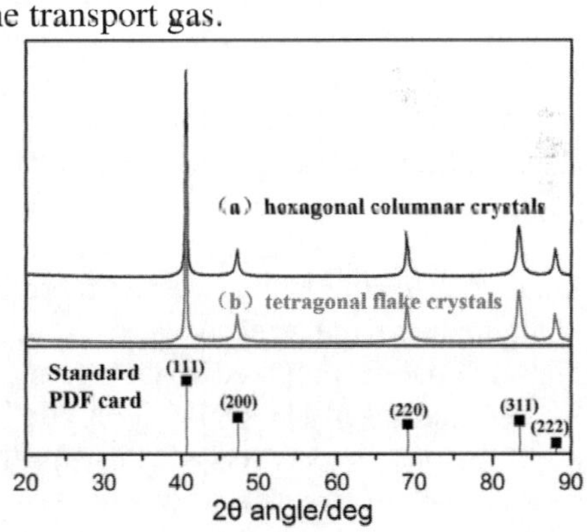

Figure 7. XRD spectrum of iridium coating prepared of $Ir(acac)_3$ crystalline powder in different shapes: (a) hexagonal columnar crystals (b) tetragonal flake crystals.

3.3. Surface morphology of the Ir coatings

Figure 7 shows XRD spectrum of iridium coating prepared of $Ir(acac)_3$ crystalline powder in different shapes. The Ir coatings both showed (111) preferred growth.

Figure 8 shows SEM image of surface morphology of Ir coating prepared of $Ir(acac)_3$ crystalline powder in different shapes. The surface of the Ir coating was consisted of many cellular protrusions, whose diameter was 10μm-40μm measured by intercept method. There was no crack or hole in the Ir coating prepared of hexagonal columnar crystals, with a smooth and uniform continuity surface. While there was obvious cracks between the cellular protrusions in the Ir coating prepared of tetragonal flake crystals.

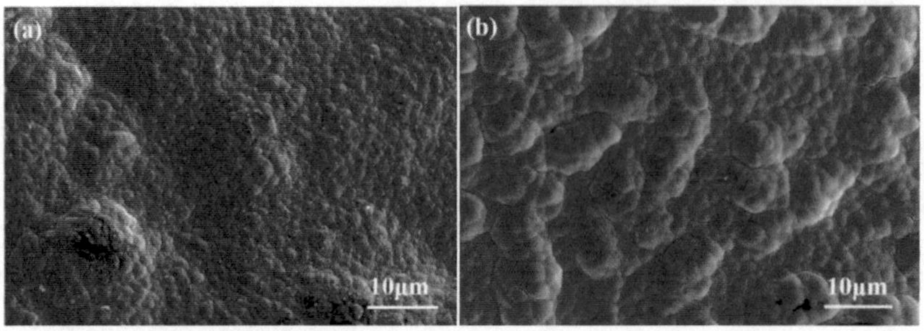

Figure 8. SEM image of surface morphology of Ir coating prepared of $Ir(acac)_3$ crystalline powder in different shapes: (a) hexagonal columnar crystals (b) tetragonal flake crystals.

The growth of the Ir coating in MOCVD progress followed the island growth mechanism. The lateral spreading of the islands allowed continuous growth of the iridium coating, and the vertical spreading of the islands increased the thickness of the iridium coating. The sublimation rate of the tetragonal flake crystals was smaller than that of hexagonal columnar crystals in the first 20 min of heating in Figure 6, which lead to the insufficient supply of precursor during the pre-deposition period. The Ir coating deposited in initial period was uneven and the stress presented in the Ir coating grew subsequently resulted in the crack on the surface of Ir coating eventually. Thus, it's better to use hexagonal columnar crystals of $Ir(acac)_3$, which has a better volatility as precursor in Ir MOCVD process.

4. Conclusions

1. The anhydrous ethanol solution with the concentration of 2g/100mL was prepared at the boiling point of anhydrous ethanol, and crystallized at 30℃ for 30min at 200r/min to obtain the hexagonal columnar crystals. The methylene chloride solution with the concentration of 1g/8mL was prepared at 20℃ and crystallized at -60℃ for 30min to obtain the tetragonal flake crystals.

2. The sublimation rate of hexagonal columnar crystals was 0.0342g/min, and the sublimation rate of tetragonal flake crystals was 0.0265g/min. The volatility of the hexagonal columnar crystals is better than that of the tetragonal flake crystals under the experimental conditions. It's due to the hexagonal columnar crystals exposed more crystal faces than the tetragonal flake crystals, increasing its contact area with the transport gas.

3. The Ir coating prepared from the hexagonal columnar crystals of $Ir(acac)_3$ had a smooth and uniform continuity surface, while the Ir coating prepared from the tetragonal flake crystals of $Ir(acac)_3$ had more obvious cracks. Associated with the volatility of the precursor, it's better to use hexagonal columnar crystals of $Ir(acac)_3$ as precursor in Ir MOCVD process.

References

[1] Mumtaz K, Echigoya J, Enoki H, Hira T and Shindo Y 1995 *J. Mater. Sci.* **30** 465-72.
[2] Hu C and Chen L 2007 *Precious metals* **28** 57-62.
[3] Tuffias R H, Williams B E and Kaplan R B 1995 *Joint Propulsion Conference and Exhibit* (San Diego, CA: the American Institute of Aeronautics and Astronautics) pp 1-9.
[4] Zhang X, Wang X, Jia Z, Hu X and Lv H 2006 *Aerospace Materials & Technology* **35** 11-15.
[5] Schneider S J 1992 *Acta Astronautica* **28** 115-125.
[6] Cai H, Hu C, Chen L and Wang Y 2009 *Chinese Journal of Rare Metals* **33(2)** 200-204.
[7] Isakova V G, Baidina I A, Morozova N B and Igumenov I K 2000 *Polyhedron* **19** 1097.
[8] Goto T, Ono T and Hirai T 2001 *Scripata Mater* **44** 1187.
[9] Sun Y M, Yan X M, Mettlach N, Endle J P, Kirsch P D, Ekerdt J G, Madhukar S, Hance R L, White J M and Vac J 2000 *Sci. Technol. A* **18** 10.
[10] Yang S, Tan C, Yu X, Liu K, Wang Z, Wang Y, Ma H, Wang F and Cai H 2015 *Surface and Coatings Technology* **265** 38-45.

IOP Conf. Series: Materials Science and Engineering **220** (2017) 012011 doi:10.1088/1757-899X/220/1/012011

A new fracturing material in oil and gas industry

Bo Cai[1], Jiangwen Xu[2], Chunming He[1], Ning Cheng[2], Yuebin Gao[1], Jin Chen[2], Jia Wang[2]

[1]PetroChina Research Institute of Petroleum Exploration & Development, Langfang city, Hebei Province, 065007, China

[2]Xinjiang Oilfield Company, PetroChina, Kelamayi city, Xinjiang Province, 834000, China

Abstract. In this study, a novel non-residual fracturing fluid was developed. This fracturing fluid system is crosslinked under acid condition and owns the advantages of fast dissolving, excellent sand carrying ability, non-residual, good anti-swelling property and low damage etc. Experimental results show that this system viscosity can reach 95% of peak viscosity in 3 minutes, the surface tension of the gel breaking liquid is 25 mN/m ,the residue content of the gel breaking fluid is 10mg/L, core permeability damage rate less than 15% which greatly reduces the damage to formation and fracture conductivity. The shear viscosity of fracturing fluid for 90min is 260 mPa.s, it has good resistance to high temperature and shearing performance and can meet the requirements of fracturing in tight reservoir. The new fluid was tested in 65 wells in the tight gas and oil reservoir in western China. Oil &gas production after stimulation using the new fluid increased 2-5 times compared with wells in similar locations.

1. Introduction

The exploration areas of oil and gas are extensive in China, but faced with complexly geological characteristics with low porosity and permeability.it becomes inevitable to adopt reservoir stimulation so as to maintain their economic development that makes strict demands on protecting reservoirs with characteristic of "low damage". Fracturing fluid system is the key to successful reservoir stimulation during the fracturing process. It has proved that fracturing fluids provide a more efficient and environmentally friendly way to develop reservoirs. We should reduce reservoir damage and improve the research of efficient fracturing technology, build a new applied technology of fracturing fluid with "low-cost, low-damage, environmental protection". In this paper, a novel non-residual fracturing fluid was developed (Figure 1).

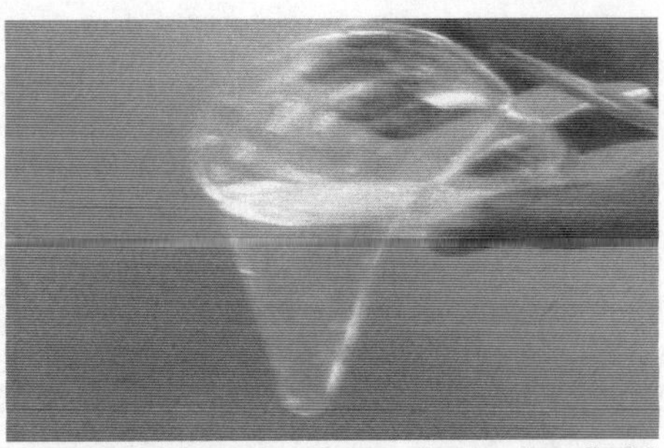

Figure 1. Picture of new fracturing fluid.

This fracturing fluid system is cross-linked under acid condition, and owns the advantages of fast dissolving, excellent sand carrying ability, non-residual, good anti-swelling property and low damage etc. Experimental results show that this system viscosity can reach 95% of peak viscosity in 3 minutes, the surface tension of the gel breaking liquid is 25 mN/m ,the residue content of the gel breaking fluid is 10mg/L, core permeability damage rate less than 15% which greatly reduces the damage to formation and man-made fracture conductivity. The new fluid was tested in 65wells in the tight gas and oil reservoir in western China. Oil &gas production after stimulation using the new fluid increased 2-5 times compared with wells in similar locations. Furthermore, it is expected to reduce direct costs of about 550 million yuan annually in petroleum industry in China, thus becoming an alternative fracturing fluid [1].

2. Main material

Fracturing fluid temperature and resistance to shear, gel, filtration and residue content based on the oil and gas industry standard SY/T5107-2005 "water base fracturing fluid performance evaluation method" to determine. Evaluation of the use of fracturing fluid formula: 0.30%A-1+0.2%B-1+0.3%KCl+0.5%C-1+0.5%D-1, the use of organic zirconium as a crosslinking agent, crosslinking ratio of 100:0.4. The main ingredients are as follows: cellulose A-1, tackifierB-1, crosslinking agent F-1, additive D-1.

3. X-ray experiment analysis

Comparing the photos of modified material focus on before or after dissolution by electron microscopy scanning, it is clear that the swelling of new material has dissolved drastically as shown in figure 2 and figure 3. The solubility of the modified material has been greatly enhanced with special characteristics of rapid dissolving [2], it is clear that the laminated structure of base-liquid material is in echelon [3]. Material gel has formed a dense-spatial-network structure by cross-linked carboxyl group, greatly improving temperature resistance and carrying capacity [4].

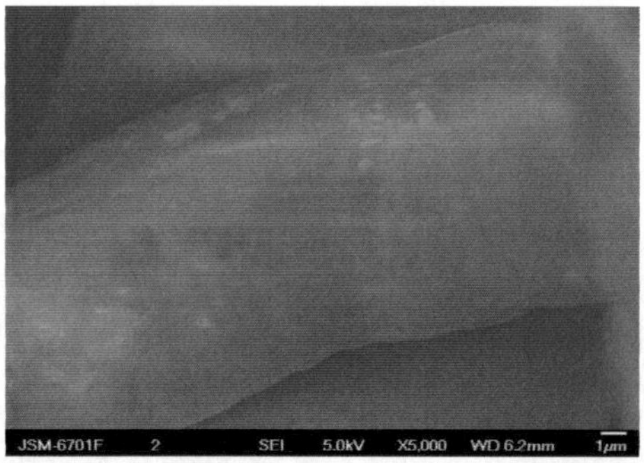

Figure 2. Fracturing material before dissolving.

Figure 3. Fracturing material after dissolving.

3.1. *Temperature factor analysis of fracturing fluid*

This paper studies the hydroxypropyl guar gum, carboxymethyl guar gum base fluid temperature resistance was seen in table 1. Table 1 shows 90 ℃ and 120 ℃ under the conditions of shear for a certain period of time, when the temperature return to the starting temperature (30 ℃), hydroxypropyl guar gum and carboxymethyl guar gum viscosity can recover more than 93%, taking into account the instrument error, that is to say the chemical bond gum molecular chain basically no fault open; when the temperature reached 160℃, the recovery rate of viscosity is below 25%, indicating guar gum molecular chemical bond will be broken at the temperature.

3.2. *Shear properties*

At 80℃, 170s^{-1}, Nevinne shear performance measurement model of clean fracturing fluid using HAAKE-RS600 rheometer, the results are shown in figure 4. The shear viscosity of fracturing fluid for 90min, 260mPa.s, it has good resistance to high temperature and shearing performance and can meet the requirements of fracturing in tight reservoir[5].

Table 1. Hydroxypropyl guar gum and carboxymethyl guar gum temperature resistance

	Temperature ℃	Time Min	Viscosity MPa.S
0.4%HPG	90	1.5	41
0.4%HPG	120	2	42
0.6%HPG	160	2	28
0.4%CMHPG	90	1.5	45
0.4%CMHPG	120	2	45.6
0.6%CMHPG	160	2	25

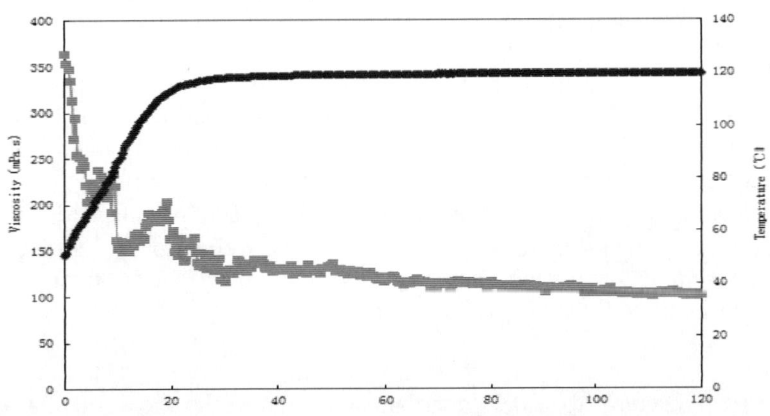

Figure 4. Temperature resistance and shear resistance about 0.3% material.

3.3. *Fracturing fluid property*

3.3.1. Filtration performance. Fracturing fluid filtration effect of fracturing fluid fracturing capacity, the greater the filter loss, the lower the efficiency of the fracturing fluid, the difficulty of the fracturing construction, the greater the damage to the reservoir. This paper evaluates the ultra-low concentration of hydroxypropyl guar gum fracturing fluid formula system of 120 ℃ filtration performance. Taking into account the filtrate loss of diesel, adding 0.5% diesel in fracturing, compared the filtration performance results show that the 5MPa pressure difference, the temperature of 120 ℃, liquid filter loss performance without adding diesel with conventional fracturing fluid, add diesel after filtration has a lower rate, at the beginning of filtration and filtration coefficient.

3.3.2. Study on the mechanism of cross linking. The effects of crosslinking agent in water complexation and repeated hydrolysis, hydroxyl bridge has complexions of polynuclear hydroxyl bridge with positive charge; modified carboxymethyl material base electronegative, polynuclear olation complex ion with modified material in carboxyl and Xian formed polar bond and coordination bond and stock exchange [6]. In this paper, using organic zirconium as crosslinking agent of fracturing fluid acid complex, organic zirconium produced polynuclear olation complex ion in water complexation and multiple hydrolysis, hydroxyl bridge, positive charge polynuclear olation complex ion with high, and high metal ions to form coordination bonds; and the thickening agent the COOH band negatively, oxygen and nitrogen with lone pair electrons, polynuclear olation complex ion is formed by polar bonding and ligand healthy cross-linked with thickening agent in carboxyl and amide groups, the crosslinking reaction is as follows: (1) organic zirconium + H+, hydroxyl zirconium hydroxyl ion; (2) zirconium ion and polynuclear hydroxyl ions by bridge; (3) polynuclear olation complex ion + crosslinking agent, crosslinking gel[7].The first step is to restrict the reaction speed of the crosslinking agent. The ligands and the formation of Zr+ organic zirconium chelate ring, with strong, slow down the process of dissociation of organic zirconium, controlled the formation rate of hydroxy and hydrous zirconium ions, thereby delaying the crosslinking reaction speed, the system with delayed crosslinking, but by adjusting the size of crosslinked pH and ion concentration of the environment. Adjust the crosslinking time [8].

4. Field test

At present, new fracturing fluid system has been successfully applied in tight gas reservoirs in 65 wells field tests show that the new material residue fracturing fluid system has simple preparation, rapid dissolution, base fluid without fisheye and controllable delayed crosslinking, strong carrying capacity and breaking characteristics of glue thoroughly, no residue etc. Site construction of low friction, the average reduction rate of 78%, construction of the largest displacement of $8m^3$/min; cracking ability and sand carrying ability, the highest single well liquid level of 1450 m^3, the largest single sand volume is $96.5m^3$, average sand ratio 30%; Flow back fluid viscosity is 1.18mPa.s, average flow back rate is 81%, discharge time is reduced from the conventional guar gum fracturing fluid of 20h to 48h, after the pressure is 3-8 times the average daily production of adjacent wells. The new fracturing fluid system transformation of S1 well non main gas bearing, before the pressure of no production, obtained high yield unimpeded flow of $13\times 10^4m^3/d$ pressure, and good control of fracture height, no water phenomenon, is 3.3 times the same wells group of guar gum fracturing fluid applications. The result shows that the fracturing fluid with no residue is better to reduce the damage to the reservoir, especially to the tight gas reservoir.

5. Conclusions

China's oil and gas reservoirs are widely distributed and the exploration potential is tremendous, but with the more and more low permeability reservoirs have become the mainly oil production potential trend. Hydraulic fracturing treatment has been playing an important role in oil and gas industry which is a well-stimulation technique in which rock is fractured by a pressurized liquid. Therefore, it is very important to apply low-damage fracturing fluid in oil and gas field. In this study, a novel non-residual fracturing fluid was developed. New fracturing fluid has good temperature and shearing performance, sand carrying capability, anti-filtration performance, can meet in tight gas reservoir reconstruction requirements. New fracturing fluid has the advantages of no residue, its anti-swelling, very good to improve traditional guar gum and its derivatives are serious problems such as a large number of fracturing fluid residue, clay swelling, greatly reduce the damage to the reservoir. The new fluid was tested in 65 wells in the tight gas and oil reservoir in western China. Oil &gas production after stimulation using the new fluid increased 2-5 times compared with wells in similar locations. It provides an effective low damage fracturing fluid for the development of the region.

Acknowledgments
This work is supported by the National Science and Technology Major Project (NO. 2016ZX05023). The authors would express their appreciation to the project for contribution of research fund. The authors would also like to thank lab staff for carrying out the experiments.

References

[1] Zhang R, Lu Y, Wang Y 2006 Study on fracturing fluid system with low damage and high elastic polymer *Drilling fluid and completion fluid* 5-10.

[2] Cai B, Wang X D, & Wang X 2011 Mathematical study of fracture face skin in hydraulic fractures *Journal of China University of Mining &Technology* **40(6)** 938-942.

[3] Li L, Zhao S, Hu H 2009 Developement in Solvent System of Celluse *Journal of Celluse Science and Technology* 69-75.

[4] Cai B, Ding Y, & Lv Y 2014 Optimizing flow coupling in complex artificial fracture network systems *Journal of China University of Min-ing&Technology* 470-474.

[5] Cai B, Ding Y, & Lv Y 2014 A Case Study of Hydraulic Fractures Optimization in Heavy Oil *SPE* 172849.

[6] Econmides M, Martin T 2008 Modern fracturing enhancing natural gas production *Houston: Gulf Publishing Co.* 116-125.

[7] Multalik P 2008 Case history of sequential and simultaneous fracturing of the Barnett shale in Parker County *SPE* 116124.

[8] Olson J, Dahi T 2009 Modeling simultaneous growth of mul-tiple hydraulic fractures and their interaction with natural fractures *SPE* 119739.

Spread prediction model of continuous steel tube based on BP neural network

Jian-wei ZHAI[1], Hui YU[1*], Hai-bei ZOU[1], San-zhong WANG[1] and Li-gang LIU[1]

[1]College of Mechanical Engineering, Yanshan University, Qinhuangdao 066004, Hebei, China

*E-mail: yuhui@ysu.edu.cn

Abstract. According to the geometric pass of roll and technological parameters of three-roller continuous mandrel rolling mill in a factory, a finite element model is established to simulate the continuous rolling process of seamless steel tube, and the reliability of finite element model is verified by comparing with the simulation results and actual results of rolling force, wall thickness and outer diameter of the tube. The effect of roller reduction, roller rotation speed and blooming temperature on the spread rule is studied. Based on BP(Back Propagation) neural network technology, a spread prediction model of continuous rolling tube is established for training wall thickness coefficient and spread coefficient of the continuous rolling tube, and the rapid and accurate prediction of continuous rolling tube size is realized.

1. Introduction

As a high efficiency continuous rolling-tube unit of hot-rolled seamless steel tube production, three-roller continuous mandrel rolling mill has been widely employed in the domestic and foreign seamless steel tube manufacturers [1-3]. In the continuous rolling process of seamless steel tube, the steel tube is extruded by roller and mandrel, which makes the metal produce complex three-dimensional deformation flow. The bottom and side walls of pass have the largest amount of reduction, which makes the metal flow towards the roll gap. However, the metal in the roll gap is in free condition without contact with the mandrel and the roller and forms spread in the roll gap. Spread decides whether pass would have overfill or underfill in the rolling process. The overfill of pass will lead to rolling fold on the steel tube outer surface and the underfill of pass will lead to hollows on the steel tube inner surface [4]. At the same time, the accurate calculation of the spread determines the accuracy of each stand's outlet area, and it is important for making roller rotation speed. The rolling process of eight-stand full floating mandrel seamless steel tube was simulated by ZHAO Zhi-yi[5], and the simulated wall thickness, outer diameter and rolling force were a good agreement with the actual results. The effect of rolling parameters on spread was studied by SHUAI Mei-rong[6] using finite element method, and she concluded that the diameter of inscribed circle of pass has an upmost effective on spread, roller diameter ranks secondly, and the rolling velocity, the temperature, and the friction coefficient have little effect on spread. The application of artificial neural network with its unique advantages on spread study of strip rolling has achieved good results [7-11], but artificial neural network is less employed in the spread of continuous tube. In this paper, a finite element model is established based on the geometric pass of roll and technological parameters in a factory to simulate the continuous rolling process of steel tube. And the effect of roller reduction, roller rotation speed and blooming temperature on the spread rule is studied. At the same time, based on BP neural network technology and according to the finite element numerical simulation results on the wall thickness coefficient and spread coefficient, a spread prediction model of continuous tube is established to predict the size of continuous tube quickly and accurately by learning and training.

2. Spread parameters of continuous tube

In the process of steel tube continuous rolling, spread variation will affect the section size of shell. The wall thickness and outer diameter at the roll gap are important parameters in researching the spread of continuous tube. The wall thickness variation can be expressed as the wall thickness coefficient: α. The outer diameter variation can be expressed as the spread coefficient: β. The formulas of α and β are as follow:

$$\alpha = \frac{S_{gi} - S_{bi-1}}{S_{bi-1}} \quad (1)$$

$$\beta = \frac{R_{gi} - R_{bi-1}}{R_{bi-1}} \quad (2)$$

Where, S_{gi} is the wall thickness of steel tube at the roll gap of the pass.$S_{bi\text{-}1}$ is the wall thickness of steel tube at the pass bottom of the upper pass. R_{gi} is the outer diameter of steel tube at the roll gap of the pass.$R_{bi\text{-}1}$ is the outer diameter of steel tube at the pass bottom of the upper pass.

3. Establishment and verification of finite element model

3.1. Establishment of finite element model

A five stands three-roller continuous mandrel rolling mill is adopted as the research object in this paper. The basic dimension of the tubular billet is Φ222×18.45mm while the dimension of shell is Φ193.28×10.24mm. The outlet dimension of extracting mill is Φ182×10.55mm. The blooming temperature is 1150℃. The material of tubular billet is adopted 45#. The inlet velocity and outlet velocity of tubular billet is 1500mm/s and 2969.9mm/s respectively.

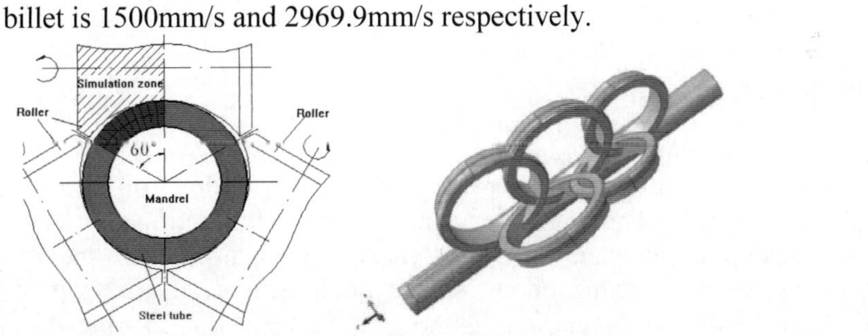

(a) The simulation zone (b) The finite element model

Figure 1. The simulation area of three-roller mandrel pipe mill and finite element model.

The simulation area of three-roller continuous mandrel rolling mill and the finite element model are shown in figure 1. The tubular billet was defined as elastic-plastic deformation body while the mandrel and the roller were defined as rigid bodies. The penalty friction model was taken. The friction coefficient between tube and roller was set to 0.3 and the friction coefficient between mandrel and tube was set to 0.07. The plastic work heat transfer coefficient is 0.9. The contact heat transfer coefficient between mandrel and tube is 8KW/(m^2·k) and the contact heat transfer coefficient between tube and roller is 20KW/(m^2·k). Radiative heat transfer coefficient and convective heat transfer coefficient are considered as comprehensive effect: 0.17KW/(m^2·k).

3.2. Verification of finite element model

After simulation, the average rolling force of each stand was obtained. By comparing simulation values with the measured rolling force as shown in figure 2 (a), the maximum error of rolling force is 7.88%. Figure 2 (b) is the radar chart of comparison between simulated and measured wall thickness values of outlet shell of extracting mill. And it is shown that the maximum absolute error of the wall thickness is 0.13mm and the relative error is 0.22%. Figure 2 (c) is the radar chart of comparison between simulated and measured outer diameter values of outlet shell of extracting mill. And it is shown that the maximum absolute error value of outer diameter is 0.15mm and the relative error is 0.16%. The simulation results are in good agreement with the measured results, so the established finite element model of three-roller continuous rolling tube is reliable.

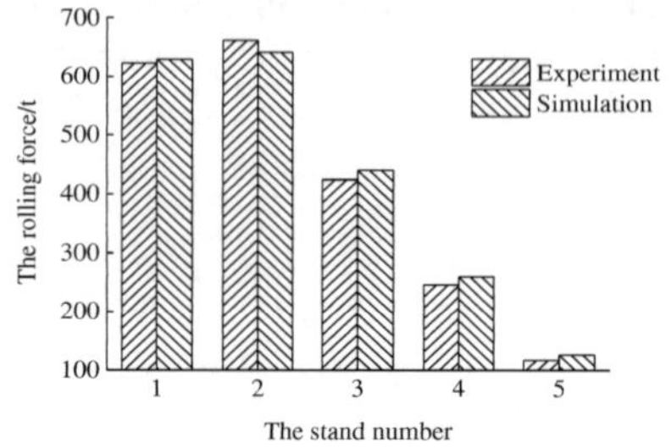

(a) Comparison of rolling force

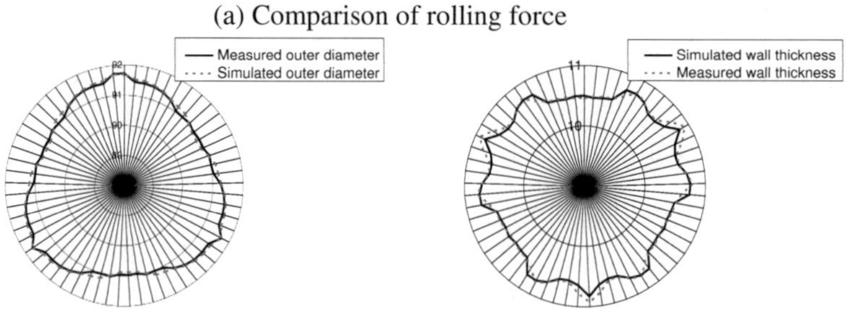

(b) Comparison of wall thickness (c) Comparison of outer diameter

Figure 2. Comparison of measured and simulated results.

4. Spread rule analysis

4.1. The influence of reduction on spread

The size of the roller reduction of three-roller continuous mandrel rolling mill is decided by the eccentricity of the roller pass. The original eccentricity of the first mill is 3mm in this paper. Keeping the blooming temperature and the velocity schedule unchanged and the eccentricity between 2.4~3.6mm, the influence of reduction on spread coefficient of the pass is studied. The spread coefficient with the variation of eccentricity is shown in figure 3. It is concluded from figure 3 that the spread coefficient of continuous tube decreases gradually with the increase of eccentricity. This is because with the increase of eccentricity the roll gap value and the volume of metal accommodated in the roll gap decrease. Therefore, the outer diameter value of continuous tube decreases. According to the law of the lowest resistance, the metal has more axial extension.

4.2. The influence of blooming temperature on spread

Keeping the roller reduction and the velocity schedule of continuous tube unchanged, the blooming temperature increases by 25℃ between 950~1150℃. Then simulate the continuous rolling process of steel tube at different blooming temperature and study the influence of blooming temperature on spread. Figure 4 shows the effect of different blooming temperature on spread coefficient of continuous tube. As can be seen from figure 4, the spread coefficient of continuous tube shows an increasing trend with the increase of blooming temperature. The reason is that with the increase of the blooming temperature the surface metal of continuous tube is oxidized and the friction coefficient increases. Therefore, the longitudinal resistance of metal flow increases and more metal flow to the roll gap which will lead to the increase of spread coefficient.

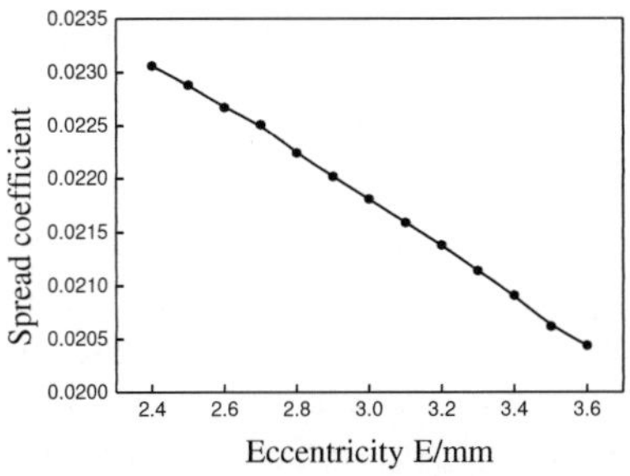

Figure 3. The effect of eccentricity on spread of continuous tube.

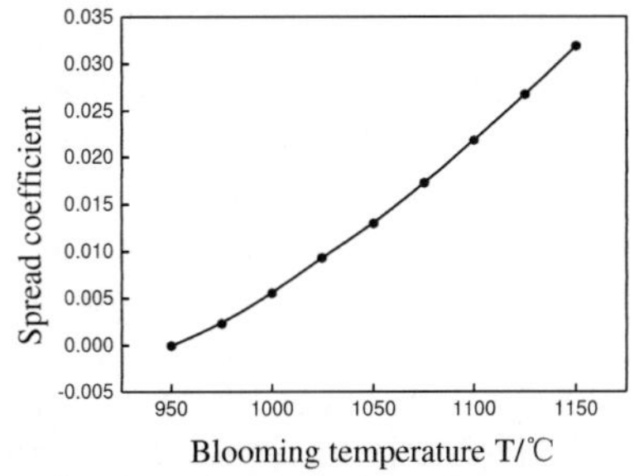

Figure 4. The effect of blooming temperature on spread of continuous tube.

4.3. The influence of velocity schedule on spread

4.3.1. The influence of retaining speed of mandrel on spread. Keeping the roller rotation speed, the blooming temperature and the roller reduction unchanged, the retaining speed of mandrel changes every 0.3m/s. The influence of retaining speed of mandrel on spread coefficient of continuous tube is obtained as shown in figure 5.Figure 5 shows that with the increase of retaining speed of mandrel the spread coefficient of continuous tube basically unchanged, which illustrates that the retaining speed of mandrel has little effect on the spread of continuous tube.

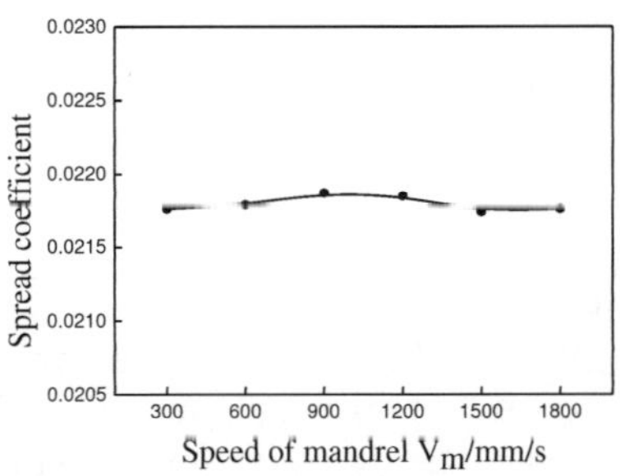

Figure 5. The effect of retaining speed of mandrel on spread of continuous tube.

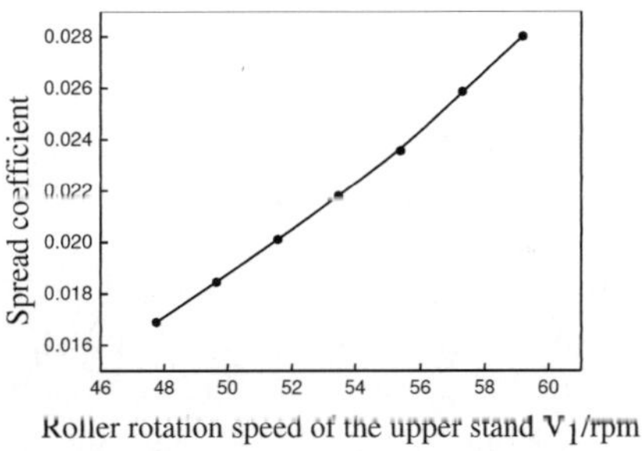

Figure 6. The effect of roller rotation difference decrease on spread of continuous tube.

4.3.2. The influence of roller rotation speed difference on spread of continuous tube. When setting the initial process of three-roller continuous tube, the tension force between stands is firstly required to be zero and then the tension force between stands is produced by the increase or decrease of roller rotation speed. The rotation speed difference between two adjacent stands is reduced by increasing the rotation speed of the upper pass' rollers. The rotation speed difference between two adjacent stands is increased by increasing the rotation speed of the next pass' rollers.

(1) The influence of roller rotation speed difference decreases between two adjacent stands on spread of continuous tube.

Keeping the retaining speed of mandrel, the blooming temperature, the reduction and the roller rotation speed of other four stands unchanged, the roller rotation speed of the upper pass between two adjacent stands is only changed. The influence of the roller rotation speed difference decrease on spread coefficient is obtained as shown in figure 6.It seems from figure 6 that with the decrease of roller rotation speed difference between two adjacent stands the spread coefficient of continuous tube increases. It is because the compressive stress increases between the two adjacent stands on the basis of original roller rotation speed with the decrease of roller rotation speed difference. Meanwhile, the longitudinal flow velocity of metal in the deformation zone becomes slow in the upper pass and more metal is forced to flow towards the roll gap. Therefore, the tendency of piling-up of steel increases and the spread coefficient of continuous tube increases.

IOP Conf. Series: Materials Science and Engineering **220** (2017) 012012 doi:10.1088/1757-899X/220/1/012012

(2) The influence of roller rotation speed difference increases between two adjacent stands on spread of continuous tube.

Keeping the retaining speed of mandrel, the blooming temperature, the reduction and the roller rotation speed of other four stands unchanged, the roller rotation speed of next pass between two adjacent stands is only changed. The influence of the roller rotation speed difference increases on spread coefficient is obtained as shown in figure 7.Figure 7 illustrates that with the increase of roller rotation speed difference between two adjacent stands the spread coefficient of continuous tube decreases. The reason is that the tensile stress increases between the two adjacent stands on the basis of original roller rotation speed with the increase of roller rotation speed difference. In addition, the metal flows along the longitudinal direction more easily and the tendency of pulling steel increases. Therefore, the spread coefficient of continuous tube decreases.

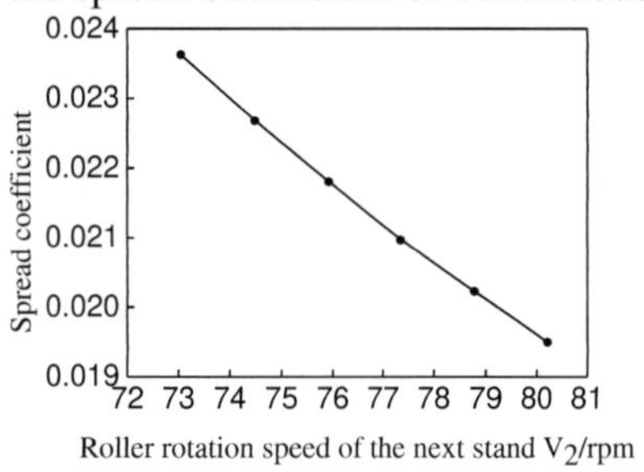

Figure 7. The effect of roller rotation speed difference increase on spread of continuous tube.

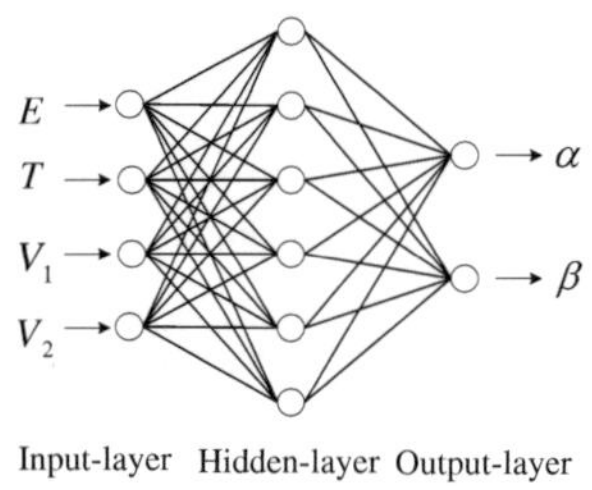

Figure 8. Neural network structure for spread prediction.

5. Establishment of BP neural network model

BP network is a multilayer feedforward artificial neural network. It employs the back-propagation algorithm to train the network model. And the steepest descend algorithm is adopted to keep adjusting the weights and thresholds of network to reduce the error to the minimum.

Design of input layer: The material and size of tubular billet in finite element model are the same. During the simulation process, the influencing factors of spread are studied including the roller eccentricity E of the first stand, the blooming temperature T, the retaining speed of mandrel V_m, the roller rotation speed V_1 of the first stand and the roller rotation speed V_2 of the second stand. Considering the little influence of retaining speed of mandrel on spread of continuous tube, there are four input units of BP neural network: $\boldsymbol{X}=\{E,\ T,\ V_1,\ V_2\}$.

Design of hidden layer: In this paper a single hidden layer is adopted. The trial and error method is adopted to determine the number of hidden layer units. Therefore, six units of hidden layer are obtained by testing and training.

Design of output layer: The output units are wall thickness coefficient α and spread coefficient β. Therefore, the output vector is: $\boldsymbol{Y}=\{\alpha,\beta\}$.Considering the large difference in magnitude, the input date and the output date should be normalized before training. The fundamental structure of spread prediction model of continuous mandrel rolling mill based on BP neural network can be determined as shown in figure 8.

Design of transfer function: In this neural network model, the transfer function of hidden layer is type S logarithmic function *logsig*, and the transfer function of output layer is type S tangent function *tansig*.

6. The implementation of spread prediction model

The date of wall thickness coefficient and spread coefficient adopted in establishing the neural network model in this paper arises from the finite element numerical simulation. After learning and training, the error curve is obtained as shown in figure 9. Figure 9 shows when the number of training steps reaches 8 steps it has been convergence and achieved a certain accuracy requirement. Another 15 groups of date were input into the model for prediction. The predicted date obtained were normalized and compared with the value of finite element simulation, as shown in figure 10. Figure 10 shows that the spread coefficient and the wall thickness coefficient obtained by learning and training fit well with simulation value and BP neural network can predict well for nonlinear spread of continuous tube. In production, if these variables are arbitrarily changed, the spread of continuous tube can be predicted accurately by this neural network model.

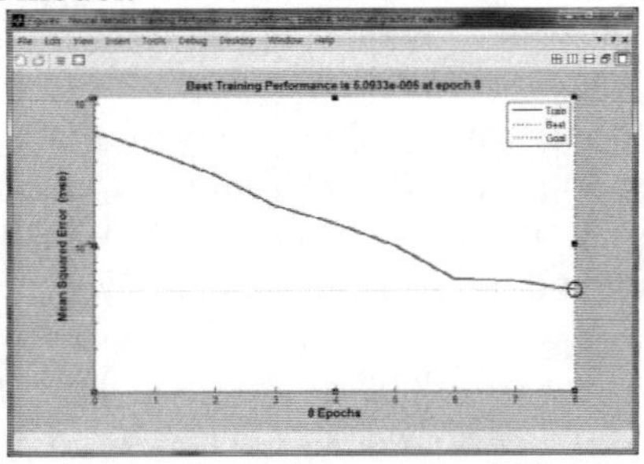

Figure 9. The deviation curve of BP network.

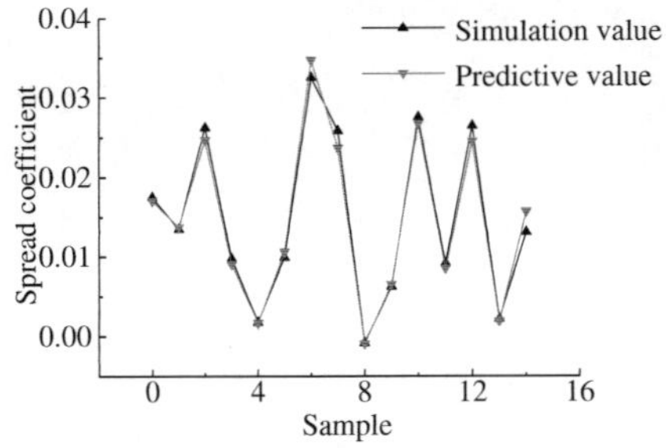

(a) The spread coefficient

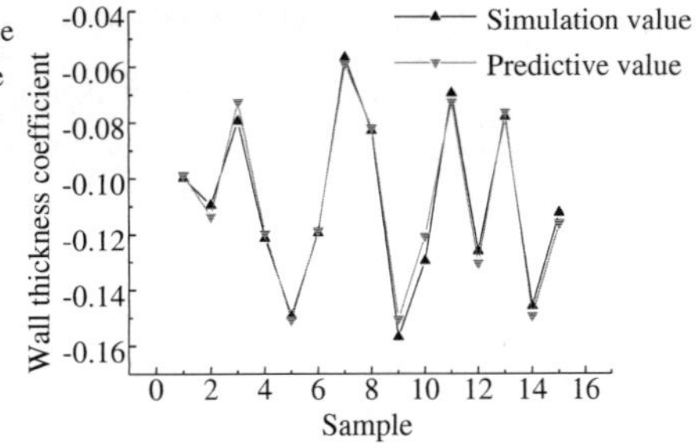

(b) The wall thickness coefficient

Figure 10. The comparison between simulation value and predictive value.

7. Conclusion

The spread of continuous tube decreases with the increase of roller reduction and increases with the increase of blooming temperature. The influence of retaining speed of mandrel on spread is little in the process of steel tube rolling. The spread coefficient increases with the decrease of roller rotation speed difference and decreases with the increase of roller rotation speed difference of the two adjacent stands.

Based on the BP neural network, the prediction model of continuous tube spread is established. The predicted values of spread coefficient and wall thickness coefficient by learning and training fit well with simulation values which illustrates that it is feasible to predict continuous tube spread by the established neural network model.

Acknowledgments

This research was supported by the National Natural Science Foundation of China (Project No.51275445) and Natural Science Foundation of Hebei Province of China (No. E2014203077).

References

[1] Yin Y D and Li S Z 2005 Progress of continous rolling technology for seamless steel tube *Journal of Anhui University of Technology* (Natural Science) **22(3)** 229-233.
[2] Zhang D T, Gao R Q, Sun S Z and Zhong X D 2015 Development history and distribution of domestic mandrel pipe mill lines *Steel Pipe* **44(2)** 6-10.
[3] Yu H, Du F S and Xu Z Q 2011 Influence of pass parameters on retained mandrel rolling process *J IRON STEEL RES INT* **18(2)** 31-37.
[4] Zhuang G and Yin X Q 2006 Analysis of defects of mandrel mill-rolled seamless steel tubes *Steel Pipe* **35(5)** 26-29.

[5] Zhao Z Y, Xie J X and Hong H P 2009 Finite element analysis of wall thickness changes of the cross section of seamless tube in hot continuous mandrel rolling *Journal of Plasticity Engineering* **16(2)** 109-112.
[6] Shuai M R, Huang Q X, Zhu Y C and Pan L 2013 Spread model for TC4 alloy rod during the three-roll tandem rolling process *RARE METAL MAT ENG* **42(5)** 0909-13.
[7] Jia C Y 2001 A method of artificial neural network to spread forecast of hot-rolling steel strip on finishing stands *Steel Rolling* **18(1)** 13-15.
[8] Gao S F, Liu H F, Xi A M and Yang X 2016 Prediction and control of thickness transverse distribution in hot rolling strip *Journal of Harbin Institute of Technology* **48(1)** 180-183.
[9] Wang A L, Yang Q, He A R and Liu H Q 2010 FES width spread model and its optimization for a rough rolling mill *Journal of University of Science and Technology Beijing* **32(4)** 515-519.
[10] Han L L, Zhang L M and Meng L Q 2006 Width variable prediction in medium plate mill by means of neural network *China Mechanical Engineering* **17(18)** 1948-50.
[11] Fu Q and Zhang J Z 2003 Width control model on finishing train based on BP neural networks *Journal of Iron and Steel Research* **15(4)** 70-72.

IOP Conf. Series: Materials Science and Engineering **220** (2017) 012013 doi:10.1088/1757-899X/220/1/012013

Composite Grounding Application of Transmission Line Tower with Flexible Graphite Grounding Material

Hongtao Liu[1], Lei Zhang[1], Jia Xiong[1], Zhenxing Cui[2] and Qi Yang[2]

[1]State Grid Corporation of Datong power supply company in Shanxi Province, China
[2]HongMen Electrical Company in Wuhan Province, China

Abstract. To solve the metal corrosion problem of transmission line tower grounding grid, a composite grounding material technique based on flexible graphite grounding is proposed. Using CDEGS software, the power frequency grounding resistances with different soils layers and different ground network size of tower are simulated. The researches show that layered soil resistance can be reduced by laying vertical grounding body and uniform soil can reduce ground resistance by increasing grounding network size.

1. Introduction

In order to meet the basic requirements of transmission line tower grounding grid design, the following problems must be solved:

1) Grounding resistance meet the requirements of standard. Grounding resistance is directly related to the grounding short-circuit of power frequency and the ground potential raised when lightning flow into the ground. Resistance reduction measures currently used in engineering mainly include increasing grounding grid area, leading an external grounding connection, installing additional vertical grounding, replacing soil and using resistance reducing agent etc. These resistance reduction measures will bring some serious problem such as difficulty of maintain, large economic losses and soil pollution.

2) Serious corrosion of grounding grid. Since the ground operation of the device in the ground brings a harsh operating conditions, especially in the presence of moisture and harmful gases where the soil is acidic. Grounding grid corrosion parameters will change, even cause the problem of open circuit between electrical equipment grounding and each part of ground network, and thus should be paid special attention[1].

Graphite grounding body appears like the shape of cable, made of high-carbon conductive graphite through advanced manufacturing process, using connected the tower with stainless alloy tip. This product belongs to nonmetallic conductor, it has some features like corrosion-resistant. It has the characteristics of high and low temperature resistance, grounding resistance stability, maintenance free, safe and reliable. The product is light-weight, with less consumption, effect significantly, conservation of land resources and construction costs, energy saving, environmental protection, security and particular suitable for using in the places where construction difficulty and high soil resistivity such as acidic soil, alkaline soil, marshes, damp areas, the beach, the mountains and the hills.

In order to save the excavation of earthwork quantity, reduce the intensity of construction, curtail construction time, increase service life and improve economic efficiency, this paper presents a composite transmission line tower grounding technique based on the flexible graphite grounding, studies the power frequency grounding resistance of different hierarchical soils and different network size parameters of tower.

2. Flexible graphite composite grounding design requirements

2.1. Different soil stratification

The "Grounding for AC Electrical Installations" standard GB50065-2011 said, transmission line tower grounding resistance should meet the appropriate standard in different soil resistivity places, and not bigger than the parameters specified in Tab.1.

Table 1. Transmission line power frequency grounding resistance requirements

Soil resistivity ρ(Ω·m)	Ground Resistance(Ω)
≤100	10
≤500	15
≤1000	20
≤2000	25
>2000	30

In reference[2], graphite grounding system installation number has a corresponding reference index according to the soil resistivity, as shown in Tab.2.

Table 2. Laying graphite grounding reference index

No.	Soil resistivity ρ(Ω·m)	Reference data of laying time / month
1	≤50	2~4
2	≤100	4~6
3	≤200	6~8
4	≤500	10~12
5	≤1000	12~16
6	>1000	20~24

In the view of the tower grounding problem for high soil resistivity areas, focusing on whether grounding resistance and the laid number of grounding bodies meets the standard when soil resistivity greater than 500Ω·m.

This paper discusses the grounding resistance of two and three layered soil conditions when put the same depth and clearance of vertical grounding system. During the discussion, ground root net held at open 10 m, rays 30 m, ground depth 0.8 m[3].

2.2. Two layers of soil

According to the selected under layers of soil in the reference[4], this paper selects the following two stratified soil to be calculated combined with actual soil stratification situation in high soil resistance region[4][5].The one is upper for gravel and base for sandy soil structure (upper soil resistivity of 5000Ω·m, subsoil resistivity of 300Ω·m), and the other is upper for gravel and base for stone soil structure (upper layer soil resistivity of 3000Ω·m, subsoil resistivity of 400Ω·m).

We set the upper soil resistivity is ρ1, lower soil resistivity is ρ2, the thickness of the layers of soil were h1 and h2, l means grounding depth. Two layers of soil thickness h1, h2 are set to 2 m and infinity respectively, soil resistivity parameter settings are shown in Tab.3.

Table 3. Two layers of soil vertical grounding body simulation parameter

Number of grounding body(Each ray)	ρ1(Ω·m)	*l*(m)	ρ2(Ω·m)
0	5000		300
1	5000	0.4、0.6、0.8、1.0、1.2、1.4、1.6	300
2	5000	0.4、0.6、0.8、1.0、1.2、1.4、1.6	300
3	5000	0.4、0.6、0.8、1.0、1.2、1.4、1.6	300
4	5000	0.4、0.6、0.8、1.0、1.2、1.4、1.6	300
5	5000	0.4、0.6、0.8、1.0、1.2、1.4、1.6	300
6	5000	0.4、0.6、0.8、1.0、1.2、1.4、1.6	300
0	3000		400
1	3000	0.4、0.6、0.8、1.0、1.2、1.4、1.6	400
2	3000	0.4、0.6、0.8、1.0、1.2、1.4、1.6	400
3	3000	0.4、0.6、0.8、1.0、1.2、1.4、1.6	400
4	3000	0.4、0.6、0.8、1.0、1.2、1.4、1.6	400
5	3000	0.4、0.6、0.8、1.0、1.2、1.4、1.6	400
6	3000	0.4、0.6、0.8、1.0、1.2、1.4、1.6	400

2.2.1. Three layers of soil. According to the three soil resistivity selected in the reference[4], in the three soil level simulation, we set the upper soil resistivity is ρ1, the intermediate layer soil resistivity is ρ2, lower soil resistivity value ρ3, three soil thicknesses of h1, h2 and h3. The layer soil's thickness h1, h2 and h3 are set to 1.5 m, 0.5 m and infinity respectively, the soil resistivity parameters are shown in Tab.4.

Table 4. Three vertical ground soil simulation parameters

Number of grounding body	Soil resistivity relationship	ρ1(Ω·m)	ρ2(Ω·m)	ρ3(Ω·m)	*l* (m)
1、2、4、6	ρ1>ρ2>ρ3	200	1100	2000	1.6
1、2、4、6	ρ1>ρ2<ρ3andρ1=ρ3	2000	200	2000	1.6
1、2、4、6	ρ1<ρ2>ρ3andρ1=ρ3	200	2000	200	1.6

2.2.2. Different grounding grid area. There are two kinds of fault currents flowing in the fault current: the vertical flow along the grounding body and the transverse flow to the soil. The current ratio of the forward flow and the current to the soil is closely related to the electromagnetic characteristics of the grounding material, soil resistivity and so on. So, first of all, the influences of grounding materials on the grounding resistance of typical grounding grid structure in different soil resistivity are analyzed, when the area of the local network is consistent. In this paper, a typical power plant, substation commonly used metal grounding material, such as brass, 45# steel and different diameters of the flexible graphite grounding material are chosen for studying the influence of the material on grounding resistance. The relevant parameters of each grounding material are shown in Tab.5, and the structure of grounding grid is shown in Fig.1.

IOP Conf. Series: Materials Science and Engineering **220** (2017) 012013 doi:10.1088/1757-899X/220/1/012013

Table 5. The relevant parameters of each grounding material

Material Parameter	Brass	45# Steel	Flexible graphite grounding material
Electrical resistivity×10^{-7}/Ω·m	0.175	1.75	325
Relative permeability (μ_r)	1	636	1
Diameter /mm	20	20	Φ20, Φ28

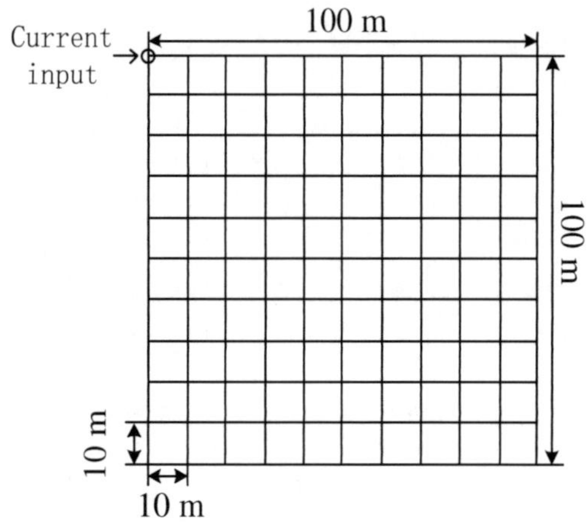

Figure 1. Grounding grid structure.

3. Calculations of the models

3.1. Different soil stratification

3.1.1. Two layers of soil. In the national standard, this section further requires the grounding resistance value be-low 25 Ω. For the layered soil which the upper one is 5000Ω·m and the lower one is 300Ω·m, the simulation have been done with CDEGS according to the parameters table in 2.1.1 section. The calculated results are shown in Fig.2.

In Fig. 2, when the depth of the grounding body is 1.2 m, the grounding resistance is greatly reduced. The more number of the grounding body is, the larger amplitude of the grounding body is. In order to meet the grounding resistance in 25 Ω or less, 20 grounding bodies must be buried at least. Taking into account the economy, the two layers soil optimal solution is 20 vertical ground bodies with 1.556 m deep.

Similarly, the gravel and rocky layered soil (upper 3000Ω·m, lower 400Ω·m) were simulated and calculated. The results are shown in Fig.3. Similarly, from Fig.3, the optimal solution for the two layers soil is 16 vertical grounding bodies with 1.443 m deep.

Therefore, in the two layer of soil, the grounding resistance has the most obvious changes when the vertical body digs from the high resistivity soil into the low resistivity soil, and the more number of grounding body is, the faster the grounding resistance reduces. Therefore, for the two soil layers, the engineer should ground through the upper soil with high resistivity and bury 16~20 vertical grounding bodies.

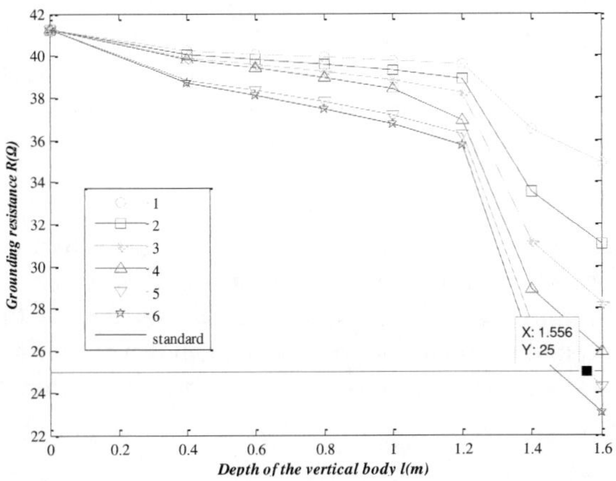

Figure 2. The grounding resistance changes with the depth of ground under different number of vertical grounding body.

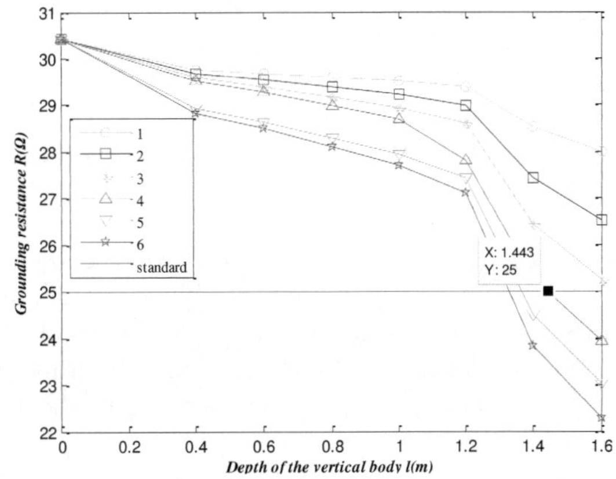

Figure 3. The grounding resistance changes with the depth of ground under different number of vertical grounding body.

3.1.2. Three layers of soil

Table 6. Grounding resistance under different soil

No.	soil resistivity Relationship	ρ_1 (Ω·m)	ρ_2 (Ω·m)	ρ_3 (Ω·m)	l (m)	The number of grounding body	Grounding resistance R(Ω)
1	$\rho_1<\rho_2<\rho_3$	200	1100	2000	1.6	0	22.1
2	$\rho_1>\rho_2>\rho_3$	2000	1100	200	1.6	0	20.6
3	$\rho_1>\rho_2<\rho_3$ and $\rho_1=\rho_3$	2000	200	2000	1.6	0	35.1
						1	33.3
						2	32.5
						4	31.1
						6	30.3
4	$\rho_1<\rho_2>\rho_3$ and $\rho_1=\rho_3$	200	2000	200	1.6	0	10.3

It can be seen from Tab.6:

1) For the soil resistivity of①②④, it can meet the standard without laying vertical grounding bodies.

2) When the soil resistivity is small in the middle, the laying of the vertical grounding body at equal distance can effectively reduce the grounding resistance. 24 or more vertical grounding bodies should be laid, each of which is at least 1.6 m.

3.2. Different grounding grid area

3.2.1. Effect of grid area on the grounding resistance. With the increase of the grid area, the difference of the grounding resistance of the flexible graphite grounding material with the brass, 45# steel material is increased. In addition, the grounding resistance decline rate becomes slow with the increase of the grid area. After the analysis, it is found that the reason is that with the continuous expansion of the grid area, the far end of the grounding body to reduce the effect of bulk flow, the effective grounding area of the ground material tends to be saturated. Under the same soil resistivity, the grounding grids made by those different materials only exist differences in the bulk resistance. With the increase of the grid area, the ratio of grounding material bulk resistance in the whole grounding resistance increased, so the grounding resistance of the material with large resistivity is bigger than other. Because the resistivity of the flexible graphite grounding material is larger than that of the brass and the 45# steel grounding material, the grounding resistance difference increases when the grounding grid area is larger.

3.2.2. Effect of grid area on touch voltage. According to the Fig.4 with the increase of the grid area, the maximum touch voltage of the flexible graphite grounding material, brass and 45# steel are reduced. At the same time, with the increase of the grid area, the maximum touch voltage of the grounding grid different between flexible graphite grounding material and other mental material are in-creased. This is due to the increase in the grid areas, the effective grounding area of different grounding materials are saturated.

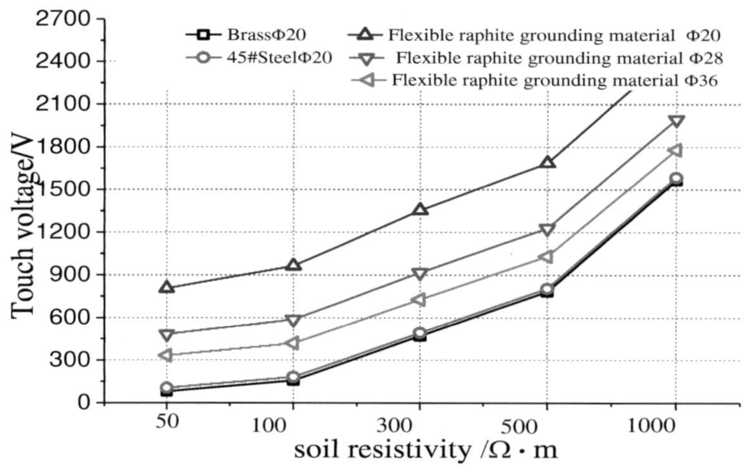

Figure 4. Effect of soil resistivity on resistance value.

3.2.3. Effect of grid area on step voltage. As shown in Fig.5, with the increase of the grid area the step voltage of each material is also increasing which is similar to the touch voltage. By comparative analysis of the application of different grounding materials in substation, except for the grounding resistance, the grounding characteristics of flexible graphite grounding material under different soil conditions and different grounding grids are different from that of brass and steel. On the one hand, the flexible graphite grounding material diamagnetic properties have not been reflected because of the simulation calculation by alternating current. On the other hand, the resistivity of the flexible graphite grounding material of the material itself is higher than that of the metal grounding materials, that the grounding resistance, touch voltage and the step voltage of the flexible graphite grounding material is poorer than that of the metal grounding materials, and the advantages of flexible graphite grounding material in touch with soil are not reflected in the simulation process.

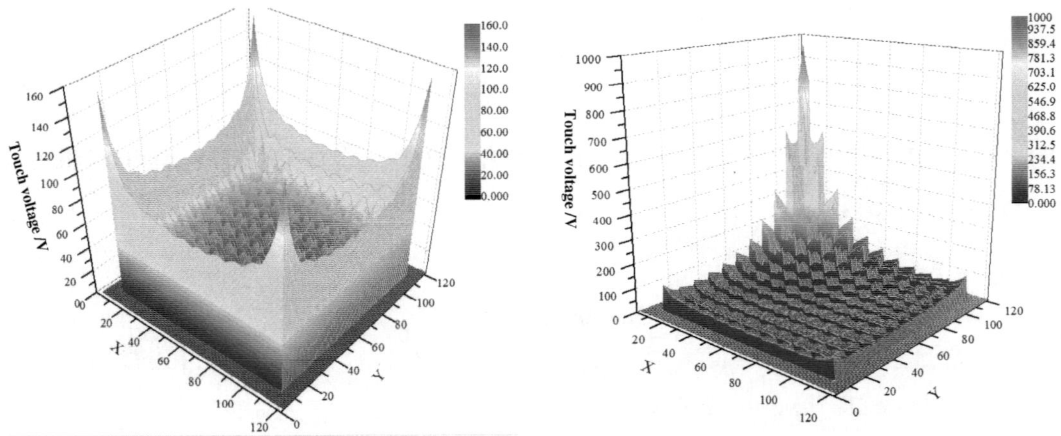

(a)Φ20 Brass (b) Φ20 Flexible graphite grounding material

Figure 5. Effect of soil resistivity on resistance.

4. Conclusion

By comparing the ground resistance of different soil layers and different grounding grid areas, this paper presents a composite grounding scheme of flexible graphite grounding system, and the following conclusions can be drawn:

1) For the two soil layers, the engineer should ground through the upper soil with high resistivity and bury 16~20 vertical grounding bodies. In the case of three layers of soil, only when the soil resistivity is small in the middle, the laying of the vertical grounding body should be buried at equal distance.

2) With the increase of the diameter of flexible graphite grounding material, the grounding resistance has also a trend of decrease.

Acknowledgments

We would express our sincerely gratitude to all those who helped us during writing this paper, thank for your valuable suggestions. And we greatly acknowledge the help in paper writing and rhetoric aspects from our fellow colleagues, they had provided many novel and rigorous ideas. In the end, specially thank you for Linsey Chen's precious advice.

References

[1] Zeng L T 2008 Calculation and design of grounding grid resistance under non-uniform soil Guangxi University.

[2] Shi J F 2000 Grounding technology and construction of the practical graphite material *Journal of Meteorology and Environment* **25** 37-38.

[3] Liu X L 2003 Grounding design of power plant and substation in high soil resistivity area *Ph.D. thesis, Zhengzhou University*.

[4] Xing T F 2013 Research on grounding test and calculation method of layered soil *Master thesis, Chengdu University of Information Technology, Sichuan.*

[5] Tang S N, Mo W Q, Zhou Y L 2006 Grounding treatment of substation in high soil resistivity area *High Voltage Engineering* **32** 121-122.

IOP Conf. Series: Materials Science and Engineering **220** (2017) 012014 doi:10.1088/1757-899X/220/1/012014

Analysis of Piezoelectric Actuator for Vibration Control of Composite plate

Ahmed R Gomaa[1*] and Huang Hai[1]

[1]School of Astronautics, Beihang University (BUAA), 37 Xueyuan Rd, Beijing, China

*E-mail: ragabeta@yahoo.com

Abstract. Vibration analysis is studied numerically in this paper for a simply supported composite plate subjected to external loadings. Vibrations are controlled by using piezoelectric patches. Finite element method (ANSYS) is used for obtaining finite element model of the smart plate structure, a layered composite plate is manufactured experimentally and tested to obtain the structure mechanical properties. Different piezoelectric patch areas and different applied gain voltage effects on vibration attenuation is studied. The numerical solution is compared with the experimental work, a good agreement achieved.

1. Introduction

The use of layered composite materials in the aeronautics and astronautics industry have been increased, plate and shell structures are most widely used elements. It had been showed that the piezoelectric materials are useful and used for smart adaptive structures [1-4]. Control and suppression of structural vibrations are necessary and also important. Recently, smart materials and structures especially piezoelectric elements are preferred and widely used for vibration control and suppression [5-7]. FEM is used to reduce the iterations done experimentally and to evaluate the behaviour of the smart structures subjected to electromechanical loads [8-10].

ANSYS is used in this work to study the vibrations of a fully simply supported composite plate subjected to external force. Vibration is controlled and suppressed successfully using the PZT piezoelectric patches bonded to the plate surface. The effect of different PZT patch area is examined experimentally and the results compared with the numerical solution, a good agreement is achieved. Also the effect of different applied voltage to the piezoelectric film is studied. 8-ply E-glass epoxy composite plate is manufactured and its mechanical properties is obtained using tensile tests, then vibration tests is applied on the plate with the PZT patches.

2. Production of the laminate specimens

E-Glass woven fiber was used as a bidirectional fabric as shown in Figure1. And laminating epoxy resin as matrix for the composite material of the laminates specimens. Industry has evolved a dozen separate manufacturing processes for fabricating composites each one offers advantages and specific benefits. Hand lay-up is a basic moulding process which is used in this work.

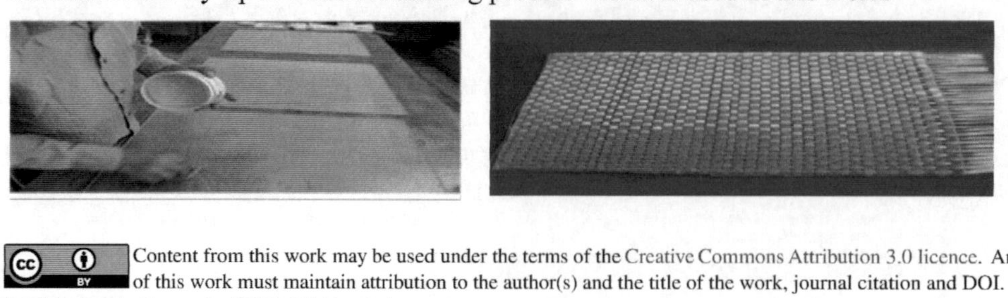

Figure 1. E-Glass woven fabric.

3. Experimental determination of mechanical properties

The first set of tests in the experimental work was performed for the mechanical characterization of the used composite laminate plates. Characterization was used to determine the mechanical properties of the composite laminate (Glass/epoxy). These mechanical properties were required as inputs to the finite element model for natural frequency detection by free vibration. The test program and specimen's configuration for laminate and specimen's configuration for laminate characterization are shown in Table.1. In order to obtain any material property, three specimens were tested and the average value was considered.

Table 1. Characterization test program

Elastic constants	No. of specimens	Length	Width	Layers	Orientation
E_{11}	3	250	15	8	0°
E_{22}	3	175	25	8	90°
ν_{12}	The same specimens from E_{11} test.				
ν_{21}	The same specimens from E_{22} test.				
G_{12}	3	250	25	8	$\pm 45^o$

The testing machine used in the present experimental work was a computerized universal testing machine (Shimadzu) of capacity 50 kN.

3.1. Tensile test procedure

Nine specimens were prepared for tensile test using material test system machine as relevant to ASTM standard code (D3039) [11] as shown in Figure 2. The mechanical properties of 8-layered woven fiber Glass/Epoxy were determined using 5 mm electrical strain gages. Unidirectional tensile tests were performed on specimens cut in longitudinal and transverse directions to obtain E_{11}, E_{22}, υ_{12} and υ_{21}. Specimens cut at angle $+45°$ to the longitudinal direction were prepared to find the shear modulus G_{12} according to Jones [12].

Figure 2. Tensile test machine and the strain gauges mounted to specimen.

3.2. Results of tensile test

3.2.1. Determination of E_{11} and υ_{12}. The longitudinal Young's modulus E_{11} and the major Poisson's ratio υ_{12} were determined from a simple tension test. The longitudinal Young's modulus E_{11} was determined from the slope of the stress-strain curve, the major Poisson's ratio υ_{12} was determined from the slope of the transverse and axial strain diagram during the same experiment as shown in Figure 3.

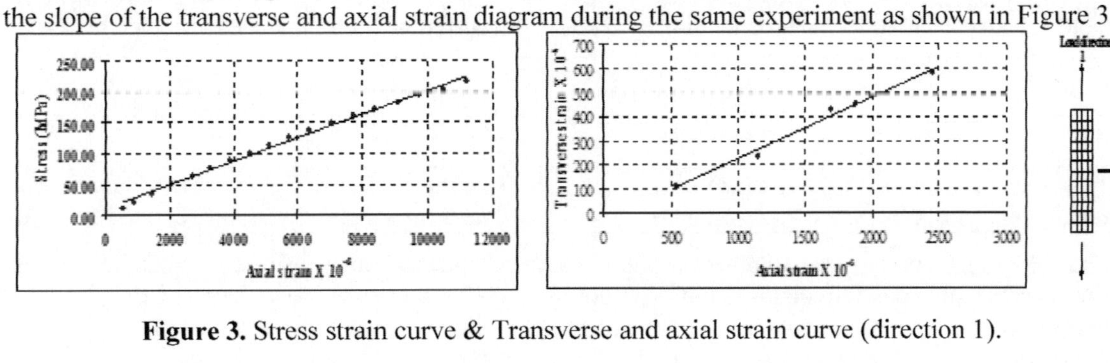

Figure 3. Stress strain curve & Transverse and axial strain curve (direction 1).

3.2.2. Determination of E_{22}, υ_{21}, and G_{12}. Similarly, the transverse Young' s modulus E_{22}, Poisson's ratio υ_{21}, and in-plane shear modulus G_{12} was determined from slope of stress strain diagrams.

From the previous experimental tensile tests, the average values of material properties of the 8-layered woven fiber Glass/Epoxy laminate are given in Table 2.

Table 2. Experimentally measured mechanical properties of 8-layer woven fiber Glass/Epoxy laminate

E_{11}	E_{22}	υ_{12}	υ_{21}	G_{12}
19 GPa	19 GPa	0.256	0.249	2.8 GPa

4. Vibration test

The second set of tests is performed on square composite laminate with simply supported boundary conditions is the vibration test. The natural frequencies of the composite plate with different piezoelectric patch areas is measured and compared with the numerical results from ANSYS, then the effect of changing of applied volt on the PZT film is studied.

4.1. Effect of piezoelectric patches area on natural frequencies

The composite laminate plate specimens used in present experiment consist from 8 layers of (0/90) woven E-Glass fiber with epoxy matrix. After the cure process, four test specimens of size (200x200x2.5) mm^3 without piezoelectric patches and twelve test specimens of the same dimension with different areas of piezoelectric patches (50x50, 110x110, 150x150) mm^2 bonded on the top face of the composite plate with special wax were prepared to study the effect of increasing the piezoelectric actuator area on natural frequencies Figure 4.

The natural frequencies were measured for all specimens with four edges simply supported boundary conditions. Ten repeated tests were conducted for each plate. The force was measured using a force transducer with charge amplifier (B&K type 2626). This output was captured by one accelerometer to measure dynamic response and was amplified using a conditioning amplifier (B&K type 2626) as shown in Fig.4 and then read using the high resolution signal analyzer (B&K type 3562A), giving the frequency response function (FRF).

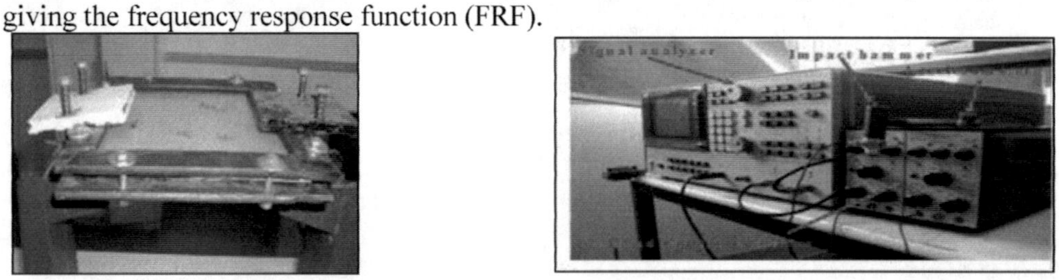

Figure 4. Vibration test measuring system with SS fixture of square laminate.

The first four measured experimental modes of vibration for different piezoelectric patch (actuator) area with 1 volt applied current on the top surface of structure are given in Table 3.

Table 3. Experimentally measured natural freq. (Hz) with different central piezoelectric patch areas

Mode	host plate only	Mid plane Piezoelectric film area		
		(50x50) mm^2	(110x110) mm^2	(150x150) mm^2
First	172.2 Hz	176.3 Hz	191.2 Hz	199.4 Hz
Second	459.5 Hz	462.8 Hz	550.5 Hz	615.2 Hz
Third	466.3 Hz	467.5 Hz	560.4 Hz	620.2 Hz
Fourth	686.4 Hz	695.3 Hz	798.2 Hz	870.1 Hz

From Figure 5, we can clearly observe that, the active control via the piezoelectric patches results in the increase of natural frequency, more predominantly for higher modes. The sensitivity of the primary frequency is high for the largest actuation area but lower for the small piezoelectric patch area, while the sensitivity of the secondary, third and fourth frequency is high for both the small and large delamination areas.

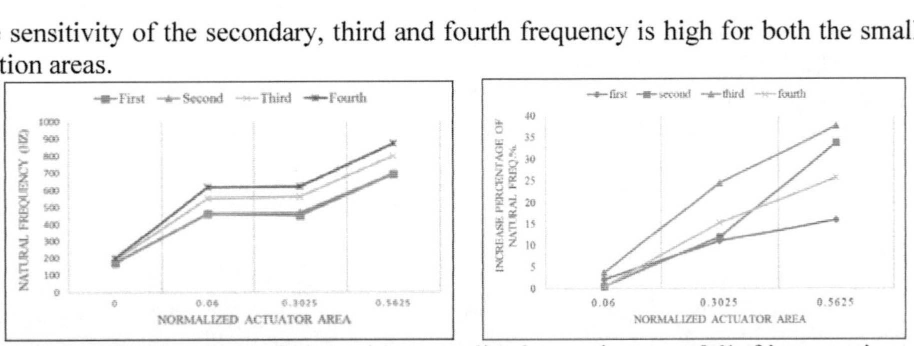

Figure 5. Natural frequencies variation with normalized actuation area &%of increase in natural freq.

It can be seen that, as expected, for all modes the larger actuation area, the more significant increase in natural frequencies, also it can be deduced that different modes have different sensitivities to the actuation areas. For instance, the natural frequency of first mode has relatively small increase (16%) for actuation area (150x150) mm^2 whereas the increase in the higher mode is significant (37.7%) relative to host plate as shown in Figure 5.

4.2. Numerical results compared with experimental work

The results from present FEM validated with the experiments conducted on eight plies (0/90) of dimensions (200x200) mm^2 square laminate with four edges simply supported boundary conditions accomplished in our laboratory. Table 4 show the comparison between frequencies predicted for the first four modes by using the finite element model (ANSYS) and those measured experimentally. It arises that the results obtained from numerical calculations were in good agreement with the experimental investigation. The maximum error between FEM and experimental test results is less than 5.41%. The present verified finite element model can be used for further analysis.

Table 4. Experimental measured natural frequencies and numerical results for SS square laminates in Hz

Actuation area mm^2	First mode		Second mode		Third mode		Fourth mode	
	EX	FEM	EX	FEM	EX	FEM	EX	FEM
Host Plate	172.2	171.73	459.5	455.37	466.3	455.37	692	690.79
50x50	176	173.78	462	470.48	467.5	470.48	695	698.83
100x100	191	186.16	551	546.86	560	546.86	798	788.31
150x150	199	195.94	615	607.85	620	607.85	870	889.28

Figure 6 shows the effect of increasing the piezoelectric patch area on reduction of the first frequency of composite plate. it's very clear that vibration attenuation of 55.2%, 66.4%, and 79.7% are achieved as the patch area are set to (50x50), (110x110), and (150x150) mm^2 respectively.

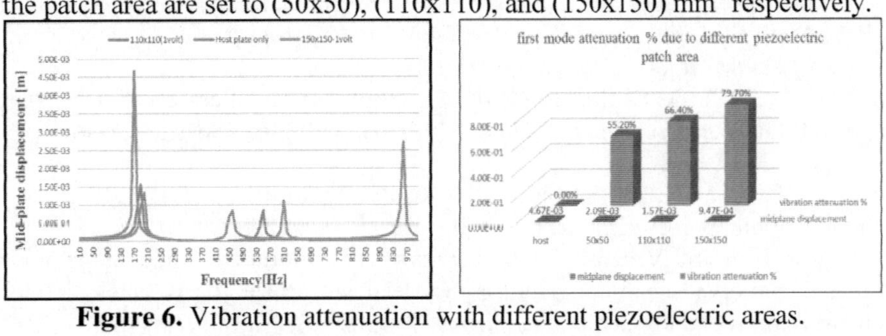

Figure 6. Vibration attenuation with different piezoelectric areas.

4.3. Effect of piezoelectric actuator applied voltage on natural frequencies

The effect of changing the applied voltage on the one specified patch area is studied also numerically. By choosing the 150X150 mm^2 area patch and apply different voltage value on the upper face of the piezoelectric film (-50v,1v,50v) we noticed that the amplitude of mid-plane displacement of the plate reduces by increasing the voltage value. Figure 7 shows the displacement amplitude versus the frequency due to -50v, 1v, and 50v respectively.

It is also noticed that the 2nd mode frequency is not effected. It's very clear that vibration attenuation of 54.8%, 79.7%, and 96.5%are achieved as the control voltage are set to -50,1, and 50 volts respectively.

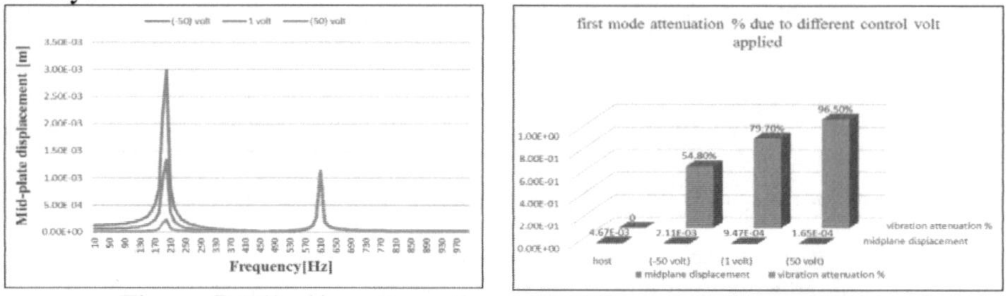

Figure 7. Vibration attenuation with different applied voltage.

5. Summary

ANSYS, FEM code, is used to investigate the response of the smart structure by control the vibration of E-Glass Epoxy composite plate. 8-ply (0/90) composite plat is manufactured and mechanically tested to measure its mechanical properties. Vibration test is performed experimentally for the plate with different piezoelectric patch areas bonded to measure the modal frequencies. The experimental results are compared with the numerical results and a good agreement is achieved. Also the effect of applied voltage on the piezoelectric patch is studied, and 96% of first mode frequency is attenuated.

References

[1] Abreu G L C M, Ribeiro J F, and Steffen V 2004 Finite element modeling of a plate with localized piezoelectric sensors and actuators *J. of the Braz. Soc. Of Mech. Sci.&Eng.* **26(2)** 117-128.

[2] Zhou Y, and Wang X 2004 Finite element modeling for piezoelectric smart composite including debonding *45th AIAA/ASME/ASCE/AHS/ASC str., Str. Dyn. & Mat. Conf.* 19-22.

[3] Lasova Z, and Zemcik R 2012 Comparison of finite element models for piezoelectric materials *Procedia Engineering* **48** 375-80.

[4] Nestorovic T, Shabadi S, Marinkovic D, and Trajkov M 2013 Modeling of piezoelectric smart structures by implementation of a user defined shell finite element *Facta Universitatis: Mech. Eng.* **11(1)** 1-12.

[5] Saviz M R, and Mohammad M 2010 Dynamic analysis of laminated cylindrical shell with piezoelectric layers under dynamic loads *J. finite elem. in analysis & design* **46** 770-81.

[6] Wang S Y, Quek S T, and Ang K K 2001 Vibration Control of Smart Piezoelectric Composite Plates *J. of Smart Materials and Structures* **10** 637-44.

[7] Araujo A L, Carvalho V S, Soares C M M, Belinha J, and Ferreira A J M 2016 Vibration analysis of laminated softed core sandwich plates with piezoelectric sensors and actuators *J. Comp. Stru.* **151** 91-98.

[8] Bendigeri C, Narayana K, and Ramachandra K 2013 Finite Element Development for Analysis of Smart Structures *International Journal of Mechanical Engineering* **6 (1)**.

[9] Gohari S, Sharifi S, and Vrcelj Z 2016 New explicit solution for static shape control of smart laminated cantilever piezo-composite-hybrid plates/beams under thermos-electro-mechanical loads using piezoelectric actuators *Journal of Composite Structures* **145** 89-112.

[10] Kulkarni P, and Kanakaraddi R K 2015 Finite element modeling of piezoelectric patches for vibration analysis of structures *Int. Res. J. of Engineering and Technology* **2(4)**.

[11] ASTM D3039M 1997 Standard test method for tensile properties of polymer matrix composite materials *American Soc. of Test & Mat.* West Conshohocken, PA 19428, USA, **15**.03.

[12] Jones M 1975 *Mechanics of Composite Materials* McGraw Hill, New York.

IOP Conf. Series: Materials Science and Engineering **220** (2017) 012015 doi:10.1088/1757-899X/220/1/012015

Simultaneously enhanced mechanical and damping properties of Mg–Zn–Y alloys reinforced with LPSO phase

Jingfeng Wang[1], Haibo Wang[1], Shun Li[1], and Shaohua Wang[1]

[1]Chongqing University, Shapingba District, Chongqing 400044, China

Abstract. The microstructure, mechanical properties and damping capacities of Mg-Zn-Y alloys were investigated and compared systematically. The results showed that strength and damping of the alloy were increasing markedly with the increase of the volume fraction of long period stacking ordered (LPSO) phase (8%, 16%, 32%, 64%, respectively) on the whole. The corporate effect of LPSO phase and solid solution atoms was beneficial to the strengths. The Mg-1.36Zn-2.28Y can be classified as high damping metals ($Q^{-1} \geqq 0.01$) at strain amplitudes surpassing 1×10^{-3}. With increasing of the LPSO phase, the critical strain amplitudes of alloys gradually decreased so that alloys can break away from pinning points more easily, thus, achieving a more superior damping performance. In addition, the strain amplitude-independent damping and strain amplitude-dependent damping of the Mg–Zn–Y alloys both increased. The damping capacities of the alloys cannot be explained by the Granato–Lücke theory exclusively.

1. Introduction

Magnesium alloys with low density, high strength, anti-dynamic impact load capacity, electromagnetic radiation shielding ability are widely used in aerospace, rail and 3C products [1-4]. However, with the rapid development of modern society, more and more sophisticated equipments, complex and diverse work environment, putting forward more urgent requirements for reducing energy consumption, vibration and noise. Thus, developing the lightweight and structure-function integration alloy (High-strength and high-damping alloy) will be the focus. However, the current research and development of structure-function integration magnesium alloy is very scarce. Most of the research are aimed at the mechanics or single function characteristics of magnesium alloys [5,6], because it is not easy obtain an alloy with good mechanical property and high damping capacity simultaneously. For instance, merchant high-strength Mg alloys, currently represented by the AZ and ZK series, do not have excellent damping capacities [7–10], and the high-damping Mg–Zr alloy does not have enough strength for engineering applications [11].

Fortunately, in our previous work, it was found that the damping and strength of the alloy at low strain gradually increased after the introduction of LPSO phase into the Mg-Cu-Mn-Zn-Y alloy [12]. This provides a new way for preparing high strength and high damping alloys. Simultaneously, study have been reported that Mg-Zn-Y alloy, producing by rapid solidification of powder metallurgy, has a very superior mechanical property [13]. Therefore, introducing the LPSO phase into the Mg-Zn-Y alloy is possible to prepare a structural-functional alloy with superior strength and damping property.

In this paper, the effect of LPSO structure on the mechanical property and damping capacity are expected to be researched by introducing LPSO phase into a new Mg-Zn-Y alloy. The LPSO structure is obtained by adding Zn and Y in the same atomic proportions into the alloys. The microstructure, mechanical property and damping capacity of these alloys are discussed and compared synthetically.

IOP Conf. Series: Materials Science and Engineering **220** (2017) 012015 doi:10.1088/1757-899X/220/1/012015

The present investigation hopes to contribute to the lightweight material designs of magnesium alloy for anti-vibration and noise-reduction applications.

2. Experimental procedure

The materials used in this study were the as-cast Mg–Zn–Y magnesium alloys with varying Zn and Y contents. The chemical compositions of these alloys were determined by X-ray fluorescence (XRF) and are listed in Table1. Mg-Zn-Y alloys were prepared by melting ultra-high purity Mg, pure Zn and Mg-30wt.%Y master alloy in an electric resistance furnace. All these master alloys were kept at 750C° for an hour, and then the impurities in it were removed. After maintained for 30 minutes, the metal solution was poured in a mould, which were pre-heated at 300C° for 2 hours. In order to prevent the oxidation, a larger flow of gas was used in the above process.

The phase compositions were by measured by a Rigaku D/MAX2500PC X-ray diffractometer (XRD) with a copper target at a 10–90° scanning angle and 1.5°/min scanning speed. Microstructure of the specimens was measured by using an Olympus optical microscope and a TESCAN VEGA 3 LMH scanning electron microscope (SEM) equipped with EDS (energy dispersive X-ray spectrometry). Tensile testing was carried out on a Shimadzu CMT-5105 material tester at a stretching rate of 1mm/min at room temperature. Damping capacity was determined by loss tangent Q^{-1}=tan φ, which φ is the lag angle between the applied strain and the response stress. The damping capacity was tested on an internal consumption meter MFP-1000 in a torsion mode. The machined dimension of the damping test specimen was 35 mm ×Φ2 mm using an electric spark cutter. Measurements were made at strain amplitudes (ε) ranging from 1×10^{-5} to 2×10^{-3} at room temperature with a vibration frequency (f) of 1 Hz to determine the strain-dependent damping capacity.

Table 1. Chemical composition and LPSOphase volume fractionof the as-cast Mg–Zn–Y alloys

Nominal Alloys	Chemical Composition (wt%)			Y/Zn Ratio	Volume Fraction of LPSOPhase (%)
	Mg	Zn	Y		
Alloy 1	Bal.	1.36	2.28	3/2	8
Alloy 2	Bal.	2.35	5.08	3/2	16
Alloy 3	Bal.	4.23	8.31	3/2	32
Alloy 4	Bal.	8.37	18.13	3/2	64

3. Results and discussion

3.1. Microstructure of the As-cast Mg-Zn-Y alloys

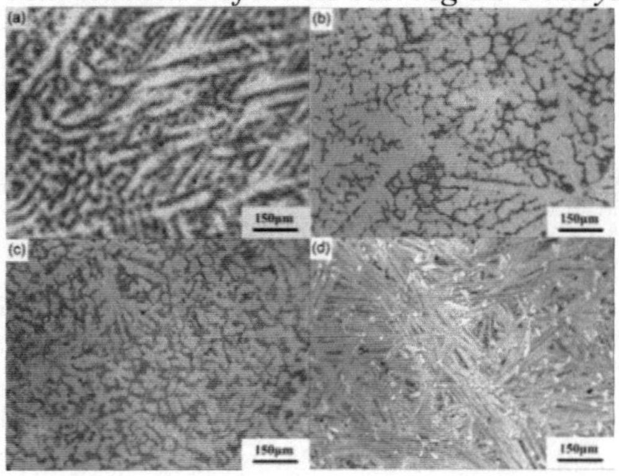

Figure 1. Microstructures of the as-cast Mg-Zn-Y alloys:(a)Alloy 1;(b)Alloy 2;(c) Alloy 3; (d) Alloy 4.

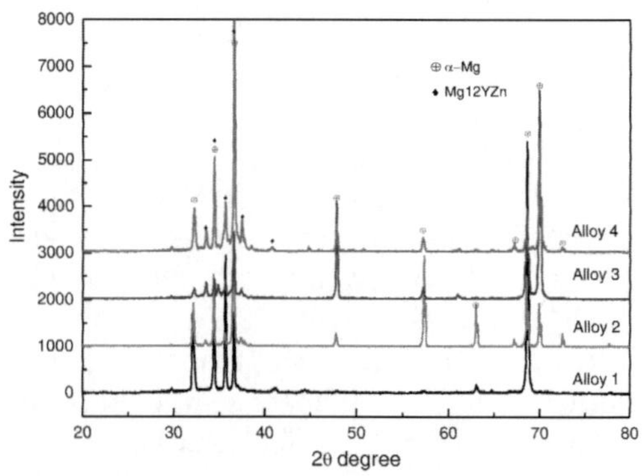

Figure 2. XRD patterns of as-cast Mg-Zn-Y alloys.

Figure 1 shows the OM images of the microstructures of the as-cast Mg-Zn-Y alloys. The microstructures of these as-cast Mg-Zn-Y alloys are very variable. With the increase of alloying elements, dendrites gradually refined, small equiaxed crystal appeared, phases gradually increased, and finally a large number of strip second phase. During this period, great amount of white particulates appear at the grain boundaries, especially in the alloy 4.

In order to reveal the phase ingredient, the crystallographic analysis was carried out by the XRD. Figure 2 shows the XRD patterns of the studied alloys. It is observed from Fig. 2 that all the experimental alloys are mainly composed of α-Mg and $Mg_{12}YZn$ phases, which is well coincided with the previous results reported by Wang et al.[14].It implies that the second phase is entirely a LPSO phase with increasing content of Zn and Y elements.

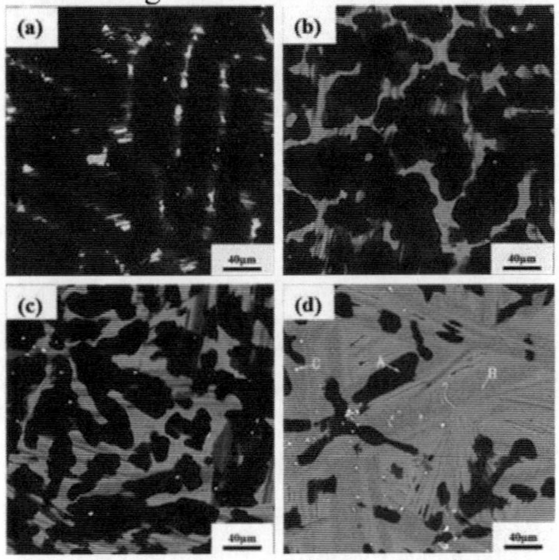

Figure 3. Scanning electron microscope images of as-cast Mg–Zn–Yalloys: (a) alloy 1; (b) alloy 2; (c) alloy 3 and (d) alloy 4.

Table 2. EDS results of the compositions

Points	Composition(at.%)		
	Mg	Zn	Y
A	98.99	--	1.01
B	88.20	5.49	6.31
C	16.60	2.47	80.93

Figure 3 illustrates the SEM microstructure of the as-cast Mg–Zn–Y alloys. These alloys exhibit a typical dendrite microstructure, which mainly comprise α-Mg dendrites and intermetallic compounds precipitating along the grain boundary. With the increase of Zn and Y contents, the overall morphologies changed greatly. The white areas gradually increased, the black areas decreased gradually, and the small white-bright particles distributed in the magnesium matrix increased by degrees. At the same time, the morphology of the alloys also has undergone great changes. In Fig. 3 (a), the white phase is less dispersed and distributed on the grain boundaries of the dendrites. When comes to Fig.3 (b) and Fig. 3 (c), the white phase is connected into a piece of mesh, which is distributed throughout the region. And the entire organization is divided into continuous white mesh and black island-like region or chicken claw-like region. Surprisingly, an elongated strip organization is found in the local white phase. As shown in Fig. 3 (d), with the further increase of Zn and Y, the white phase of the alloy occupies in most of the region, the black phase is divided into small pieces, and dispersed in the entire organization. Similarly, elongated strips or rod-like structures distributed in different directions are also found in the white phase. According to the principle of scanning backscattering, the larger the atomic coefficient is, the higher the brightness is. Therefore, it can be deduced that the white areas in the figures are mainly composed of Zn and Y elements with relatively large atomic coefficients, and they are likely to be $Mg_{12}YZn$ phase. Those black or dark colors region, which may mainly consist of Mg elements, are likely to be α-Mg phase.

In order to verify the above inference, we carried out EDS energy spectrum analysis of the alloy 4, and the EDS results of the compositions of the corresponding positions in Fig. 2 are shown in Table 2. It can be concluded that almost all the Zn and Y atoms are concentrated on the precipitates, no Zn element was found in the magnesium matrix. 98.99% Mg atoms and 1.01% Y atoms were found in point A, this implies that the black area is α-Mg matrix, in which a small amount of Y atoms are dissolved. Mg, Zn and Y three elements are found in point B, and the atom ratio of Y to Zn is 6.31/5.49=1.15, which is very close to the Y/Zn atom ratio in the $Mg_{12}YZn$ LPSO. Obviously, it can be concluded that the second phase is the $Mg_{12}YZn$ LPSO phase. This means that the gray phase with continuous network distribution is the LPSO phase. And there are also Mg, Zn and Y three elements in point C, and the contents of Mg, Zn and Y atoms are 16.60%, 2.47%, 80.93%, respectively. Where Y atom is much more than Mg and Zn, which is likely to be the Y atoms enriched in the magnesium and LPSO phases. In short, these are consistent with the results of XRD, where the phase compositions in alloy 4 are identified as α-Mg and LPSO-phase.

3.2. Mechanical properties of the As-cast Mg-Zn-Y alloys

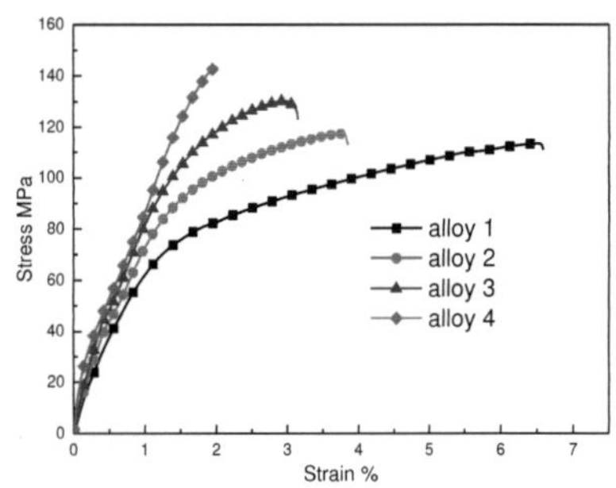

Figure 4. Tensile stress-strain curves of as-cast Mg-Zn-Y alloys.

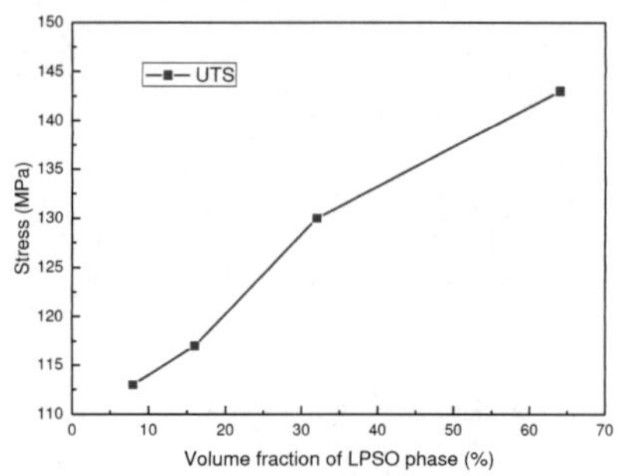

Figure 5. Relationship between the UTS and the volume fraction of LPSO phase.

Table 3. Mechanical properties of as-cast Mg-Zn-Y alloys

Alloy No.	UTS/MPa	YS/MPa	Elongation/%
1	113	51	5.4
2	117	53	2.8
3	130	60	2.1
4	143	48	1.4

Figure. 4 illustrates the stress-strain curves of the as-cast Mg-Zn-Y alloys. And the values of the mechanical properties are listed in Table 3, relationship between the UTS and the volume fraction of LPSO phase are shown in Fig. 5. As a whole, the LPSO phase is beneficial to the improvement of the strength of the alloy, but not to the improvement of the elongation. It can be seen that the ultimate tensile strengths (UTS, σ_b) range from 113MPa to 143MPa, and the yield tensile strengths (YTS, σ_s) vary between 48MPa and 60MPa.Alloy 1 showed the most excellent plasticity, but worst in tensile strength, while alloy 4 is just the opposite.

As we all know, the strength of alloy is often affected by the secondary phase, solid atoms, grain size and texture. As we mentioned above, we found that the grain size changed little, and new phase is not discovered. Since the alloy is obtained by a conventional manufacturing method, so no significant texture is produced. At the same time, the EDS results show that the alloy contains only a small amount of solid solution atoms (Zn and Y), so the second phase enhancement will be the main factor. The precipitation strengthening effect of LPSO phase is conducive to the improvement of strength. However, as the content of LPSO phase increases, the elongation of the alloys decline gradually. This may because while the LPSO phase acts as a strengthening phase to enhance the alloy strength, its reticular formation around the grain boundary is not conducive to the plastic deformation of the alloy. Thus not easy to coordinate the plastic deformation between the grain and the second phase, but easily lead to uneven stress distribution and local stress concentration, resulting in brittle fracture material, which exacerbated the plasticity.

3.3. Damping capacities of the As-cast Mg-Zn-Y alloys

The strain amplitude dependence of damping capacities at 1 Hz and room temperature of as-cast Mg-Zn-Y alloys are shown in Fig. 6. As we can see, the strain amplitude increases from 1×10^{-5} to 2×10^{-3}. It is obvious that, from the whole graph, the damping capacity of the alloys increase with the strain amplitude. It can be seen that in the small strain amplitude, the damping capacity value of every sample is small and varies little with the strain amplitude. While the strain attains a certain critical value, the damping capacity initiates to increase rapidly with the increase of strain amplitude. Obviously, there is a close relationship between the size of the final damping value and the critical strain amplitude (Which is mainly determined by the binding energy between the dislocation and the solute atoms, which should relate to the solute concentration [11]).

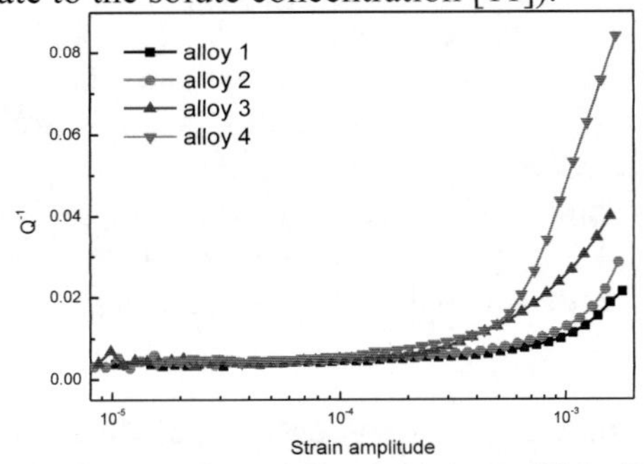

Figure 6. Damping-strain amplitude curves of Mg-Zn-Y alloys.

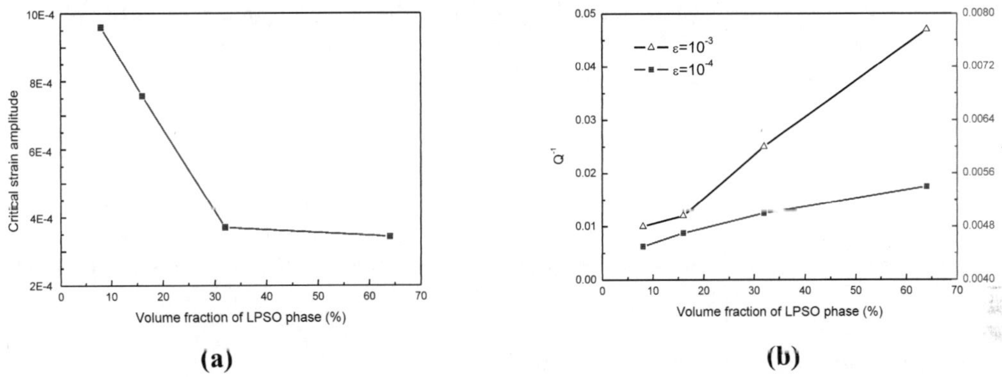

Figure 7. (a) Relationship between the critical strain amplitude and the volume fraction of LPSO phase; (b)Relationship between the damping capacities and the volume fraction of LPSO phase.

By calculating the average strain amplitude value at the inflection point of the curve, the corresponding critical strain amplitude of alloys are 9.58×10^{-4}, 7.57×10^{-4}, 3.7×10^{-4}, 3.43×10^{-4}, respectively. As shown in Fig.7(a), with the increasing content of the LPSO phase, the critical strain amplitudes of alloys gradually decrease so that alloy can break away pinning points easier, having more time to perform high energy dissipation, therefore, it will tend to show higher damping properties at high strain amplitudes. Figure.7(b) shows the relationship between the damping capacities and LPSO content of the alloys under high strain amplitude ($\varepsilon=10^{-3}$) and low strain amplitude ($\varepsilon=10^{-4}$), respectively. We find that the damping properties at the low and high strain amplitudes have the same trend. That is, with the increase of the content of LPSO phase, the growth rate at low strain is lower than that of high strain. Low strain amplitude damping capacity increased from 0.0045 to 0.0054, an increase of 20%; while high strain amplitude damping capacity increased from 0.01 to 0.047, an increase of 370%.

In general, the damping values of the four alloys are almost the same at low strain amplitudes, but the alloys with more LPSO phase have lower critical strain amplitude, which may lead to more time for energy dissipation. As a result, exhibits more excellent damping performance at high strain amplitude. Obviously, we can conclude that LPSO phase has a great effect on the damping properties of the alloys. However, it is not clear that why these alloys has such excellent damping property.

The classic "Granato-Lücke" theory is widely used to decipher the dislocation damping mechanism of Mg and its alloys[9,15-17]. However, some papers have pointed out that the alloy containing LPSO phase does not meet the G-L dislocation theory [12,18]. This means that the high damping properties of these alloys do not originate from dislocation motion, there may be other damping mechanism. As we know, this alloy just contain LPSO phase and Mg matrix. Owing to their different thermal expansion coefficients, they may reduce the amount of entangled dislocations in the alloy during the solidification. Meanwhile, there is a large difference between the hardness of the two (nearly 100Hv [19]), resulting in a great friction between the two factors, thus, consuming more energy and exhibiting a high damping performance. These deductions can well explain that the damping capacity increases with LPSO phase. Therefore, this new mechanism may provide a new way to prepare high-damping alloy, but further studies will need to provide more sufficient evidence. As mentioned above, the alloy containing more LPSO phase also has a good mechanical property, so it may provide a new idea for the preparation of structural and functional integration materials.

4. Conclusions

(1) The main phase composition for these 4 alloys consist of α-Mg and LPSO phase. With the increase of LPSO phase, grain size refined and its morphology also undergone a very big change.

(2) With the increase of LPSO phase, the critical strain amplitudes of alloys gradually decrease so that alloys can separate pinning points easier, behaving better in both low strain amplitude level and high strain amplitude level. The LPSO phase is conducive to damping capacities on the whole.

(3) Both the damping capacities and mechanical properties increase with the content of LPSO phase at the same time, which breaks the contradiction that they cannot be simultaneously improved.

Acknowledgements
This work was financially supported by National NaturalScience Foundation of China (51571044), National Basic Research Program of China (2013CB632201), Basic Research Project of Chongqing (cstc2015jcyjBX0081), and National Key Research and Development Program (2016YFB0301102).

Reference

[1] Lan A and Huo L 2016 *Mater. Sci. Eng. A* **651** 646–56.
[2] Liu D 1994 *Ordnance Mater. Sci. Eng.* **17** 67–70.
[3] Wei Y 2009 Chongqing University.
[4] Zhang J, Perez R and Lavernia E 1993 *J. Mater. Sci.* **28** 2395–404.
[5] Deng K, Li J, Nie K, Wang X and Fan J 2015 *Mater. Sci. Eng. A* **624** 62–70.
[6] Kim W 2009 *Scripta Mater.* **61** 1040–43.
[7] Watanabe H, Mukai T, Sugioka M and Ishikawa K 2004 *Scripta Mater.* **51** 291–5.
[8] Zhang Z, Zeng X and Ding W 2005 *Mater. Sci. Eng. A* 392 150–5.
[9] Wang J, Gao S, Zhao L, Liang H, Hu Y and Pan F 2010 *Trans. Nonferrous Met. Soc. China* **20** s366–70.
[10] Göken J and Riehemann W 2002 *Mater. Sci. Eng. A* **324** 134–40.
[11] Wan D,Wang J and Yang G 2009 *Mater. Sci. Eng. A* **517** 114–7.
[12] Wang J, Lu R, Wei W, Huang X and Pan F 2012 *J. Alloys Compd.* **537** 1–5.
[13] Abe E, Kawamura Y, Hayashi K and Inoue A 2002 *Acta Mater.* **50** 3845.
[14] Lu R, Wang J, Chen Y, Qin D, Yang W and Wu Z 2015 *J. Alloys Compd.* **639** 541–6.
[15] Granato A and Lücke K 1956 *J. Appl. Phys.* **27** 798–9.
[16] Granato A and Lücke K 1956 *J. Appl. Phys.* **27** 583–5.
[17] Feng H, Yang Y and Chang H 2014 *Mater. Sci. Eng. A* **609** 7–15.
[18] Qin D, Wang J, Chen Y, Lu R and Pan F 2015 *Mater. Sci. Eng. A* **624** 9–13.
[19] Chino Y, Mabuchi M, Hagiwara S, Iwasaki H, Yamamoto A and Tsubakino H 2013 *Acta Mater.* **61** 6338–51.

IOP Conf. Series: Materials Science and Engineering **220** (2017) 012016 doi:10.1088/1757-899X/220/1/012016

Study on Damping Capacities of ZK60-2.8Nd Magnesium Alloy

X L Liu[1*], W Y Zhang[1] and Y Z Wu[1]

[1]Hunan Institute of Technology, Hunan Hengyang, 421002, China

*E-mail: lxl@hnit.edu.cn

Abstract. Damping properties of ZK60-2.8Nd magnesium alloys as-cast, solution and extrusion were investigated by dynamic mechanical thermal analyzer. The results show that ZK60-2.8Nd magnesium alloy extrusion and solution exhibit almost the same strain independent damping value as that of alloy as-cast, the critical strain amplitude is about 5×10^{-5}. Solution treatment effects the distribution of solution atoms, and dynamic re-crystallization during the hot extrusion increases movable dislocation density ρ, which makes the dislocations easier to move on the basal plane. Within the strain range of $5\times10^{-5}\leq\varepsilon\leq4\times10^{-4}$, the strain dependent damping value of ZK60-2.8Nd magnesium alloy as-solution is the highest, and that of the alloy as-cast is the least due to the grain sizes increase of the alloy as-solution and dynamic re-crystallization of the alloy as-extrusion leading the increase of movable dislocation density. Within scope of the experiment, the maximum damping value of alloy as-cast, solution and extrusion is 0.03371, 0.01546 and 0.01392, respectively.

1. Introduction

Lightweight structure design has been an increasingly outstanding status in modern industries, and magnesium alloys have been widely used with their high strength and good damping performance [1]. However, pure magnesium exhibits extraordinary high damping properties but poor mechanical properties [2]. At present, there are many high strength magnesium alloys reported on processing technology, mechanical properties and strengthening mechanism, etc. [3-7]. For many years, there are more and more reported on damping properties of high strength magnesium alloys, including Mg-Al alloy such as AZ31, AZ61, AZ80, Mg-Zn alloy such as ZK60, Mg-Re alloy such QE22, WE43 [8-12]. However, there are various factors effecting on damping properties of magnesium alloy including alloy elements, grain size and grain orientation, heat treatment and deformation, strain amplitude and temperature. Meantime, damping mechanism of the magnesium is rather complex [13-18]. The rare earth neodymium (Nd) can reinforce magnesium alloy at room temperature and high temperature which makes the magnesium alloy containing Nd present good application prospect [19]. Therefore, in this investigation, the magnesium alloy containing Nd is to be prepared to test the damping performance, which would be useful for the further analysis of the damping mechanism and obtaining high strength and high damping magnesium alloy.

2. Experiments

The testing alloys were prepared from commercial Mg (99.96%), Zn (99.91%), Mg-29.98%Nd and Mg-30.96%Zr master alloys in an electric resistance furnace under the protection of mixed SF_6 and CO_2 protection atmosphere with the ratio of 100:1. The ingots with a cylindrical mould of ϕ 60 mm × 135 mm were on solution treatment at 420℃ for 24h, and then quenched into water at room temperature. Then hot extrusion samples were conducted at 360℃ with extrusion ratio 16:1. The chemical compositions of these Mg alloys are given in Table 1.

Table 1. Chemical compositions of test alloys (mass fraction, %)

	Mg	Zn	Nd	Zr
ZK60-2.8Nd	91.21	5.50	2.75	0.54

Microstructure observation was conducted on a POLYVAR-MET optical microscope. The phase analysis was carried out by D/MAX2500PC X ray diffractometry (XRD) with copper target.

The damping capacity was determined by $Q^{-1} = \tan\varphi$. The specimens for damping measurements with dimensions of 38 mm×6 mm×1 mm were machined out by an electric spark cutting method. The damping tests were carried out using a TA Q800DMA.

3. Results and discussion

3.1. Microstructures

Fig.1 shows the XRD patterns of the alloy as-cast, solution and extrusion. There are three phases in three states of ZK60-2.8Nd alloy including α-Mg, MgNdZn and MgZn phases.

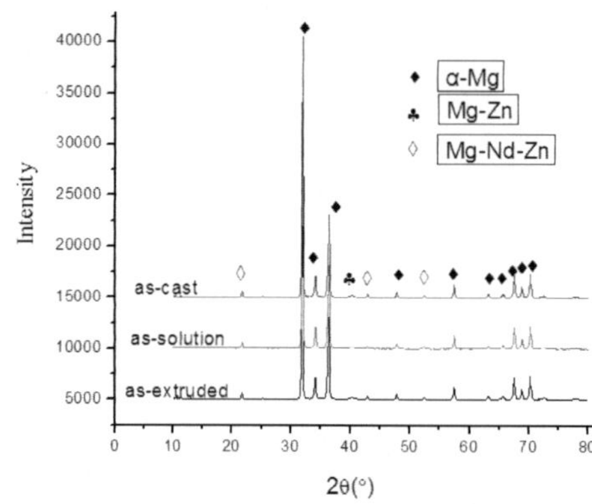

Figure 1. XRD patterns of ZK60-2.8Nd magnesium alloys.

Fig.2 shows the microstructure of the testing alloys as-cast, solution and extrusion. The microstructure of the as-cast alloy is comprised of dendrites of magnesium matrix phases separated by the continuous or discontinuous networks of eutectic compounds. The average grain sizes are measured to be about 35 μm for ZK60-2.8Nd alloy, shown in Fig.2 (a). Considering the Fig. 1 showed the XRD patterns of the alloys as cast, the continuous or discontinuous networks of eutectic compounds may be MgNdZn and MgZn phases.

It is found that the continuity of eutectic networks is seemingly weakened after solution treatment, and the average grain sizes increase a little, about 40 μm, seen in Fig.2 (b).

The microstructures of the ZK60-2.8Nd alloy as-extruded are shown by the optical micrographs in Fig.2 (c). It shows the average grain sizes of 10 μm, grain refinement much more than that of alloy as-cast and solution.

Grain sizes of three states of ZK60-2.8Nd alloy are listed in Table 2.

Table 2. Grain size of three states of ZK60-2.8Nd alloy (μm)

	As-cast	As-solution	As- extrusion
Grain size	35	40	10

3.2. Damping capacities

Strain amplitude dependence of damping capacity in ZK60-2.8Nd magnesium alloys as-cast, solution and extrusion are shown in Fig.3.

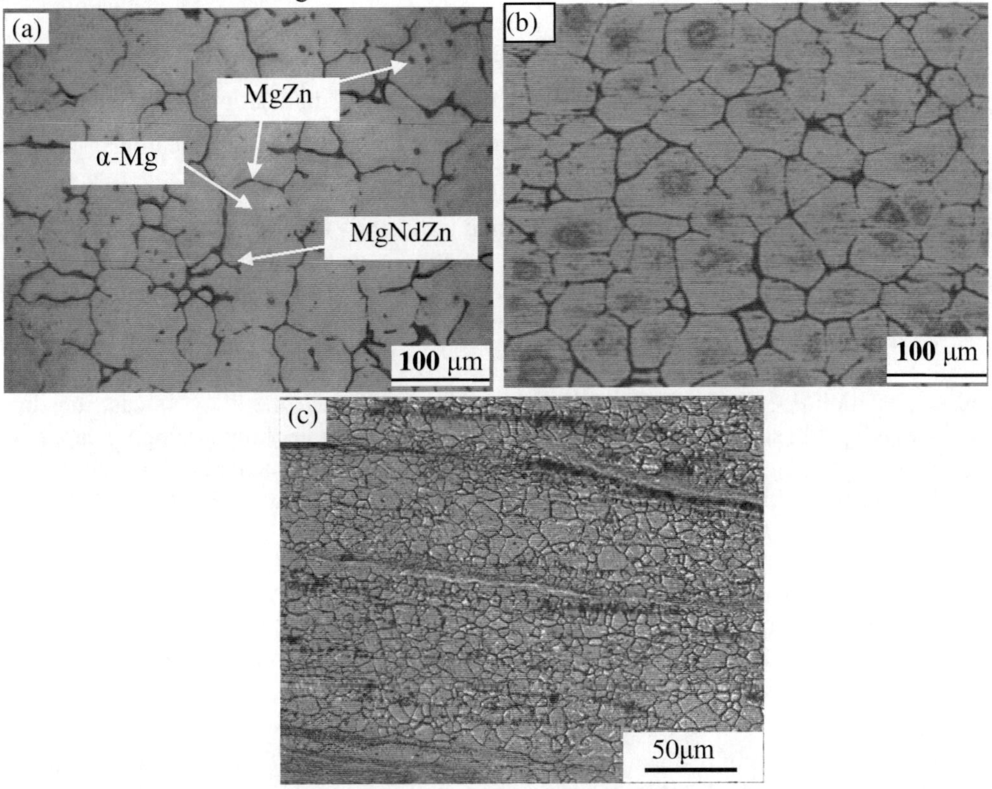

Figure 2. Microstructures of ZK60-2.8Nd magnesium alloys (a) as-cast; (b) solution; (c) extrusion.

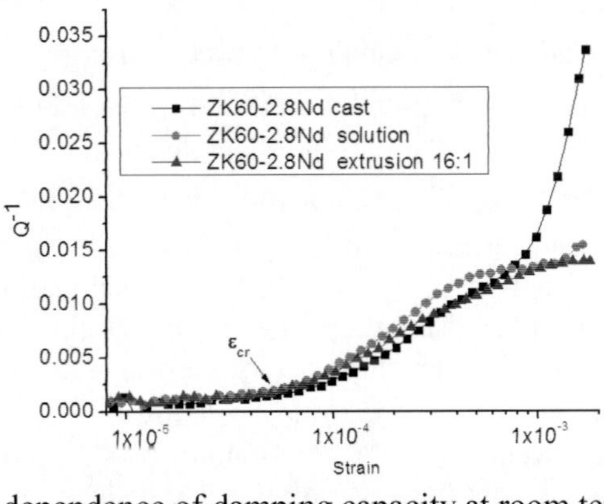

Figure 3. Strain amplitude dependence of damping capacity at room temperature for ZK60-2.8Nd magnesium alloys as-cast, solution and extrusion.

It can be seen that all the damping curves present almost the similar characteristics that the damping capacity can be divided into two parts as follows [14, 15]:

$$Q^{-1} = Q_I^{-1} + Q_H^{-1} \tag{1}$$

$$Q_I^{-1} \sim \rho L_C^{4} \tag{2}$$

$$Q_H^{-1} = (C_1/\varepsilon)\exp(-C_2/\varepsilon) \tag{3}$$

$$C_1 = \frac{\rho F_B L_N^3}{6bEL_C^2}, \quad C_2 = \frac{F_B}{bEL_C} \tag{4}$$

Where, Q_I^{-1} and Q_H^{-1} represent the strain independent damping and strain dependent damping, respectively; ρ is the dislocation density; L_C is the mean length of dislocations between the weak pinning points consisted of the soluble atoms or vacancies surrounding the dislocations; F_B is the binding force between a dislocation and a weak pinning point, E is the modulus, L_N is the mean length of dislocation segments between strong pinning points, and b is the magnitude of Burger vector.

Fig.3 shows that the strain independent damping Q_I^{-1} of alloy as-cast, solution and extrusion are all almost the same, which keep almost in horizontal lines exiting in lower strains on the damping strain curves; the critical strain amplitude (ε_{cr}) is about 5×10^{-5}. But when the strain amplitude is more than the critical strain amplitude, the strain dependent damping capacities increase rapidly with increase of the maximum strain amplitude, and the extrusion alloy rises faster than the alloy as-cast, but lower than solution alloy. When the strain is about $3\times10^{-4} \sim 4\times10^{-4}$, they show very high damping capacity of 0. 01. Especially, the damping value of alloy as-cast increases rapidly far ahead of the others, up to 0.03371 within the testing strain range, more than twice of that of alloy solution and extrusion, 0.01546 and 0.01392, respectively, seen in Fig.3.

General, solid solution treatment, as the same as the extrusion, could add the number of weak pinning point to the dislocation to make the movable dislocation segment much shorter, and decrease the strain independent damping capacity[13,18]. However, in this investigation, MgNdZn and MgZn phase included in the ZK60-2.8Nd magnesium alloy help decrease solid solution atoms in the magnesium matrix so that the alloy solution and extrusion posses the same Q_I^{-1} as that of alloy as-cast, owing to the MgNdZn phase. Certainly, solid solution treatment makes the grain sizes increase a little so as to increase the mean length of dislocation segments between strong pinning points. Although hot extrusion refined the grain size, sufficient dynamic re-crystallization improved movable dislocation density ρ, which made the dislocations easier to move on the basal plane. That is why the strain dependent damping capacity Q_H^{-1} of alloy as-cast is the least among the three states alloy in a certain strain range. According to the formula (3) and (4), the strain dependent damping capacity Q_H^{-1} increase with the increase of dislocation segments between strong pinning L_N and the dislocation density ρ. From figure 1 and table 2, the grains size of alloy as-solution are 40 μm more than that of alloy as-cast. Grain boundaries act as the strong pinning to dislocations, which offers more dislocation segments between strong pinning L_N, the strain dependent damping capacity Q_H^{-1} of alloy as-solution increase accordingly more than that of alloy as-cast. Although the grains size of alloy as-extrusion are 10 μm less than that of alloy as-cast, hot extrusion brought out the emergence of Dynamic re-crystallization to reduce dislocation tangles, which increase the movable dislocation density ρ, the strain dependent damping capacity Q_H^{-1} of alloy as-extrusion increase accordingly more than that of alloy as-cast. At higher strain amplitude, the alloy as-cast shows the maximum damping capacity owing to the coarser grain size, the continuous or discontinuous networks of eutectic compounds MgNdZn phases, and bigger density ρ of movable dislocation.

Two critical strain amplitude points in the damping strain spectrum has been studied on many alloys by some researchers [16, 17]. Present authors found it is associated with the medium intensity pinning points such as MgZn phases in the Mg-Zn magnesium alloys [18]. So it is MgZn phases in the ZK60-2.8Nd magnesium alloy to lead to the second strain amplitude points. Maybe it exits other reasons which would be further studied.

4. Conclusions

(1) The strain independent damping capacity of ZK60-2.8Nd magnesium alloys as-cast, solution and extrusion are almost the same when the strain amplitude is lower than 5×10^{-5}.They exhibits high strain dependent damping capacities ($Q_H^{-1} \geqslant 0.01$) when the strain amplitude is higher than 4×10^{-4}, the maximum damping value of alloy as-cast, solution and extrusion is 0.03371, 0.01546 and 0.01392, respectively, within scope of the experiment.

(2) The strain dependent damping capacity Q_H^{-1} of ZK60-2.8Nd magnesium alloy as-solution is the highest, and that of the alloy as-cast is the least among the three states alloy within the strain range of $5\times10^{-5} \leqslant \varepsilon \leqslant 4\times10^{-4}$ due to the grain sizes increase of the alloy as-solution and dynamic re-crystallization of the alloy as-extrusion leading the increase of movable dislocation density.

Acknowledgments
Foundation item: Project (2015JC3106) supported by China Hunan Provincial Science and Technology Department.

References

[1] Powell B R, Krajewski P E and Luo A A 2010 *Mater. Design & Manufacturing* **80** 114-73.
[2] Kojima Y, Aizawa T, Kamado S and Higashi K 2003 *Materials Science Forum* **419** 3-20.
[3] Chen X H, Huang X W, Pan F S, Tang A T and Wang J F 2011 *Trans. Nonferrous Met. Soc. China* **21** 754-60.
[4] Chen T J, Wang W, Zhang D H, Ma Y and Hao Y 2013 *J. Alloys & Compounds* **546** 28-40.
[5] Zhao X F, Li S B, Wang Q F, Du W B and Liu K 2013 *Trans. Nonferrous Met. Soc. China* **23** 59-65.
[6] Wang Y X, Wei F J and Yang Y S 2012 *Trans. Nonferrous Met. Soc. China* **22** 1322-28.
[7] Guo Q, Yan H, Chen Z, Wu Y and Chen J 2007 *Acta Metallurgica Sinica* **43** 619-24.
[8] Göken J, Swiostek J, Letzig D and Kainer K U 2005 *Materials science forum* **482** 387-90.
[9] Wei L Y, Dunlop G L and Westengen H 1995 *Metall. Mater. Trans.* **A 26** 1705-16.
[10] Chen X H, Mao J J, Pan F S, Peng J and Wang J F 2010 *Trans. Nonferrous Met. Soc. China* **20** 1305-10.
[11] Lambri O A and Richemann W 2005 *Scripta Materialia* **52** 93 97.
[12] Lambri O A, Riehemann W, Salvatierra L M and García J A 2004 *Mater.Sci. Eng.* **A 373** 146-57.
[13] Liu C M, Ji R F, Zhou H T and Chen M A 2005 *Chinese Journal of Nonferrous Metals* **15** 1319-25.
[14] Granato A and Lucke K 1956 *J. Appl. Phys.* **27** 583-93.
[15] Granato A and Lucke K 1956 *J. Appl. Phys.* **27** 789-805.
[16] Hutchison T S and Rogers D H 1962 *J. Appl. Phys.* **33** 792-99.
[17] Hu X S, Wu K, Zheng M Y, Gan W M and Wang X J 2007 *Mate. Sci and Eng* **S 452–453** 374-79.
[18] Liu X L, Liu C M, Zhang W Y, Luo J H and Zeng S M 2012 *Advanced Materials Research* **557-559** 1624-28.
[19] Huang X, Wang Q, Zeng X, Zhu Y, Lu C and Ding W J 2004 *Journal of the Chinese Society of Rare Earths* **22** 361-4.

Influence of temperature on Cole-Cole dielectric model of oil-immersed bushing

K Wang[1], X J Chen[2*], X W Xu[1], G Q Liu[1], D X Zou[1] and W D Liu[2]

[1]Yunnan Electric Power Research Institute, Yunnan Power Grid Co., Ltd., Kunming 650217, China
[2]College of Electrical Engineering, Southwest Jiaotong University, Chengdu 610031, China

*E-mail: xjchen_1994@163.com

Abstract. In this paper, 72.5 kV oil-immersed bushing was produced in laboratory. The frequency-domain dielectric response tests of oil-immersed bushings were carried out at different test temperatures. The experimental data were fitted by using the modified double relaxation Cole-Cole dielectric model. The influence of temperature variation on the dielectric response test of the oil-immersed bushing and the Cole-Cole dielectric model parameters were analysed. The results showed that with the increase of the test temperature, the spectrum of the real and imaginary of the complex permittivity are shifted to the high frequency direction; the parameters of the dielectric model are significantly affected by temperature.

1. Introduction

Insulation performance of oil-immersed bushing will directly affect the safety and stability of the transformer operation. The frequency domain dielectric response method is one of the methods used for lossless insulation diagnosis [1-2]. The modified Cole-Cole model [3-4] can better explain the dielectric properties of the dielectric in the frequency domain, and more suitable for field applications.

In this paper, 72.5 kV oil-immersed bushing was produced in laboratory. The frequency-domain dielectric response tests of oil-immersed bushings were carried out at different test temperatures. The experimental data were fitted by using the modified double relaxation Cole-Cole dielectric model. The influence of temperature variation on the dielectric response test of the oil-immersed bushing and the Cole-Cole dielectric model parameters were studied.

2. Experiment

2.1. Pretreatment of experimental materials

The internal insulation materials of the oil-immersed bushing model are the paper (thickness is about 0.1 mm) and the 45#oil. The electrode screen is 0.03 mm aluminum foil. The center duct is a hollow aluminum tube. In order to reduce the influence of the moisture content, the following pretreatment of paper and oil was carried out: The paper was cut according to the pre-calculated size; the cut paper was dried at 110 °C/50 Pa for 72 h; so that the water content was less than 0.5%. The oil was degassed and dried at 80 °C until the moisture content was about 11 ppm.

2.2. Preparation of bushing model

According to insulation structure and design specifications of 72.5 kV high voltage bushing, the bushing model was prepared. The structure of capacitance core is shown in Figure 1. In order to eliminate the influence of moisture in the air during the winding process, the capacitor core was dried in a drying oven at 110 °C for 24 h after the bushing was completed. The dried oil-immersed bushing capacitor core was sealed in an epoxy tube and immersed in insulating oil for 48 h under 40 °C.

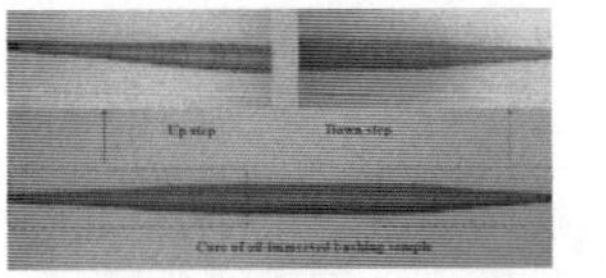

Figure 1. Structure of capacitance core.

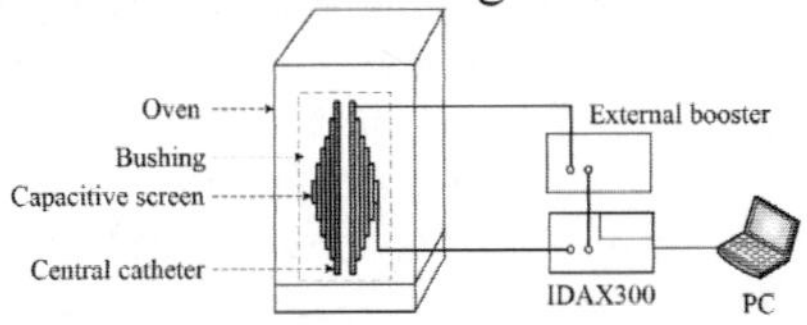

Figure 2. Experiment wiring diagram.

2.3. Experiment process

The dielectric response test was carried out at 15 °C, 25 °C, 35 °C, 45 °C, and 60 °C, respectively. The bushing model was placed in the oven and allowed to stand at a preset temperature for 24 h. IDAX300 is used to test the frequency domain dielectric response of the oil-immersed bushing model. The test frequency ranges 0.001 Hz~1000 Hz. The moisture content of the bushing model is 0.53%. Figure2 shows the experimental wiring diagram of this experiment.

3. Analysis of experimental results based on the double relaxation Cole-Cole model

3.1. Experimental results

Figure3 is the spectrum diagram of the real part and the imaginary part of the complex permittivity of the oil-immersed bushing model. The results show that with the increase of the test temperature, the spectrum of the real and imaginary of the complex permittivity are shifted to the high frequency direction.

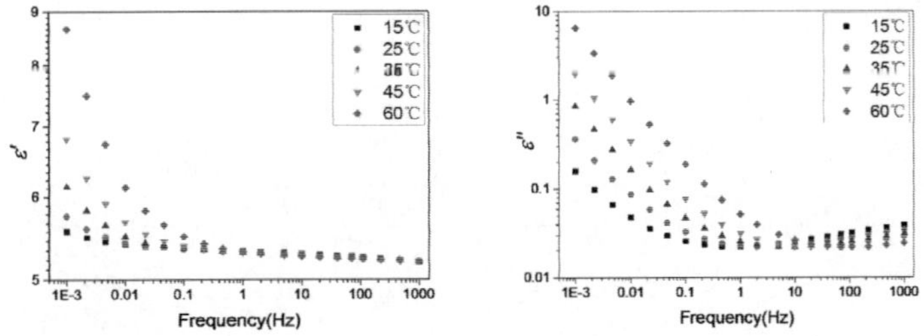

(a) ε' **(b)** ε''

Figure 3. Complex permittivity.

3.2. Establishment of double relaxation Cole-Cole model

It can be seen from Figure3 that the spectrum of the imaginary part of the complex permittivity has a transition in the vicinity of 10 Hz, which indicates that the frequency-domain dielectric response characteristics in the test band are at least include two relaxation process. Therefore, in this paper, the dielectric properties of oil-immersed bushing in the frequency domain were fitted by α relaxation process and β relaxation process. The equation (1) is the modified double-relaxation Cole-Cole model equation for oil-immersed bushing.

$$\varepsilon(\omega)=\varepsilon_\infty+\frac{\sigma_{dc}}{j\varepsilon_0\omega}+\frac{\Delta\varepsilon_1}{1+(j\omega\tau_1)^{\alpha_1}}+\frac{\Delta\varepsilon_2}{1+(j\omega\tau_2)^{\alpha_2}} \tag{1}$$

Where:ε_∞ is the high frequency dielectric constant; $\Delta\varepsilon_1$, $\Delta\varepsilon_2$ are the relaxation strength of α relaxation process and β relaxation process respectively; α_1, $\alpha_2$are the distribution parameters ofα relaxation process and β relaxation process;τ_1, τ_2 are the relaxation time of α relaxation process and β relaxation process; σ_{dc} is the direct current conductivity of oil-immersed bushing, pS·m^{-1}.

The formula (1) is divided into real and imaginary parts, then the real and imaginary parts of the complex permittivity of the oil-immersed bushing can be expressed as the following formulas:

$$\varepsilon' = \varepsilon_\infty + \mathrm{Re}[\frac{\Delta\varepsilon_1}{1+(j\omega\tau_1)^{\alpha_1}} + \frac{\Delta\varepsilon_2}{1+(j\omega\tau_2)^{\alpha_2}}] \quad (2)$$

$$\varepsilon'' = \frac{\sigma_{dc}}{\varepsilon_0\omega} - \mathrm{Im}[\frac{\Delta\varepsilon_1}{1+(j\omega\tau_1)^{\alpha_1}} + \frac{\Delta\varepsilon_2}{1+(j\omega\tau_2)^{\alpha_2}}] \quad (3)$$

Where: ε_∞, $\Delta\varepsilon_1$, $\Delta\varepsilon_2$, α_1, α_2, τ_1, τ_2, σ_{dc} are unknown parameters, and the least squares method can be used to fit results of the experiment so that the unknown parameters can be obtained.

3.3. *Analysis of results based on double relaxation Cole-Cole model*

The fitting curve of the real and imaginary parts of the complex permittivity of the bushing model are shown in Figure 4. At the same time, the unknown parameters in equation(2) and equation(3) are obtained by fitting and are shown in Table 1.

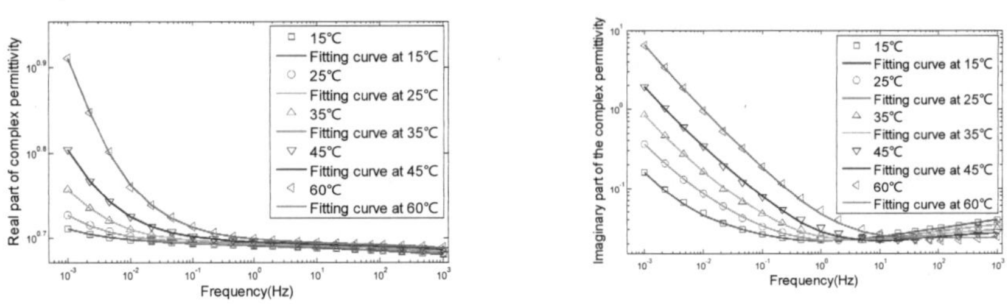

(a)ε' (b) ε''

Figure 4. Complex permittivity curve.

Table 1. Parameters of Cole-Cole model of oil-immersed bushing at different temperatures

Parameters	15 °C	25 °C	35 °C	45 °C	60 °C
σ_{dc}/(S/m)	6.70E-09	1.64E-08	4.05E-08	9.42E-08	3.24E-07
$\Delta\varepsilon_1$	0.180 497	0.282 454	0.503 608	1.005 245	2.342 287
τ_1/s	39 343.57	7 856.422	488.568 2	756.377 1	447.442 6
α_1	0.476 818	0.561 242	0.577 932	0.607 662	0.668 282
$\Delta\varepsilon_2$	0.497 421	0.745 339	0.464 511	0.290 165	0.396 576
τ_2/s	4.90E-09	3.59E-09	3.43E-09	3.27E-09	1.70E-09
α_2	0.223 823	0.251 027	0.254 668	0.268 09	0.308 281
ε_∞	3.339 657	3.007 481	3.174 107	2.871 314	2.884 781

4. Discussion

4.1. *Influence of temperature on DC conductivity*

The fitting relation curve shown in Figure5 and the fitting relation shown in equation (4). It can be seen from Figure5 and equation (4) that the oil-immersed bushing increases exponentially with the increase of the temperature *T*, which meets the study that the DC conductivity and temperature *T* satisfy the Arrhenius equation.

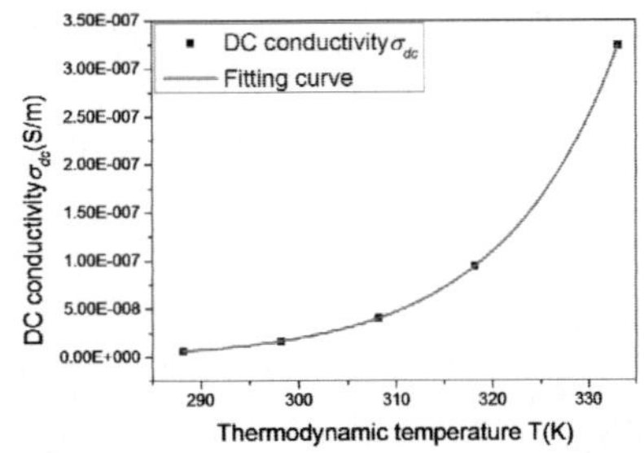

Figure 5. The fitting curves of DC conductivity and temperature.

$$\sigma_{dc} = 48658.56\exp(-\frac{1.1833\times10^{-19}}{1.38\times10^{-23}T}) \tag{4}$$

4.2. Influence of temperature on relaxation time constant

The fitting curve of the relaxation time and the temperature T is shown in Figure6.The fitting relation shown in equation (5) is obtained by fitting the relationship between the relaxation time and the thermodynamic temperature in Table 1. From Figure 6, it can be seen that the relaxation time τ_1 decreases exponentially with the increase of the test temperature and the relaxation time τ_2 decreased with the temperature rise and in the process of change there is a slow down in the process of declining.

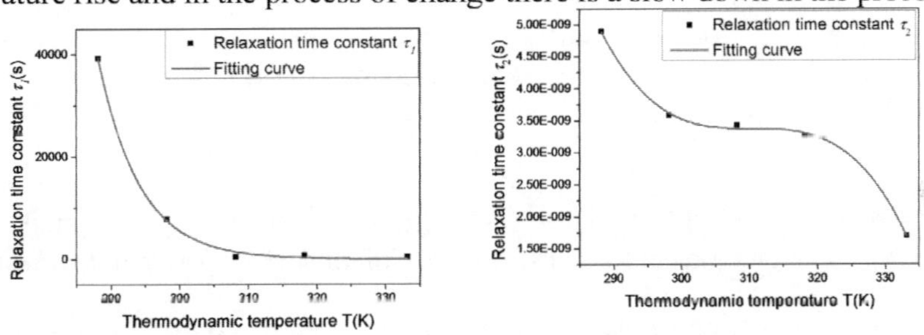

(**a**)Relaxation time τ_1 (**b**) Relaxation time τ_2

Figure 6. The fitting curve of relaxation time and temperature.

$$\begin{aligned} \tau_1 &= 1.8452\times10^{25}\exp(-0.16518T) \\ \tau_2 &= 4.173\times10^{-6} - 4.032\times10^{-8}T + 1.299\times10^{-10}T^2 - 1.396\times10^{-13}T^3 \end{aligned} \tag{5}$$

4.3. Influence of temperature on relaxation strength

The relaxation strength and temperature in Table 1 are fitted to obtain the fitting curve of the relaxation strength and temperature T shown in Figure 7, and the fitting relation is obtained as shown in equation (6). Figure 7 shows that the dielectric relaxation strength $\Delta\varepsilon_1$ of the Cole-Cole model increases exponentially with the increase of temperature and $\Delta\varepsilon_2$ showed a tendency to increase first, then decrease and then increase with the increase of test temperature.

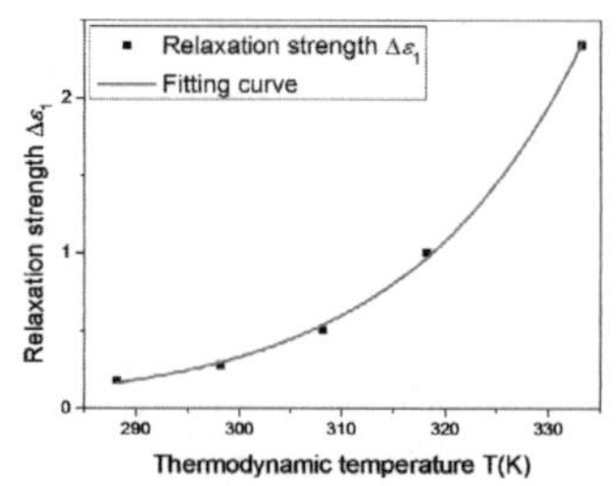

(**a**)Relaxation strength $\Delta\varepsilon_1$

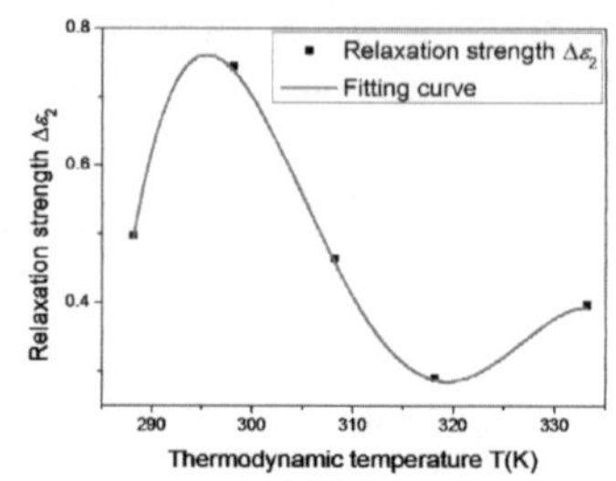

(**b**) Relaxation strength $\Delta\varepsilon_2$

Figure 7. The fitting curves of relaxation strength and temperature.

$$\begin{aligned} \Delta\varepsilon_1 &= 6.94813\times10^{-9}\exp(0.05895T) \\ \Delta\varepsilon_2 &= -20248.256+257.315T-1.225T^2+0.00259T^3-2.047\times10^{-6}T^4 \end{aligned} \tag{6}$$

5. Conclusion

1) With the increase of the test temperature, the spectrum of the real and imaginary of the complex permittivity are shifted to the high frequency direction.

2) the DC conductivity and thermodynamic temperature satisfy Arrhenius relation; the relaxation time τ_1 decreased exponentially with the increase of test temperature; τ_2 decreased with the temperature rise and in the process of change there is a slowdown; the relaxation strength $\Delta\varepsilon_1$ exponentially increased with the increase of test temperature; $\Delta\varepsilon_2$ showed a tendency to increase first, then decrease and then increase with the increase of test temperature.

References

[1] Wang D Y, Zhou L J, Jiang J F, et al. 2015 Modelling the electrode polarization of oil gap under low frequency excitation and the expressions of dielectric parameters *Proceedings of the CSEE* **35** 5950-6.

[2] Saha T K, Purkait P and Müller F 2005 Deriving an equivalent circuit of transformers insulation for understanding the dielectric response measurements *IEEE Transactions on Power Delivery* **20** 149-57.

[3] Wolny S, Adamowicz A and Lepich M 2014 Influence of temperature and moisture level in paper-oil insulation on the parameters of the Cole-Cole model *IEEE Transactions on Power Delivery* **29** 246-50.

[4] Yang L J, Gao S H, Gao J, et al. 2016 Characteristic parameters extracted from modified Cole-Cole model and moisture content assessment methods study on frequency-domain dielectric spectroscopy of oil-paper insulation *Transactions of China Electro Technical Society* **31** 310-7.

IOP Conf. Series: Materials Science and Engineering **220** (2017) 012018 doi:10.1088/1757-899X/220/1/012018

Numerical study of solid particle erosion in butterfly valve

Bo Liu[1*], Jiangang Zhao[1] and Jianhua Qian[1]

[1]China Nuclear Power Technology Research Institute, Shenzhen, 518031, China

*E-mail: lbsc88@sina.com

Abstract. In the actual operation of butterfly valve, the butterfly valve is found severe erosion wear. A solid particle erosion analysis of butterfly valve based on the erosion theory is researched in this study. A CFD model has been built to simulate the flow erosion. Different parameters of butterfly valve including inlet velocity, particle mass fraction and solid particle diameter are separately analysed. The results show that erosion rate increase with the increase of inlet velocity, particle mass fraction and solid particle diameter. The peak erosion rate is up to 4.63E-5 ($kg/m^2/s$) and erosion of valve disc mainly occurs around the upstream edge and the cylinder face.

1. Introduction

Butterfly valve has advantages of simple structure, small volume, light weight and low cost, which has been widely applied in the industry. In the nuclear power station, butterfly valve is usually used to the circulation water filtration system and is found serve erosion wear frequently which would lead to functional failures and threaten the cooling water supply of nuclear power station. So it is essential to study the influencing parameters on flow erosion of butterfly valve.

Erosion wear of valve is an important issue for the valve industry which causes the concern of researchers [1]–[3]. Solid particle erosion is one of main reasons of erosion wear [4]–[6] and studies of solid particle erosion of valves with different influencing parameters [7]–[9] are highly performed. However, Solid particle erosion of butterfly valve is rarely studied at present.

Since erosion flow features of butterfly valve, especially for erosion rate are highly difficult to obtain by experiment. In this study, a numerical model of butterfly valve based on the solid-liquid two-phase flow theory has been built to simulate the flow erosion in FLUENT and erosion rates under different parameters including inlet velocity, particle mass fraction and solid particle diameter are discussed separately.

2. Theoretical models

The Computational Fluid Dynamics (CFD) model of butterfly valve is simulated by FLUENT. The k-ε model and discrete phase model are used in numerical analysis and theories of these models are described as follows.

2.1. k-ε model

There are three k-ε models in FLUENT including standard k-ε model, RNG k-ε model realizable k-ε model. The standard k-ε model has characters of robustness, economy reasonable accuracy for a wide range of turbulent flows which leads to widely using in industrial flow simulations. In this study, the standard k-ε model [10] is usually used to simulate turbulent flow.

The turbulent viscosity equation can be written as

$$\mu_t = \rho C_\mu \frac{k^2}{\varepsilon} \tag{1}$$

where ρ is the fluid density, C_μ is a constant, k is the turbulent kinetic energy, ε is the dissipation rate of k.

The turbulent kinetic energy and its rate of dissipation can be obtained from the following transport equations:

$$\frac{\partial}{\partial t}(\rho k) + \frac{\partial}{\partial x_i}(\rho k u_i) = \frac{\partial}{\partial x_j}\left((\mu + \frac{\mu_t}{\sigma_k})\frac{\partial k}{\partial x_j}\right) + G_k + G_b - \rho\varepsilon - Y_M + S_k \tag{2}$$

$$\frac{\partial}{\partial t}(\rho \varepsilon) + \frac{\partial}{\partial x_i}(\rho \varepsilon u_i) = \frac{\partial}{\partial x_j}\left((\mu + \frac{\mu_t}{\sigma_\varepsilon})\frac{\partial \varepsilon}{\partial x_j}\right) + C_{1\varepsilon}\frac{\varepsilon}{k}(G_k + C_{3\varepsilon}G_b) - C_{2\varepsilon}\rho\frac{\varepsilon^2}{k} + S_\varepsilon \tag{3}$$

in which G_k is the generation of turbulence kinetic energy due to the mean velocity gradients, G_b is the generation of turbulence kinetic energy due to buoyancy, Y_M is the contribution of the fluctuating dilatation in compressible turbulence to the overall dissipation rate, $C_{1\varepsilon}$, $C_{2\varepsilon}$ $C_{3\varepsilon}$ are constants, σ_k and σ_ε are the turbulent Prandtl number for k and ε S_k and S_ε are user-defined source terms.

2.2. Discrete phase model

2.2.1. Equations of motion for particles. Fluent predicts the trajectory of a discrete phase particle by integrating the force balance on the particle, which is written in a Lagrangian reference frame. This force balance equates the particle inertia with the forces acting on the particle can be written as

$$\frac{d\vec{u}_p}{dt} = F_D(\vec{u} - \vec{u}_p) + \frac{\vec{g}(\rho_p - \rho)}{\rho_p} + \vec{F} \tag{4}$$

where $\vec{F}$ is and additional acceleration term, $F_D(\vec{u} - \vec{u}_p)$ is the drag force per unit particle mass, $\vec{u}$ is the fluid phase velocity, $\vec{u}_p$ is the particle velocity, ρ_p is the density of the particle.

2.2.2. Particle erosion and accretion theory [11]. Particle erosion and accretion rates can be monitored at wall boundaries. The erosion rate is defined as

$$R_{erosion} = \sum_{p=1}^{N_{particles}} \frac{\dot{m}_p C(d_p) f(\alpha) v^{b(v)}}{A_{face}} \tag{5}$$

in which $\dot{m}_p$ is the particle mass flow, $C(d_p)$ is a function of particle diameter, α is the impact angle of the particle path with the wall face, $f(\alpha)$ is a function of impact angle, v is the relative particle velocity, $b(v)$ is a function of relative particle velocity A_{face} is the area of the cell face at the wall.

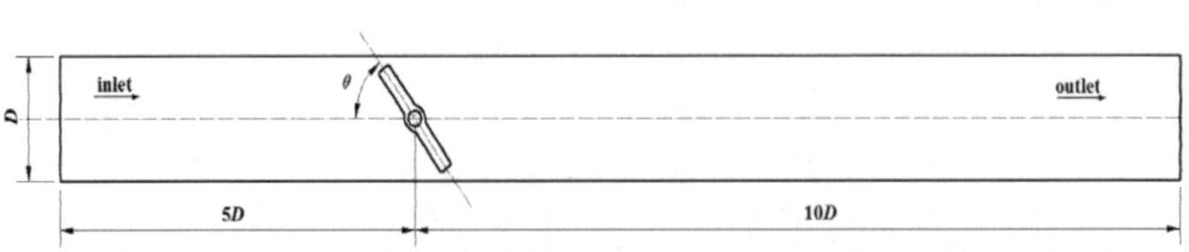

Figure 1. Geometry model of butterfly valve.

3. Numerical simulations

3.1. Geometry model
To accurately simulate the solid particle erosion of butterfly valve, the geometry model is built at real size and the upstream length and downstream length of butterfly valve are five times and ten times of pipe diameter, respectively. The geometry model of butterfly valve is shown in Figure 1.

3.2. Mesh model
Figure 2 shows the mesh model of butterfly valve at the opening angle 45°. Both tetrahedral element and hexahedral element are used to the mesh model. Around the valve disc, the velocity gradient has a drastic change when fluids flow through the butterfly valve. Therefore, this zone's mesh is smaller than other zones and is mainly comprised of tetrahedral element. Finally, the mesh model owns 289,8802 elements and 56,4394 nodes, respectively.

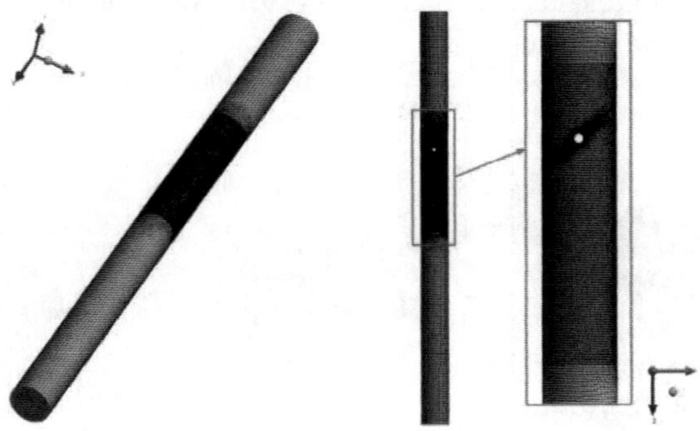

Figure 2. Mesh model of butterfly valve.

3.3. Boundary conditions and solution method
Velocity inlet and pressure outlet boundary conditions are used for the inlet and outlet of the numerical model and valve opening angle is set to 45°. Inlet velocity, particle mass fraction solid particle diameter are variable to separately study the effects of different factors. All solid walls including piping wall and valve disc wall have no-slip boundary conditions fluid gravity is considered in simulation model.

The standard k-ε model and discrete phase model are used to simulate turbulent flow and particle trajectory. The SIMPLE algorithm is adopted to pressure-velocity coupling. The QUICK scheme is applied for volume fraction. The PRESTO! scheme is used for pressure.

4. Numerical results and discussion
The solid particle erosion model has been calculated in FLUENT. Different parameters including inlet velocity, particle mass fraction and solid particle diameter are discussed successively. At inlet, velocity is separately set to 10m/s, 20m/s and 30m/s to observe the effect of inlet velocity. Particle mass fraction is separately set to 1%, 2%, 3%, 4% and 5%. Solid particle diameter is separately set to 0.05mm, 0.07mm, 0.1mm, 0.12mm and 0.15mm. Simulation results of solid erosion rate with different values of inlet velocity, particle mass fraction and solid particle diameter are shown in Figure 3-7.

Figure 3 shows that erosion rate of valve disc obviously increases with the increase of inlet velocity when the solid particle diameter and particle mass fraction keep at a certain valve. With the increase of the particle mass fraction, the erosion rate rapidly increases and the peak erosion rate is up to 4.63E−5 ($kg/m^2/s$) with inlet velocity v_{in}=30 m/s, particle mass fraction η=5% and solid particle diameter d_p=0.1mm. Effects of solid particle diameter and inlet velocity can be observed in Figure 4. Erosion rate regularly becomes larger with the increase of solid particle diameter shown in Figure 4.

Figure 5-7 vividly show the distribution of solid particle erosion and the erosion intensity of valve disc is highly enhanced with the increase of inlet velocity shown in Figure 5. Erosion mainly occurs around the upstream edge and the cylinder face of valve disc which can be seen in Figure 5. The same trend can be found in particle mass fraction and solid particle diameter shown in Figure 6-7.

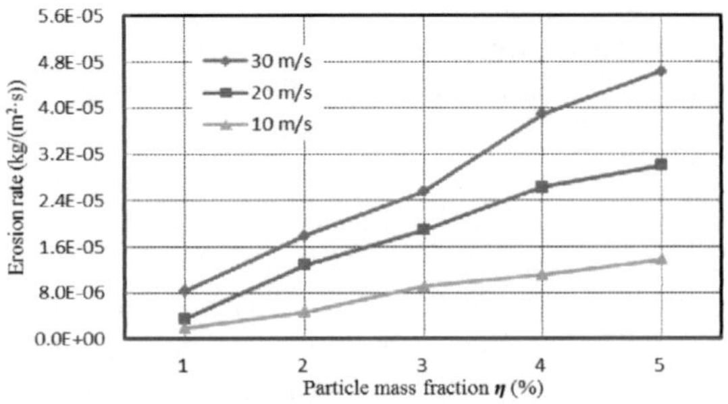

Figure 3. Erosion rate of butterfly valve at different inlet velocity v_{in} and particle mass fraction η with solid particle diameter d_p=0.1 mm.

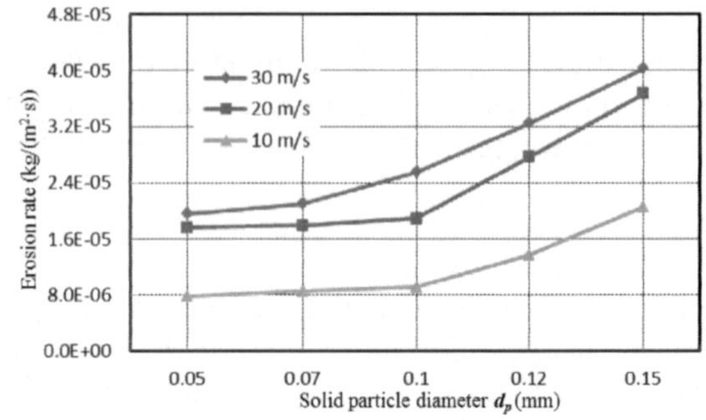

Figure 4. Erosion rate of butterfly valve at different inlet velocity v_{in} and solid particle diameter d_p with particle mass fraction η=3%.

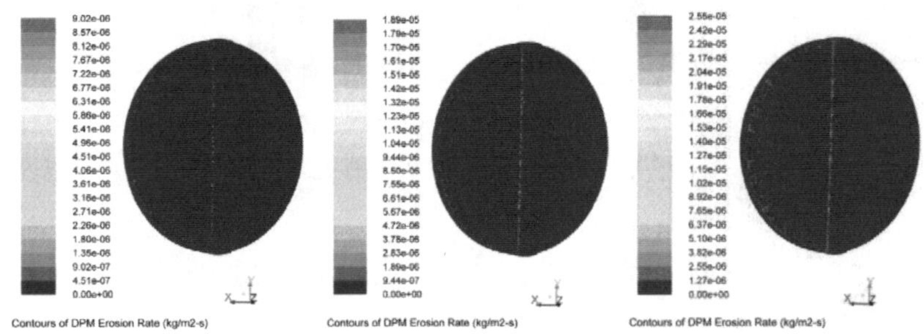

v_{in}=10 m/s v_{in}=20 m/s v_{in}=30 m/s

Figure 5. Erosion rate at different inlet velocity v_{in} with solid particle diameter d_p=0.1 mm and particle mass fraction η=3%.

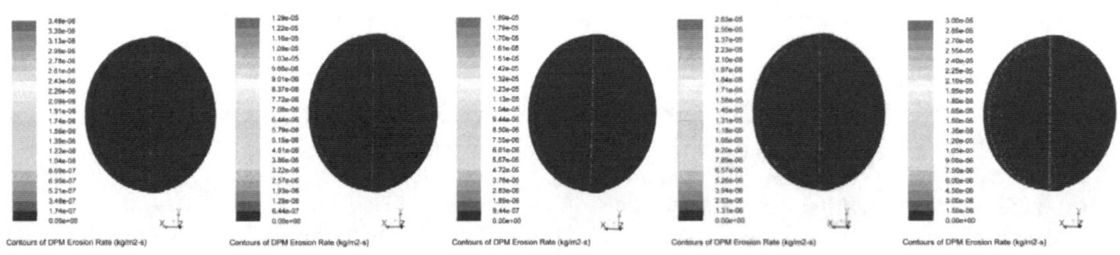

η=1% η=2% η=3% η=4% η=5%

Figure 6. Erosion rate at different particle mass fraction η with inlet velocity v_{in}=20 m/s and solid particle diameter d_p=0.1mm.

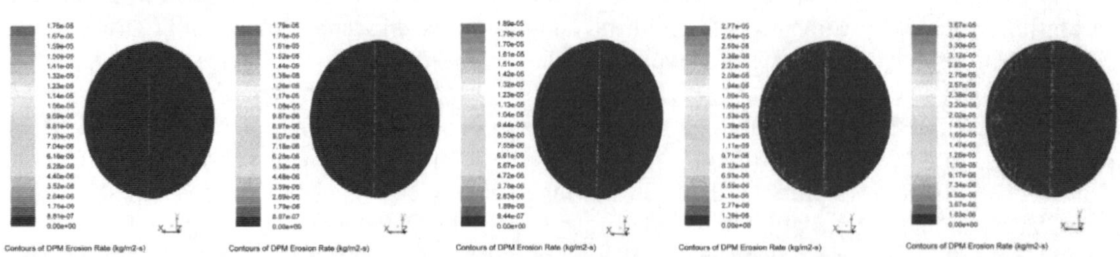

d_p=0.05mm d_p=0.07mm d_p=0.1mm d_p=0.12mm d_p=0.15mm

Figure 7. Erosion rate at different solid particle diameter d_p with inlet velocity v_{in}=20 m/s and particle mass fraction η=3%.

5. Conclusions

In this paper, a solid particle erosion analysis of butterfly valve based on the erosion theory is investigated and a CFD model of butterfly valve has been built to simulate the solid particle erosion. Different parameters of butterfly valve including inlet velocity, particle mass fraction and solid particle diameter are separately analyzed. From the study above, the following conclusions can be drawn:

(i) With increase of inlet velocity, particle mass fraction and solid particle diameter, erosion rate of valve disc increases, regularly. The peak erosion rate is up to 4.63E-5 ($kg/m^2/s$).

(ii) The erosion intensity is highly enhanced with the increase of inlet velocity, particle mass fraction and solid particle diameter.

(iii) Erosion of valve disc mainly occurs around the upstream edge and the cylinder face. To improve the erosion of butterfly valve, it can use the corrosion resistance of material at the specific location of valve disc and efficiently modified the parameters of valve.

Acknowledgments

The authors gratefully acknowledge the anonymous reviewers and associate editors for their suggestions to improve the article. This research was supported by China Nuclear Power Technology Research Institute.

References

[1] Zhang S, Wu C L, Zhang C H, Guan M and Tan J Z 2016 Laser surface alloying of FeCoCrAlNi high-entropy alloy on 304 stainless steel to enhance corrosion and cavitation erosion resistance *Opt. Laser Technol.* **84** 23-31.

[2] Hutli E, Nedeljkovic M S, Bonyár A and Légády D 2017 Experimental study on the influence of geometrical parameters on the cavitation erosion characteristics of high speed submerged jets *Exp. therm. fluid sci.* **80** 281-292.

[3] Ou G F, Rao J, Zhang L T, Zheng Z J and Ye J 2013 Numerical investigation of cavitation erosion/solid particle erosion in high differential pressure control valves in coal liquefaction *Tribology* **33** 149-61.

[4] Atkinson M, Stepanov E V, Goulet D P, Sherikar S V and Hunter J 2007 High pressure testing sand erosion in 3D flow channels and correlation with CFD *Wear* **263** 270-7.

[5] Zhu X, Liu S, Tong H, Huang X B and Li J 2012 Experimental and numerical study of drill pipe erosion wear in gas drilling *Eng. Fail. Anal.* **26** 370-80.

[6] Pang Y X, Tang Y, Liang L, Zhu Z M, Xu Y, Liu H C and Li B 2012 Flow field simulation and experimental research on interactive erosion and cavitation wear in three phases *Journal of Mechanical Engineering* **48** 115-20.

[7] Fang X, Yao J, Yin X Z, Chen X and Zhang C H 2013 Physics-of-failure models of erosion wear in electrohydraulic servovalve, and erosion wear life prediction method *Mechatronics* **23** 1202-14.

[8] Wheeler D W, Wood R J K, Harrison D and Smith E 2006 Application of diamond to enhance choke valve life in erosive duties *Wear* **261** 1087-94.

[9] Zhu H, Pan Q, Zhang W, Feng G and Li X 2014 CFD simulations of flow erosion and flow-induced deformation of needle valve: Effects of operation, structure and fluid parameters *Nuc.l Eng. Des.* **273** 396-411.

[10] ANSYS Inc. 2013 *ANSYS Fluent Theory Guide* Southpointe AZ.

[11] Edwards J K, McLaury B S and Shirazi S A 2000 Evaluation of alternative pipe bend fittings in erosive service *In proceedings of ASME FEDSM'00: ASME 2000 Fluids Engineering Division Summer Meeting.*

IOP Conf. Series: Materials Science and Engineering **220** (2017) 012019 doi:10.1088/1757-899X/220/1/012019

The Effect of Nano-ZnO Surface Wettability on Modulating Protein Adsorption

Qian Hu[1], Yadan Ding[1], Hong Shao[1], Tie Cong[1], Xiaoguang Yang[2,3] and Xia Hong[1]

[1]Key Laboratory of UV-Emitting Materials and Technology (Northeast Normal University), Ministry of Education, Changchun 130024, P. R. China
[2]College of Chemistry, Northeast Normal University, Changchun 130024, P. R. China
[3]National Engineering Laboratory for Druggable Gene and Protein Screening, Northeast Normal University, Changchun 130024, P. R. China

Abstract. Although surface wettability plays a major role in regulating protein adsorption and nanostructured ZnO has shown great potential in various biomedical fields, few reports have examined the influence of nano-ZnO surface wettability on protein adsorption. Herein, we explored the adsorption behavior of bovine serum albumin (BSA) on the superhydrophilic, hydrophilic, hydrophobic and superhydrophobic nano-ZnO surfaces. The adsorption amount of BSA increased with increase of hydrophilicity because of increased adsorption sites on the hydrophilic surface. The protein adsorption was proved to occur along with the desorption and conformational changes by well-fitted kinetic adsorption curves with the Spreading Particle Model and Fourier transformation infrared spectral analysis. The rates of BSA adsorption and desorption increased with hydrophobicity of the ZnO surfaces, which was considered to be related with the energy barrier created by water bound to the ZnO surfaces via hydrogen bonding. The rate of conformational change varied in a complex way, which might be influenced by the surface wettability of ZnO and some other factors. The present work may open up a new avenue to design nano-bio interfacial materials for advanced biological study and clinical applications.

1. Introduction

As an attractive semiconductor nanomaterial with good biocompatibility and excellent optoelectronic properties, nanostructured ZnO has shown great promise in various biomedical applications, including fluorescent biological imaging[1], resonant Raman scattering-based biological detection[2], field effect transistor-based biosensing[3], and so on[4,5]. It is imperative to study the interfacial properties between ZnO and biological components (*e.g.*, proteins, genes, and cells) to provide guidance for the relevant applications. As one of the most important surface properties of solid that affect the interaction between the solid surface and biological components, wettability of various solid surfaces has been extensively investigated[6]. Although the wettability of ZnO has also been studied[7], few reports have examined its influence on protein adsorption, whose regulation is crucial for biomedicine and tissue engineering. A systematic investigation into the effects of the wettability of ZnO on protein adsorption and the establishment of the adsorption model will shed light on the understanding of the relevant interaction mechanisms, and provide theoretical and technical basis for the development of efficient nano-biomedical platform.

In this work, the macroscopic ZnO surfaces, constructed by the ZnO crystal seed layer (ZnO SL) or the ZnO nanowire array (ZnO NA), were used as the model surfaces to study the biological adsorption

IOP Conf. Series: Materials Science and Engineering **220** (2017) 012019 doi:10.1088/1757-899X/220/1/012019

process. Their wettability was adjusted between superhydrophilicity and superhydrophobicity via ultraviolet light (UV) irradiation and dark storage. The adsorption kinetics of bovine serum albumin (BSA), an abundant protein in blood plasma and traditional blocking agent[8], on these surfaces was studied in detail with the Spreading Particle Model (SPM). And the involved rate constants were calculated and discussed.

2. Experimental section

2.1. Materials

Zinc acetate dehydrate ($Zn(CH_3COO)_2{\cdot}2H_2O$, ≥ 99.0%) and hexamethylene tetramine ($(CH_2)_6N_4$, ≥ 99.0%) were purchased from Sinopharm Chemical Reagent Co., Ltd. Monoethanolamine ($HOCH_2CH_2NH_2$, ≥ 99.0%) was bought from Tianjin Guangfu Fine Chemical Research Institute. Ethylene glycol methyl ether ($CH_3OCH_2CH_2OH$, ≥ 99.0%) was purchased from Tianjin Tiantai Fine Chemicals Co., Ltd. Fluorescein isothiocyanate-labeled bovine serum albumin (BSA, ≥ 95.0%) was bought from Beijing Zhongke Universal Technology Co., Ltd. Phosphate saline buffer (PBS, powder) was bought from Beijing Dingguo Biotechnology Co., Ltd. All the reagents were of analytical grade and used directly as received without further purification.

2.2. Preparation of the ZnO SL and ZnO NA

The ZnO SL was prepared using sol-gel dip coating method. Firstly, $Zn(CH_3COO)_2{\cdot}2H_2O$ (36.2 g) was dissolved in $CH_3OCH_2CH_2OH$ (220 mL) and $HOCH_2CH_2NH_2$ (10 mL). Then, the cleaned glasses were immersed into the above solution for 3 s, and lifted at a constant speed. The films were heated in air to 100 °C and kept for 30 min, and then heated to 450 °C to anneal for another 30 min. The heating rate was 2 $°C{\cdot}min^{-1}$. Subsequently, the films were cooled down to room temperature naturally.

The ZnO NA was prepared by hydrothermal method. The as-prepared ZnO SL was suspended in an aqueous solution of $Zn(CH_3COO)_2$ (0.02 M) and hexamethylene tetramine (0.02 M) in a Teflon-lined autoclave, and then reacted at 90 °C for 2 h. After cooling down to room temperature, it was removed from the solution, rinsed with deionized water and dried. Subsequently, it was suspended in an aqueous solution of $Zn(CH_3COO)_2$ (0.01 M) and hexamethylene tetramine (0.01 M) at 90 °C for another 4 h to perform secondary growth.

In order to obtain the surfaces with different wettability for protein adsorption, part of the obtained surfaces were exposed to UV for more than 4 h, and the rest of the surfaces were stored in dark.

2.3. Protein adsorption on the ZnO surfaces

BSA solution was prepared in PBS (pH = 7.4). The adsorption experiments were carried out in weighing bottles of 5 mL at room temperature. First, 2 mL of BSA solution was added into each weighing bottle. Then, the ZnO surfaces with similar size were carefully placed into the solution and incubated at room temperature. The adsorption kinetic curves of BSA on the ZnO surfaces were obtained with the solution concentration of 800 μg/mL, and the incubation time from 5 to 1800 min. The solution concentrations before and after the adsorption experiment were determined by fluorescence intensity of the solution. And the adsorption quantities on the ZnO surfaces were calculated by equation 1.

$$q = (C_0 - C_1) \times V / S \quad (1)$$

where C_0 is the initial concentration of BSA solution, C_1 is the concentration of the BSA solution at a certain time, V is the solution volume, and S is the area of the surface.

2.4. Characterization

X-ray diffraction (XRD) patterns were obtained using a Rigaku D/MAX-2400 X-ray diffractometer with Cu-Kα radiation (λ = 0.1541 nm). Morphologies of the ZnO SL and NA were characterized by XL-30 Field Emission Scanning Electron Microscope (SEM). The contact angle (CA) of each surface was measured by Krüss Drop Shape Analyzer-DSA100. The fluorescence intensities of BSA-FITC solution were measured by LS-55 Perkin Elmer luminescence spectrometer. The excitation wavelength for BSA-FITC was 497 nm. Fourier transformation infrared (FT-IR) absorption spectra of adsorbed BSA on the ZnO surfaces were obtained with reflection mode using Nicolet IS10 spectrometer.

3. Results and discussion

The crystal structures of as-prepared ZnO SL and NA were investigated using XRD technique. As shown in Figure 1, all of the diffraction peaks can be indexed to hexagonal wurtzite structured ZnO (JCPDS card no. 36-1451). It is worth noting that the intensity of (002) diffraction peak of the ZnO NA surpasses others, indicating that the ZnO NA was growing preferentially along the [0001] direction, and the outer top surface is the (001) plane of the ZnO NA.

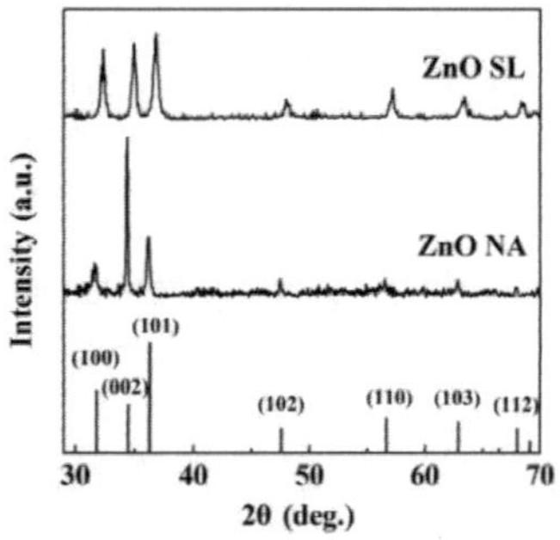

Figure 1. The XRD patterns of the ZnO SL and the ZnO NA. And the standard diffraction pattern of hexagonal wurtzite structured ZnO.

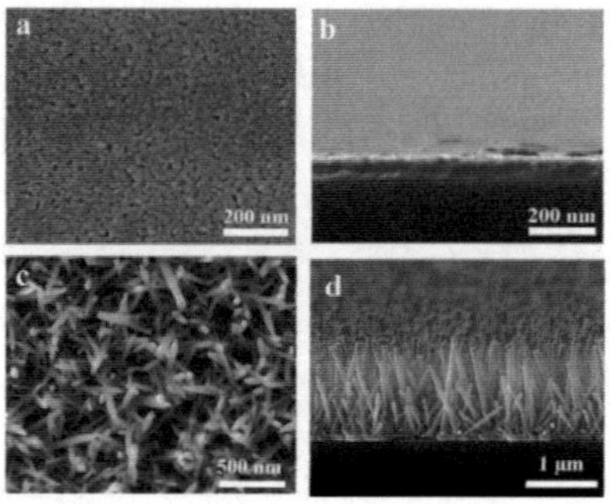

Figure 2. SEM images of (a, b) the ZnO SL and (c, d) the ZnO NA. (a, c) Top view and (b, d) cross-sectional view.

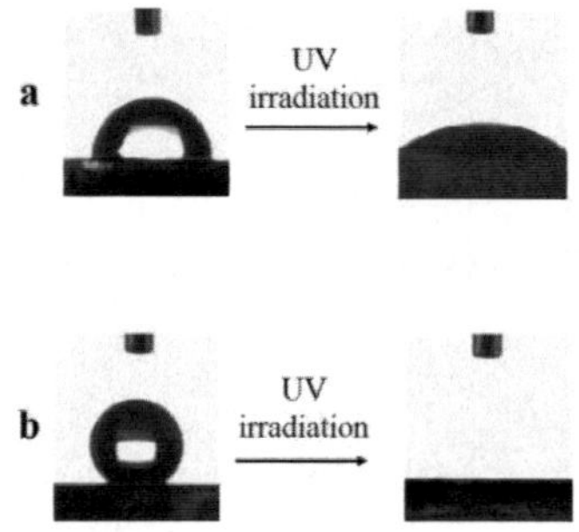

Figure 3. Photographs of water droplet on the surface of (a) the ZnO SL and (b) the ZnO NA before (left) and after (right) UV irradiation.

In order to get more detailed information about the surface structure and morphology of the ZnO SL and NA, SEM observation was carried out. Figure 2a is the top view of the ZnO SL. It can be seen that the ZnO SL consists of numerous small crystals with a diameter of 22.3 ± 4.0 nm. The thickness of the film is about 48.9 ± 1.9 nm, as shown in the cross-sectional view (Figure 2b). Figure 2c and 2d are the top view and cross-sectional view of the ZnO NA, respectively. The diameters of the nanowires are about 30.9 ± 3.2 nm, and the lengths are about 1.1 ± 0.2 μm. The anisotropic structure of the ZnO NA was consistent with the XRD results.

The morphology would profoundly affect the wettability of the ZnO surfaces. For the relatively flat ZnO SL, a photoinduced switching between hydrophobicity and hydrophilicity is shown in Figure 3a. The water CA decreases from 105.6 ± 1.3° to 28.1 ± 1.1° after UV irradiation. It results from the photogenerated oxygen vacancies on the ZnO surfaces under UV irradiation, which could be occupied quickly by the hydroxyl groups in air, and thus improve the surface hydrophilicity[8,9]. For the rough ZnO NA, the photoinduced wettability switching was enhanced greatly. The ZnO NA was superhydrophobic with the CA of 157.4 ± 2.2°, while it turned to superhydrophilicity with the CA of about 0° after UV irradiation (Figure 3b). The distinction between the wettability of the ZnO SL and the ZnO NA is ascribed to the difference in the surface structures of ZnO. On the one hand, the anisotropic ZnO NA oriented with the well-faceted (001) end protruding out has the lower surface free energy than the randomly oriented ZnO SL. On the other hand, air could enter the gaps between the ZnO NAs, and increase the hydrophobicity of the surface due to the shrink of the water droplet at the gas-liquid interface[10]. After UV irradiation, more hydroxyl groups were adsorbed on the ZnO surface, and water would enter and fill the gaps between the ZnO nanowires due to the three-dimensional capillary effect of the rough surface[11]. Therefore, after UV irradiation, superhydrophilicity was observed with the ZnO NA, while hydrophilicity was observed with the ZnO SL.

The adsorption kinetics of BSA on the superhydrophilic, hydrophilic, hydrophobic and superhydrophobic ZnO surfaces were measured in order to study the protein adsorption process on the ZnO surfaces. It can be clearly seen from Figure 4 that the adsorption amount increases with the increase of hydrophilicity. It is because that the solution on the hydrophilic, especially superhydrophilic surfaces of ZnO tends to spread and infiltrate into the gaps between the ZnO crystal seeds or nanowires due to the three-dimensional capillary effect. Hence the contact area between the protein solution and the ZnO surfaces would be extended, and more adsorption sites would be provided for BSA. On the hydrophobic surface, the spread and infiltration of the solution would be reduced, and less adsorption sites would be available for BSA. And this phenomenon is amplified on the superhydrophobic surface because the air trapped within the gaps of the nanowires greatly prevents intimate contact of the surface with the protein solution and thus strongly inhibits BSA adsorption.

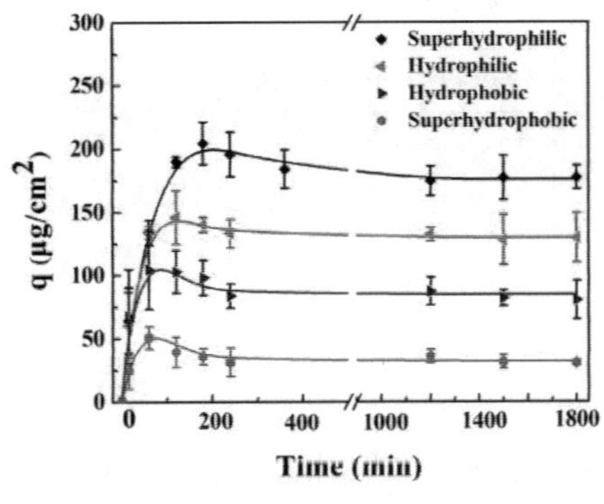

Figure 4. Kinetic curves of BSA adsorbed onto the ZnO surfaces with different wettability. Solid lines embedded are the theoretical values fitted according to the SPM.

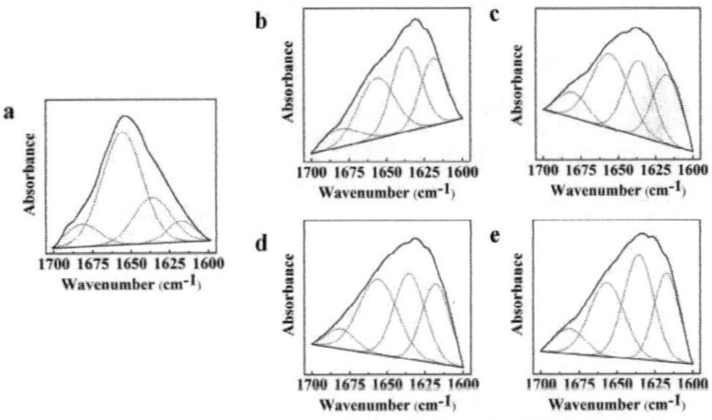

Figure 5. FT-IR spectra of (a) BSA in PBS solution, BSA adsorbed onto the (b) superhydrophilic, (c) hydrophilic, (d) hydrophobic and (e) superhydrophobic ZnO surfaces accompanied with curve-fitting.

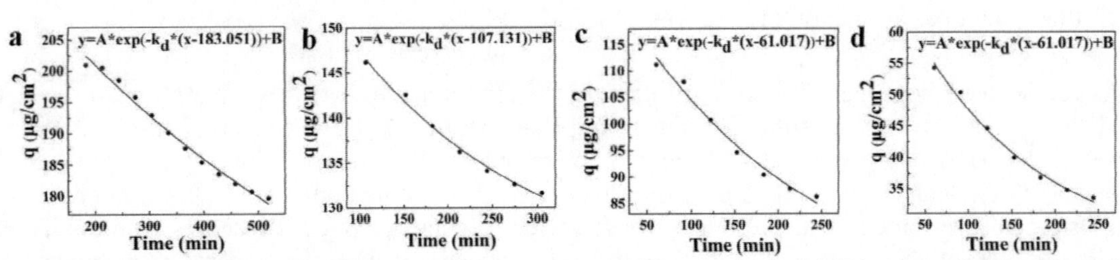

Figure 6. Desorption kinetics for the (a) superhydrophilic, (b) hydrophilic, (c) hydrophobic and (d) superhydrophobic ZnO surfaces. The corresponding fitting formulas are shown above the curves.

It can also be seen from Figure 4 that the adsorption amount (q) of BSA on the superhydrophilic surface reached a maximum within the first 180 min, and then decreased slightly to attain equilibrium. The similar variation tendency was observed for the other ZnO surfaces. It indicates that the adsorption process was predominant at first, and then replaced by the desorption process after a certain time. The adsorption-desorption equilibrium was attained finally. The distinction in the time to reach the maximum adsorption amount and the adsorption-desorption equilibrium on the ZnO surfaces with different wettability might be related to the difference in the rate of adsorption and desorption of BSA.

In addition, the conformational change of BSA which could cause irreversible adsorption was also a factor that might be considered.

FT-IR spectra were recorded to study the conformational changes of BSA adsorbed on these ZnO surfaces. The amide I region (~1700-1600 cm^{-1}), which was largely ascribed to C=O stretching vibration[12,13], was employed for the secondary structure analysis of BSA. Four peak positions centered at 1681 ± 1 cm^{-1} (turns), 1656 ± 1 cm^{-1} (α-helixs), 1636 ± 1 cm^{-1} (β-sheets) and 1617 ± 1 cm^{-1} (side chains) were used in the curve fitting (Figure 5). As the absorption band at 1617 ± 1 cm^{-1} was assigned to the side chains of BSA[14], not the amide carbonyl group in the main chain, the peak areas of the other bands were computed to obtain the contents of the secondary structures. As shown in Table 1, BSA adsorption on the ZnO surfaces caused obvious changes in the contents of α-helixs and β-sheets, and slight changes in the contents of turns compared with those of BSA in PBS solution. It indicates the occurrence of conformational changes of BSA upon adsorption on the ZnO surfaces.

Table 1. Secondary structures (%) of BSA in solution and adsorbed on the ZnO surfaces

	α-Helixs	β-Sheets	Turns
BSA in PBS solution	65.2	24.7	10.1
BSA adsorbed onto the superhydrophilic ZnO surface	41.9	46.2	11.9
BSA adsorbed onto the hydrophilic ZnO surface	49.2	38.3	12.5
BSA adsorbed onto the hydrophobic ZnO surface	48.0	43.1	8.9
BSA adsorbed onto the superhydrophobic ZnO surface	40.1	48.5	11.4

Since conformational changes existed in addition to the adsorption and desorption processes, Spreading Particle Model (SPM) was adopted to describe the adsorption kinetics of BSA on the ZnO surfaces. In SPM[15], the kinetic equation about the adsorption of protein molecules is a function of time (t). The simplest form of the equation describes the surface coverage of protein molecules as follows: first, the protein molecules adsorb on the surface reversibly at a rate k_aC. Then, the protein molecules either undergo desorption or conformational change to get adsorbed on the surface irreversibly. The rates of the two events are k_d and k_s, respectively. The expression for the total adsorption amount as a function of time was obtained as:[16,17]

$$q = A_1 \cdot \exp(A_2 \cdot t) + A_3 \cdot \exp(A_4 \cdot t) + A_5 \tag{2}$$

As can be seen from the solid lines in Figure 4, the kinetic curves of BSA adsorption were all well fitted with equation 2, suggesting that the adsorption kinetics of BSA conforms to the SPM. The values of the parameters A_1, A_2, A_3, A_4 and A_5 were obtained by curve fiiting. As $A_1 \cdot \exp(A_2 \cdot t)$ and $A_3 \cdot \exp(A_4 \cdot t)$ gradually approach zero as time increases, A_5 in equation 2 is closely related with the equilibrium adsorption amount. The value of parameter A_5 (plateau value) decreases with the increase of surface hydrophobicity according to Table 2. It is consistent with the experimental phenomena, confirming that the superhydrophilic ZnO surface is conducive to protein adsorption, but the superhydrophobic ZnO surface inhabits its adsorption. The parameters A_1, A_2, A_3 and A_4 are related with the initial stage of adsorption. A_2 and A_4 can be used to calculate the rate constants k_aC, k_d and k_s. They are described by equation 3 and 4[16].

$$-(A_2 + A_4) = k_a \cdot C + k_d + k_s \tag{3}$$

$$A_2 \cdot A_4 = k_a \cdot C \cdot k_s \tag{4}$$

Since the descending portion of the curve in Figure. 4 indicates the dominant desorption process of BSA, the desorption rate constant k_d can be estimated by fitting this region with equation 5[16].

$$q = A \cdot \exp(-k_d \cdot t) + B \tag{5}$$

where A and B are constants. The fitting results and the calculated desorption rate constant k_d are shown in Figure 6 and Table 2, respectively. And thus, k_aC and k_s can be obtained with equation 3 and 4. They are also shown in Table 2.

Table 2. Values of kinetic rate constants and plateau value

	Rate constants (min^{-1})			Plateau value
	k_aC	k_d	k_s	($\mu g/cm^2$)
Surperhydrophilicity	0.0108	0.00134	0.0095	175.61
Hydrophilicity	0.0219	0.00453	0.0170	129.95
Hydrophobicity	0.0270	0.00484	0.0219	84.36
Superhydrophobicity	0.0275	0.00695	0.0200	31.95

As shown in Table 2, both the adsorption rate constant k_aC and the desorption rate constant k_d increase with the increase of the hydrophobicity of the ZnO surfaces. It might be related with the energy barrier induced by water bound to the ZnO surfaces via hydrogen bonding interaction[18]. As the amount of hydroxyl groups on the ZnO surfaces decreases with the increase of hydrophobicity, the hydrogen bonding between the ZnO surfaces and water becomes weaker, and the energy barrier created by water for protein adsorption gets lower. Thus the adsorption of the protein to the ZnO surfaces becomes easier, which leads to the increased adsorption rate k_aC of BSA on the ZnO surfaces with increased hydrophobicity. In the meantime, lower energy barrier also makes it easier for BSA to desorb from the surface and go back to the bulk solution through the interfacial region. Therefore, the desorption rate k_d of BSA also increases with the increase of the surface hydrophobicity to achieve dynamic equilibrium. As for the conformational change rate constant k_s, it does not change monotonically with the increase of the hydrophobicity. It indicates that the conformational change of BSA is a complex process. It might be affected by not only the surface wettability, but also some other factors, such as surface free energy[19], specific geometrical figures[20], and surface functional groups[21].

4. Summary and conclusions

In summary, nano-ZnO surfaces with the wettability between superhydrophobicity and superhydrophilicity were fabricated. The adsorption amount of BSA increased with the increase of hydrophilicity of the ZnO surface due to the increased adsorption sites on the hydrophilic surface. The obvious conformational changes were caused after BSA was adsorbed on the ZnO surfaces. BSA adsorption kinetics on nano-ZnO was approximated by SPM, confirming the simultaneous occurrence of the protein adsorption, desorption and conformational changes. The rates of adsorption and desorption increased with the hydrophobicity of the ZnO surfaces. It might be related with the energy barrier caused by the water bound to the ZnO surfaces via hydrogen bonding. The rate of conformational change varied in a complex way, which might be determined by the synergistic effect of the surface wettability of ZnO and some other factors. The ZnO surfaces are promising to control the protein adsorption. This work may have instructive significance for understanding biophysicochemical interactions at the nano-bio interface and optimization of the nanobiomaterial selection, design and application.

Acknowledgments

This work was supported by the National Natural Science Foundation of China (Grant Nos. 51272040 and 11604043), the 111 project (No. B13013), and Program for New Century Excellent Talents in University of Ministry of Education of China (Grant No. NCET-12-0815).

References

[1] Ye D X, Ma Y Y, Zhao W, Cao H M, Kong J L, Xiong H M and Mohwald H 2016 *ACS Nano.* **10** 4294.
[2] Hong X, Chu X , Zou P, Liu Y and Yang G 2010 *Biosens. Bioelectron.* **26** 918.

[3] Arya S K, Saha S, Ramirez-Vick J E, Gupta V, Bhansali S and Singh S P 2012 *Anal. Chim. Acta.* **737** 1.
[4] Aydin Sevinç B and Hanley L 2010 *J. Biomed. Mater. Res., Part B* **94** 22.
[5] Talebian N, Nilforoushan M R and Zargar E B 2011 *Appl. Surf. Sci.* **258** 547.
[6] Song W and Mano J F 2013 *Soft Matter* **9** 2985.
[7] Feng X, Feng L, Jin M, Zhai J, Jiang L and Zhu D 2004 *J. Am. Chem. Soc.* **126** 62.
[8] Bhogale A, Patel N, Sarpotdar P, Mariam J, Dongre P M, Miotello A and Kothari D C 2013 *Colloids Surf., B* **102** 257.
[9] Sun R D, Nakajima A, Fujishima A, Watanabe T and Hashimoto K 2001 *J. Phys. Chem. B* **105** 1984.
[10] Cassie A B D and Baxter S 1944 *Trans. Faraday. Soc.* **40** 546.
[11] Shalumon K T, Anulekha K H, Nair S V, Chennazhi K P and Jayakumar R 2011 *Int. J. Biol. Macromol* **49** 247.
[12] Byler D M and Susi H 1986 *Biopolymers* **25** 469.
[13] Roach P, Farrar D and Perry C C 2006 *J. Am. Chem. Soc.* **128** 3939.
[14] Maruyama T, Katoh S, Nakajima M, Nabetani H, Abbott T P, Shono A and Satoh K 2001 *J. Membr. Sci.* **192** 201.
[15] Lundström I 1985 *Prog. Colloid Polym. Sci.* **70** 76.
[16] Goyal D K and Subramanian A 2010 *Thin Solid Films* **518** 2186.
[17] Wang Y, Deng H, Huangfu C, Lu Z, Wang X, Zeng X, He H and Rao H 2015 *Surf. Interface Anal.* **47** 245.
[18] Zheng J, Li L, Tsao H K, Sheng Y J, Chen S and Jiang S 2005 *Biophys. J.* **89** 158.
[19] Kwok S C H, Wang J and Chu P K 2005 *Diam. Relat. Mater.* **14** 78.
[20] Song W and Chen H. 2007 *Chinese Sci. Bull.* **52** 3169.
[21] Kidoaki S and Matsuda T 1999 *Langmuir* **15** 7639.

IOP Conf. Series: Materials Science and Engineering **220** (2017) 012020 doi:10.1088/1757-899X/220/1/012020

Modeling and simulation of the debonding process of composite solid propellants

Tao Feng[1*], Jin-sheng Xu[1], Long Han[1], and Xiong Chen[1]

[1]Department of Institute of Mechanical Engineering, Nanjing University of Science and Technology, Nanjing, 210094, China

*E-mail: 791688971@qq.com

Abstract. In order to study the damage evolution law of composite solid propellants, the molecular dynamics particle filled algorithm was used to establish the mesoscopic structure model of HTPB(Hydroxyl-terminated polybutadiene) propellants. The cohesive element method was employed for the adhesion interface between AP(Ammonium perchlorate) particle and HTPB matrix and the bilinear cohesive zone model was used to describe the mechanical response of the interface elements. The inversion analysis method based on Hooke-Jeeves optimization algorithm was employed to identify the parameters of cohesive zone model(CZM) of the particle/binder interface. Then, the optimized parameters were applied to the commercial finite element software ABAQUS to simulate the damage evolution process for AP particle and HTPB matrix, including the initiation, development, gathering and macroscopic crack. Finally, the stress-strain simulation curve was compared with the experiment curves. The result shows that the bilinear cohesive zone model can accurately describe the debonding and fracture process between the AP particles and HTPB matrix under the uniaxial tension loading.

1. Introduction

The composite solid propellant as a kind of energetic materials are widely used in the rockets and various strategic and tactical missile, its mechanical performance has a great influence on the survivability and operational capability of the missile. Many experiments have confirmed that the debonding between the particle and matrix interface is a key factor that affects the propellant macro mechanical properties[1]. Therefore, the study of the mesoscopic damage initiation, evolution, gathering and its effects on the mechanical properties of composite solid propellants is quite important and meaningful.

HTPB(Hydroxy-terminated polybutadiene) propellant is a three-phase composites, consisting of HTPB matrix, filled solid particles and the particle/binder interface. The nonlinear macroscopic mechanical response of the propellants dependents on its mesoscopic structure and the multi-scale physical process under loading. At present the meso-experiment of the propellant is difficult to carry out due to the limitations of method, so the numerical simulation method is widely applied to study the influence of mesoscopic structure on macroscopic mechanical behaviour of the HTPB propellants. Matous[2] generated the representative volume element of the propellants' mesoscopic model and used the cohesive elements to simulate the initiation and evolution of the interface debonding between the particles and the matrix. Li[3] adopted the cohesive element model to simulate propellant/linear interface and found that the failure mode experiences damage initiation, damage evolution and failure. Zhi[4-5] investigated the nonlinear mechanical behavior of solid propellants through mesomechanical methods and simulated and discussed the effects of the packing volume fraction and particle size and distribution on the debonding process of solid propellants. Han[6] employed the inversion analysis method to obtain the optimized interface parameters and then studied the change rule of the mesoscopic interface performance of HTPB composite solid propellant with different loading rate.

Thus, this paper carried out the uniaxial tension experiment to study the mechanical properties of composite solid propellants. The mesoscopic model was generated by the molecular dynamics method and the cohesive zone model was also employed to simulate the damage evolution process of the interface. Then the Hooke-Jeeves optimization algorithm was used to identify the parameters of cohesive zone model and the optimized parameters can accurately describe the debonding process between the AP particles and HTPB matrix under the uniaxial tension loading.

2. Mesoscopic model

2.1. Packing model and boundary

The components of solid propellants which are studied in this paper are listed in Table 1. In order to simulate the mesoscopic damage evolution process of the propellant interface, with the particle size distribution shown in Fig1, the molecular dynamics method was employed to generate the representative volume element(RVE) of the propellants.

Because of the very high Young's modulus, the AP particles' deformation is quite small, so the mesh type of AP particle must be CPE4. Meanwhile, as the incompressible hypothesis, the mesh type of the matrix should be CPE4H to model the mechanical response of HTPB with large deformation. According to the periodic boundary conditions and the mutual constraints of the RVE deformation, the boundary of the RVE need to keep straight when loading. Due to this fact, the left border and the lower boundary of the RVE are fixed and the right boundary is coupled the displacement in the x axis. The upper boundary, as the load recipient, undertakes the constant displacement load. In this study, the given speed of displacement load is 0.1667 mm/s and the corresponding strain rate is 0.003333 /s. The mesh model and boundary conditions are shown in Fig 2.

Table 1. Components of the HTPB propellant

Component	HTPB	AP	Al	Auxiliary
Mass fraction/%	8.0	69.5	18.5	4
Volume fraction/%	23.9	63.8	12.3	-

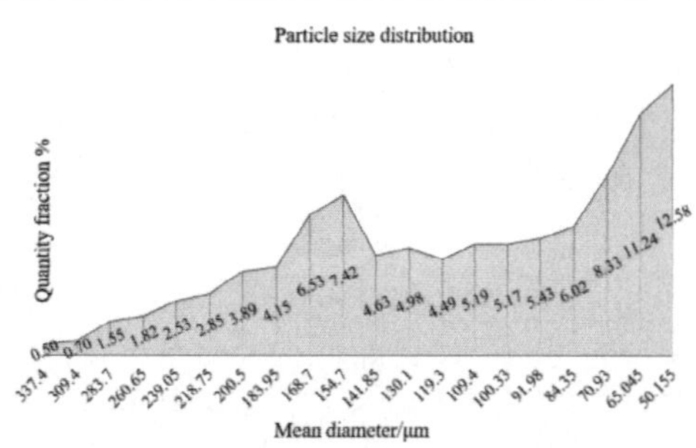

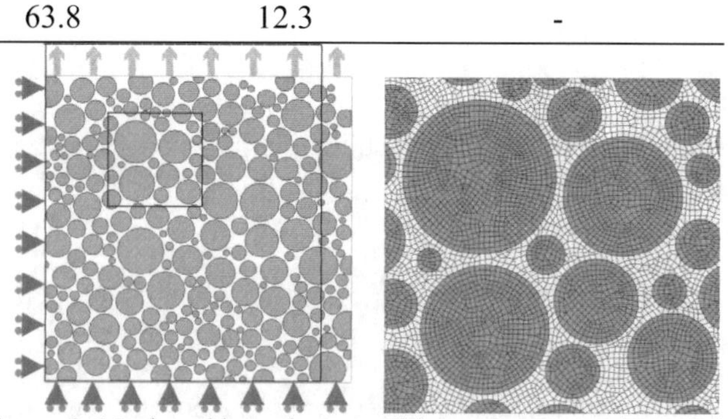

Figure 1. Particle size distribution of propellants. **Figure 2.** Mesh and boundary conditions of mesoscopic model.

2.2. Cohesive zone model(CZM)

Cohesive technology is widely deployed in simulating the damage between particles and matrix which was proposed by Dugdale[7] and Barenblatt[8]. The interface elements open when damage occurs and lose their stiffness at failure so that the continuum elements are disconnected, and in this paper, a damage model called cohesive zone model that consists of two ingredients a damage initiation criterion and a damage evolution law is accepted.

So far, several types of CZM models have been proposed, they can be distinguished by their own cohesive low[9]. The typical bilinear cohesive zone model is given in Fig 3, it relates cohesive surface traction, t ,to displacement jump, δ ,at the interface where a debonding may occur. The interface begins to damage when the maximum contact stress is equal to the maximum stress, and complete failure occurs at the maximum displacement δ_f . The slope of the traction-displacement curve before damage initial, K , is refereed as the interface stiffness. K , t_0 and δ_f are the basic parameters to confirm a bilinear CZM model.

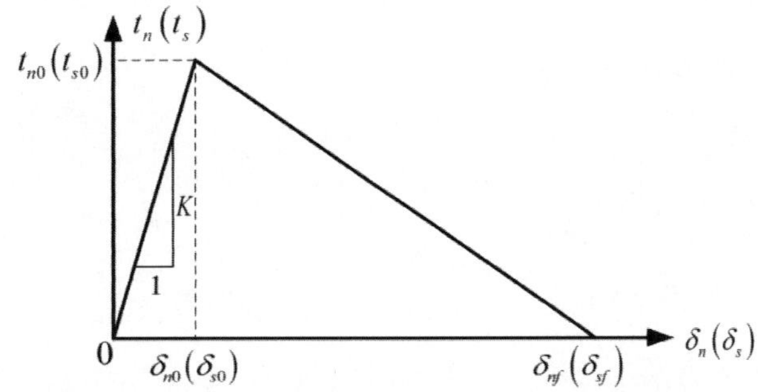

Figure 3. The schematic diagram of the bilinear cohesive zone model.

In this paper, the maximum stress criterion is used as the damage initiation and can be expressed by:

$$\max\left(\frac{\langle t_n\rangle}{t_{n0}},\frac{t_s}{t_{s0}}\right)=1 \tag{1}$$

where the symbol is the Macaulay bracket, the subscript "n" and "s" represents the normal and the shear direction.

The traction-separation law can be expressed by:

$$\begin{cases} t_n=(1-D)\dfrac{\delta_n}{\delta_{n0}}t_{n0} \\ t_s=(1-D)\dfrac{\delta s}{\delta_{s0}}t_{s0} \end{cases} \tag{2}$$

where, D is the damage factor, can be expressed by:

$$D=\begin{cases} 0 & (\delta\le\delta_0) \\ \dfrac{\delta_f(\delta-\delta_0)}{\delta(\delta_f-\delta_0)} & (\delta>\delta_0) \end{cases} \tag{3}$$

3. Material properties and mechanical parameters

3.1. Material properties

The components of solid propellants are listed in Table 1. The main material properties of solid propellants are the Young's modulus E and the Poisson's ratio v of the filled particles AP and Al. They are all considered as isotropic elastic solids, and can be set as follow: $E_{AP}=32450$ MPa , $v_{AP}=0.14$; $E_{Al}=68300$ MPa , $v_{Al}=0.33$.

Some researchers considered the matrix of solid propellants as a kind of elastic material. This simplification is inaccurate. In fact, the material usually shows a viscoelastic behavior just like rubber. In this paper, the universal stress relaxation texts on HTPB solid propellants matrix samples are conducted to reveal the viscoelastic response. The applied load is 100% strain for 1200 s. it is assumed that the matrix is a linear viscoelastic material which is defined by a Prony series[10]:

$$E(t) = E_{\infty} + \sum_{i=1}^{n} E_i \exp\left(-\frac{t}{\tau_i}\right) \tag{4}$$

where the E_{∞}, E_i, and $\tau_i, i = 1, 2, ..., N$ are material constants.

The experimental data are fitted by a nonlinear curve fitting method. The result shows that the Prony series conforms well with the experimental data, as shown in Fig.4

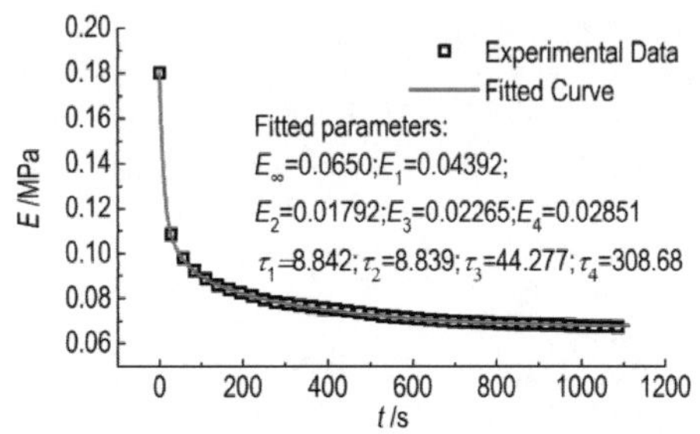

Figure 4. Stress-relaxation curve of the matrix.

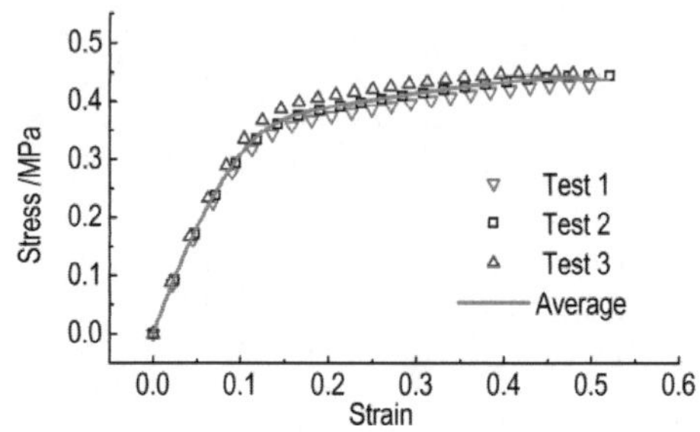

Figure 5. The uniaxial tensile stress-strain curve.

3.2. Mechanical parameters

The interface meso-mechanical parameter values have a vital influence on the accuracy of the numerical calculation results. At present, the approximate cohesive parameters are usually estimate according to the experience and literature and it is often inaccurate. In order to solve the disadvantages, many researchers raised the inversion optimization method based on Hooke-Jeeves algorithm to acquire the more exact parameters.

It's necessary to conduct the uniaxial tensile test of composite solid propellants for the inversion analysis of the interface parameters. Taking no account of the temperature and strain rate dependency, the applied load is 10 mm/min and the corresponding strain rate is 0.003333 /s at room temperature. The engineering stress-train curve of the propellant is given in Fig.5

There are three basic parameters in a typical bilinear cohesive zone model and they have different influence on the stress-stain curve's tendency respectively. In view of this feature, the step-by-step inversion method was determined and the objective function that characterize the coincidence degree of the test and the simulation curve was build.

$$R = \frac{1}{N}\sum_{i=1}^{N}\left(\sigma_{\varepsilon_i}^{sim}\left(K, t_0, \delta_f\right) - \sigma_{\varepsilon_i}^{exp}\right)^2 \tag{5}$$

When the objective function value is less than the selected tolerance R_{lim}, the current interface parameters are believed to be accurate. The initial and the ultimate value of the cohesive interface parameters are listed in Table 2.

Table 2. Cohesive parameters of the interface

Value	$K/(\text{MPa/mm})$	t_0 / MPa	δ_f /mm
Initial	500	0.50	0.01
Ultimate	400	0.45	0.02

4. Simulation and analysis

4.1. Simulation result

The mesoscopic model generated in the previous section was calculated by the finite element software ABAQUS to simulate the uniaxial tension of HTPB propellants. The Mises stress nephograms of the propellant at 5%, 15%, and 50% strain respectively are given in Fig. 6.

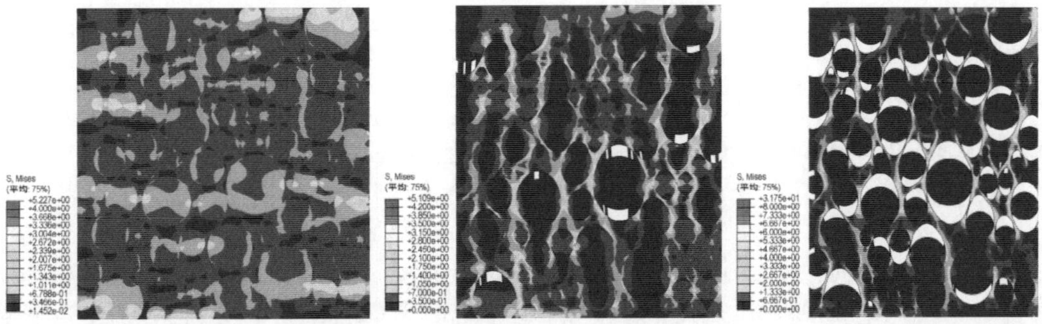

Figure 6. Mises stress distribution in the propellant when at 5%、15% and 50% strain respectively.

Figure 6 shows that when the strain is 5%, the stress and strain distribution in solid propellant is uneven, it is caused by the inconsistent deformation among the particles, matrix and interface. As an elastomer, the modulus of particles is quite larger than the matrix, so the deformation of the matrix is much bigger than the particles and this certainly makes the particle/matrix interface the weakness in the propellants. With the load increases to 15% strain, the local strain near the large particles region is very huge and this leads to a very obvious stress concentrain, the stress in the larger particle/matrix interface reaches the cohesive strength fist and as a result, these interfaces begin to debond. When the strain reaches 50%, more and more interfaces have debonded and the micro holes develop, gather and the macro fracture form, eventually, the composite solid propellants break.

4.2. Results analysis

In simulation results, the pulling force on the upper boundary of mesoscopic model can be determined by summating all the constraint reaction of each node, and then the emulational stress-strain curve of the solid propellants can be obtained through corresponding calculation. The emulational stress-strain curve and the mean experiment curve that acquired in the previous section were given in Fig. 7.

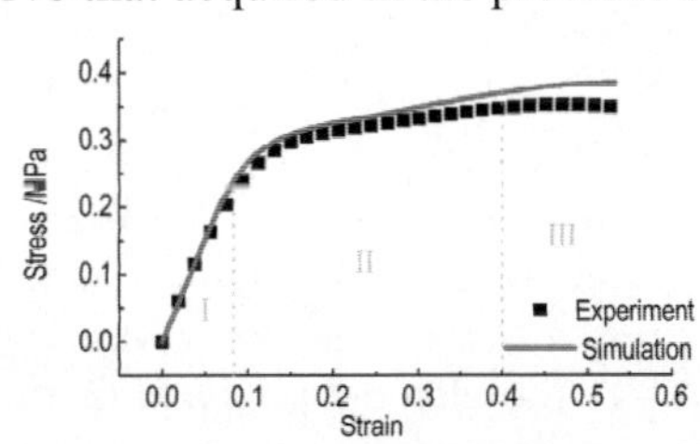

Figure 7. Engineering stress-strain curve of HTPB propellant.

Combining the stress nephograms in Fig.6 and the engineering stress-strain curve in Fig.7, it' s found that in stage Ⅰ($\varepsilon < 5\%$), the particles and matrix bond in a good condition and there is no damage inside the mesoscopic structure of the representative volume element, so the propellants show the elastic properties and the stress grows linearly with the strain increasing. Then coming to the stage Ⅱ, in this stage the large particles and matrix interfaces begin to debond and it leads to a decrease of the load bearing capability, this phenomenon can also be observed in Fig 7 that the stress-strain curve begins to appear the nonlinearity and enter a "platform area". Finally in the stage Ⅲ, more and more interfaces undertake the serious debonding and the micro holes appear, then the holes develop and gather resulting in macro crack and the stress of the propellants drops rapidly.

Comparing the simulation curve and experiment curve in Fig.7, it's clear that the simulation curve obtained by the optimized inversion method matches the actual experiment curve well, but there also exist a little gap between them. In can be thought that the main causes of the difference are: (1) the mesoscopic model established in this paper is two-dimensional and there is a certain deviation with the three-dimensional state of the actual solid propellants; (2) there are many kinds of initial defects such as micro-hole and micro-crack in the propellants and it's ignored in the mesoscopic model; (3) all the solid particles are considered as regular circular, however there may be irregular particles in the propellants and the stress concentration and interface debonding are more likely to occur near these particles.

From the above analysis, we can see that the interface debonding process is an important impact factor in the propellants' nonlinear property. The mesoscopic model and the inversion optimization interface parameters are reliable and can be used to simulate the debonding process and the mechanical response of the solid propellants.

5. Conclusion

(1) Based on the periodicity assumption, the molecular dynamics method was employed to establish the representative volume element of the composite solid propellants mesoscopic filling model. Meanwhile, the cohesive element method was employed to simulate the mesoscopic damage initiation, evolution, gathering of the particle/matrix interface and its effects on the propellants' mechanical properties.

(2) In order to obtain the exact parameters of cohesive zone model which can characterize the mechanical properties of the particle/matrix adhesive interface, the uniaxial tensile test of composite solid propellants was performed. Combining the parameter optimization inversion method based on "Hooke-Jeeves" algorithm with the experimental data, the relatively accurate parameters are acquired.

(3) The interface debonding process between particles and matrix in the solid propellants was simulated by means of the finite element platform ABAQUS. It's found that while the interface stress reaches the cohesive strength, the debonding occurs and with the increase of load, the debonding damage develops, gathers and macro fracture eventually forms. It was proved that the particle/matrix interface debonding is an important cause to the nonlinearity of the propellant stress-strain curve. Besides, the mesoscopic model and simulation method taken in this paper can be used to accurately simulate the macroscopic mechanical behavior of the composite solid propellants.

References

[1] Chang W, Ju Y, Wang P 2012 Research on Correlation Between Dewetting and Mechanical Property of HTPB Propellant *Acta Armamentarii* **32(3)** 261-266.

[2] Matouš K, Inglis H M, Gu X, et al. 2007 Multiscale modeling of solid propellants: From particle packing to failure *Composites Science & Technology* **67(7)** 1694–1708.

[3] Li G, Xing Y, Ji Z, et al. 2011 Finite element analysis of microscale interfacial debonding in composite solid propellants *Acta Material Compositae Sinica* **28(3)** 229-235.

[4] Zhi S, Sun B, Zhang J 2013 Numerical Simulation of Solid Propellant Mesoscopic Damage Using Surface-Based Cohesive Approach *Journal of Propulsion Technology* **34(2)** 273-279.

[5] Zhi S, Sun B, Zhang J 2012 Multiscale modeling of heterogeneous propellants from particle packing to grain failure using a surface-based cohesive approach *Acta Mechanica Sinica* **28** 746-759.

[6] Han L, Xu J, Zhou C 2016 Inverse identification of the rate-dependent micro interface parameters of HTPB/IPDI composite propellant *Chinese Journal of Energetic Materials* **10** 928-935.

[7] Dugdale D S 1960 Yielding of steel sheets containing slits *Journal of the Mechanics and Physics of Solids* **8(2)** 100-104.

[8] Barenblatt G I 1959 The formation of equilibrium cracks during brittle fracture. General ideas and hypotheses. Axially-symmetric cracks *Journal of Applied Mathematics and Mechanics* **23(3)** 622-636.

[9] Chandra N, Li H, Shet C, et al. 2002 Some issues in the application of cohesive zone models for metal–ceramic interfaces *International Journal of Solids and Structures* **39(10)** 2827-2855.

[10] Jung G D, Youn S K 1999 A nonlinear viscoelastic constitutive model of solidpropellant *International journal of solids and structures* **36(25)** 3755-3777.

IOP Conf. Series: Materials Science and Engineering **220** (2017) 012021 doi:10.1088/1757-899X/220/1/012021

Synthesis and Characterization of Hybrid CF/MWCNT$_S$/Epoxy Resin Composite System

Hany Fouda[1*], Lin Guo, Yonghai Yue[1], Ke Chen[1] and Karim Elsharkawy[1]

[1]School of Chemistry and Environment, Beijing University of Aeronautics and Astronautics, Beijing, China

*E-mail: hanyfouda@buaa.edu.cn

Abstract. In the present investigation, two methods were used for addition multiwall carbon nanotubes (MWCNT$_S$) into carbon fiber (CF)/epoxy resin composite system. The mechanical properties of the prepared samples were compared to show the best method for addition of MWCNT$_S$ from point of view of mechanical properties. The introduction of carbon nanotubes (CNTs) into fiber reinforced polymer composites has been achieved mainly via two routes: mixing CNTs entirely throughout the matrix (matrix modification) or attaching CNTs onto reinforcing fibers (interface modification). In all previous references the addition of CNTs occur through one route from the two routes but in this research, we introduced MWCNT$_S$ into CF/epoxy resin composite through the two routes at the same time. Three CF composite samples were prepared CF/epoxy resin composite (C1), CF/ 1wt% MWCNT$_S$ /epoxy resin composite (C2) in which MWCNT$_S$ added via one route (epoxy resin system) and the third sample was CF/ 1wt% MWCNT$_S$ / epoxy resin composite (C3) in which MWCNT$_S$ added via two routes (epoxy resin and CF fabric). The result shows that the mechanical properties of C3>C2>C1, for example, the flexural strength of C3 higher than C2 by 19% and C2 higher than C1 by 51% respectively. This is because addition MWCNT$_S$ via two routes increase the ability of good mixing of CNT$_S$ with epoxy resin and good dispersion of CNTs into the CF fabric surface and this leads to improve the interface bonding between the CF and epoxy so improve the mechanical properties.

1. Introduction

Many types of filler can be used with resin matrix composite materials but the most important one is Carbon fiber CF and this is because its excellent engineering properties, such as high specific stiffness and strength, performance to weight ratio, high thermal stability, high conductivity, corrosion resistance and self-lubrication. Most obviously, CF used in the structural application that needs weight reduction of the equipment as vehicles industries due to its high strength to weight ratio. CF reinforced resin matrix composites have a lot of applications as in aerospace applications, turbomachinery, for wind turbines applications in automotive energy systems, fuel cells, compressed gas storage and transportation, antistatic and electromagnetic shielding applications. the carbon fibers reinforced polymer matrix composites is widely used in automobile industries, aerospace, and marine during the last few decades due to their good engineering properties such as high specific strength and stiffness, lower density [1-5]. In The composite material system carbon fibers give strength and stiffness while resin matrix provides the chemical and thermal resistance of the composite [4].

CF structures have crystallized graphitic basal planes with a non-polar surface. CF is chemically inert this is because of the presence of high-temperature carbonization/graphitization step during the manufacturing process, CF has weak bonding with the matrix surface due to lipophobicity, and excessive smoothness and fewer adsorption characteristics of CF. due to the inertness of CF, improving at the fiber surface is needed in order to improve the interfacial bonding between fiber/matrix for effective stress transfer at the interface [6-7]. Liu and Kumar have studied the existing progress of carbon fiber structure, fabrication, and characterization including the addition of nanotubes in precursor fiber to improve the mechanical properties [8]. However, the fundamental mechanical properties of these composites such as longitudinal and transverse strength and toughness limited by the weak interfacial bonding between reinforcing CF surface and polymer materials. It is a long existing critical and important topic needed to discover for ensuring the continued development of CF reinforced polymer composites for different composites applications [9-14].

According to this, many studies focused on probing an understanding the physicochemical interaction at the fiber/matrix interface. For strong interfacial adhesion, adjusted level of hydrogen bond and Vander Waals forces between the CF and matrix are required during composite fabrication. In addition, the fiber/matrix interfacial adhesion energy should be higher than the cohesion energy of the matrix. The modifications to CF structure made a big difference in improving the mechanical properties of high performance polymeric composites but the focus to control the fiber/matrix interfacial properties is still a major goal [13-14].

From the high-performance thermosetting resins, epoxy resins are the most important with different applications. Epoxy resins have a lot of advantages and show unique properties. One of the most common and important polymers is Epoxy resin due to its wide applications and wide variety of industries. They are composed of polymeric molecules that are converted to a solid by polymerization. The most important advantage of epoxy resins are the ability to be transformed from a low-viscosity liquid (or thermoplastic state) into a hard, tough thermoset. Epoxy systems physically consist mainly from two components: a resin and a hardener. The hardener is the responsible for the chemical reaction, which turns the epoxy resin into a solid, cross-linked network of molecules. Epoxy resin is defined as a thermoset polymer structure with high adhesion properties and cohesive strength because, when cured, it is irreversibly rigid and relatively unchanged by heat. However, the term epoxy can also be used to indicate an epoxy resin thermoplastic or cured state [11].

Many researches about the Composites have been done. When epoxy resins are mixed with high-strength carbon fibers, the product obtained is used in structural applications requiring high strength, high stiffness and low weight. They are of relatively low density and they can be tailored to have stacking sequences to provide high strength and stiffness in the directions of high loading [13-14]. Carbon nanotubes (CNTs) have been widely used in enhancing composites for its large specific surface area, excellent mechanical properties, as well as good compatibility with the polymer [13].

Introducing CNTs to the surface of fibers to enhancement the interfacial properties of composites considered as a hot topic. So far, several methods have been developed to introduce CNTs to the fiber surface, such as chemical grafting and chemical vapor deposition. Both methods could effectively introduce CNTs on the surface of CFs with strong adhesion, but the excessive chemical treatments and the use of high temperature have hindered their practical application. Electrophoretic deposition is another efficient way to introduce and incorporate CNTs in composites. However, it was woven carbon fabric that was mostly used to receive CNTs, in which case CNTs were only deposited on part of the fiber surface rather than all surfaces. Besides, the complicated process also limits its application in industrial scale [15-18].

With comparison to above methods, sizing or coating fibers with CNT-containing sizing agent exhibit excellent advantages for their simplicity and low-cost and show a great potential application in industrial manufacturing of CNT/CF hybrid fibers [9-10].Generally, we can say that CNT_S are used with CF/polymer composites to produce hybrid CF/CNT_S/ polymer composites in order to improve the mechanical properties of CF/polymer composites [15-19]. Many experimental studies applied before to introduce CNT_S into the CF/Epoxy resin system and from these researches we observed that the addition of CNT_S occur via one route (epoxy resin or CF mat)[20]. Jianwei Zhang, Dazhi Jiang, Su Jua, Hua-Xin Peng, improved mechanical properties of composite material based on carbon fiber (CF) reinforced epoxy resin by adding a small quantity of multi-walled carbon nanotubes (MWCNTs). 1 wt. % MWCNTs were well dispersed in the epoxy resin and fiber filament wind process were used for manufacturing the CF prepregs. The hybrid MWCNT/CF/epoxy composites were fabricated by laying up technique. Investigation applied on tensile properties, interlaminar shear strength (ILSS) and microstructures for the prepared composites [15].

In this work, a simple and clear method for addition of CNT_S to the CF composite system was produced. This technique depends on adding of CNT_S to the CF/epoxy resin system through the two routes of the composite system (epoxy polymer and CF mat). In this research MWCNTs was introduced into CF and epoxy polymer separately in the first step. Then the MWCNTs/CF and MWCNTs/epoxy polymer were mixed together to get the CF/$MWCNT_S$/Epoxy Resin Composite. $MWCNT_S$ was introduced into the CF composite system by divided the amount of $MWCNT_S$ into two equal parts the first part added to epoxy polymer using ultrasonic bath sonicator and the second part of $MWCNT_S$ sprayed into the surface of CF surface using spray gun then manufacture hybrid CF/$MWCNT_S$/Epoxy resin composite using 1% weight ratio from $MWCNT_S$.

2. Experimental work

2.1. Materials

In this research the used fiber was T300 PAN carbon fiber manufactured from apolyacrylonitrile precursor with moderate mechanical properties (strength; modulus) and with fiber yarn contains 3000 filaments purchased from Kunshan Samson composite material Co., Ltd, China. The polymer used in this study is epoxy resin system with two components A epoxy (EPON 828) cured with B polyamide (versamide 125) hardener. The used resin system purchased from Aladdin Company, China. The resin system was prepared in proportions of 3:1 by weight. $MWCNT_S$ with a particle size 20-45 nm obtained from Aladdin Co., Ltd, China.

2.2. Epoxy system preparation

The used resin system is epoxy resin consists of two parts; part A (EPON 828) polymer and part B (versamide125) as curing agent. The calculated amount of part A were carefully weighed and then added the required amount of part B, then we stirring the mixture for 30 min using magnetic stirrer to prevent air trapping with low velocity rate then put in ultrasonic bath sonicator for 30 min to achieve good mixing.[13]. In case of addition $MWCNT_S$ to CF based composite, $MWCNT_S$ was added first to part A in suitable beaker with stirring for 30 min for to achieve homogenous mixture and for good distribution, then put in ultrasonic bath sonicator for 3 hours to achieve good distribution of $MWCNT_S$ in epoxy, second step by adding part B to the mixture in ultrasonic bath sonicator for 30 min and in this case the epoxy called modified epoxy resin, the last step by use the matrix contain $MWCNT_S$ with CF to produce composite material [21].

2.3. Composite sample preparation

In this paper the hand lay-up technique was used to fabricate the composite material samples. For good impregnation of carbon fiber with epoxy matrix roller and brush were used. First the epoxy resin was prepared as explained in paragraph 2.2. The CF mats treated first with acetone in order to eliminate the sizing. In the case of sample C1 CF mat was cut in to adaptable size of 250×250 mm. the brush was used coating the epoxy resin on the surface of CF. Then put and aligned the epoxy-coated fiber tape together layer above layer then the prepared composite was covered on both the sides with iron sheets and kept between the fixed and movable die of the compression molding machine under 5MPa pressure for 24 hours. In the case of sample C2 the same procedure in step A was used but by using modified epoxy resin instead of normal epoxy resin. In case of sample C3 the half amount of $MWCNT_S$ was added to epoxy resin as illustrated in section 2.2 to prepare modified epoxy resin and the other amount of $MWCNT_S$ was dissolved in 100 ml ethanol and the mixture put in ultrasonic bath soincator for 1 hour then this mixture put in spray gun and sprayed on the surface of CF mat then the

CF mat put in oven at 40 ^{o}C for 1 hour to evaporate the ethanol. The both modified CF mat and epoxy were ready to use the same procedure in case of sample C1 prepare the composite sample C3.

2.4. Mechanical testing
The tensile test machine used in this research was universal testing machine (WDW-100), purchased from Fangrui Technology Co., Ltd. Changchun, China) which used to measure all mechanical properties for the prepared samples.

2.4.1. Interlaminer shear strength test results. According to the standard (ASTM D 2344) Short Beam Shear Test Method can be considered as interlaminar shear test method that can be used to determine shear strength of unidirectional lamina. It is used mainly as quality control test and a materials screening. The test is relatively simple and a test can be done fast. We calculated and reported the average value for five specimens of each sample.

2.4.2. Flexural test (the three-point bending test). Flexural strength can be defined as the maximum stress developed when a specimen with bar shape is subjected as single beam to a bending force perpendicular to the bar. The Flexural test was processed according to ASTM D790 with specimen dimensions 12.7mm width, 200mm length and with span to thickness ratio = 32:1. For each sample five specimens were tested and the reported value is the average value for the five specimens.

2.4.3. The tensile test. According to ASTM D3039 the sample specimen with 12.7mm width, 203mm length and with thickness depend on a number of layers was used for the tensile test. We calculated and reported the average value for five specimens of each sample. The Standard Flat shape for tensile specimens is shown in figure 1.

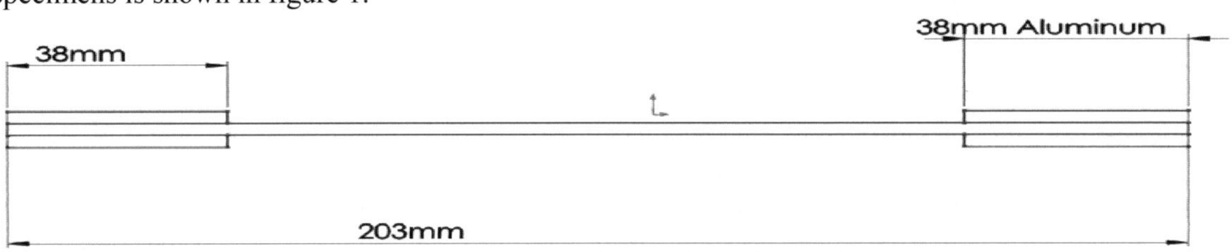

Figure 1. Standard Shape of tensile specimens.

2.5. Samples abbreviations
The abbreviations and formulations of the prepared samples used in this research explained in table 1.

Table 1. Details of sample composition

Sample abbreviations	Composition
C1	6layers[CF+ epoxy resin]
C2	6 layers [CF+ 1wt% $MWCNT_S$+epoxy matrix] $MWCNT_S$ added via one route.
C3	6 layers [CF+ 1wt% $MWCNT_S$+epoxy matrix] $MWCNT_S$ added via the two routes.

3. Result and discussion

3.1. Tensile test result
From figure 2 the tensile properties of prepared composite samples can be seen. The figure shows that there are three tensile strength values for three composite samples based on CF and have the same

number of CF layers. The composite sample C1is based on CF/1wt% MWCNT$_S$/Epoxy resin in which CNTS added via one route and the third composite sample C3 is based on CF/1wt% MWCNT$_S$/Epoxy resin in which CNT$_S$ added via two routes. The results show that the tensile strength of sample C2 is higher than the tensile strength of sample C1 by 9 %. The enhancement achieved due to addition of 1wt % of MWCNT$_S$ in to CF/epoxy resin can be explained due to a stronger resin matrix caused by MWCNT$_S$ reinforcement and enhanced matrix fiber interfacial interactions based on CNT$_S$ incorporation. The results show also that C3 is higher than C2 by 10%.This is because adding of MWCNT$_S$ through two routes in the same time (via epoxy resin and CF mat) increase the possibility for good mixing and good dispersion of CNT$_S$ in to the CF/epoxy resin composite system and this lead to stronger of resin matrix and improve matrix-fiber interfacial interactions so increase the tensile strength.

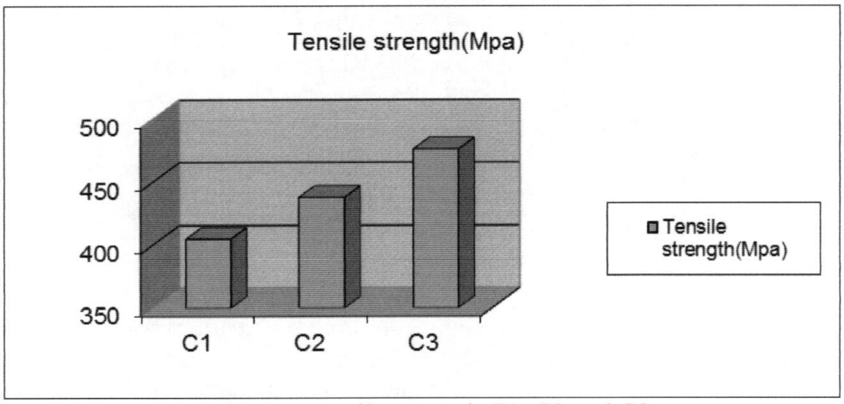

Figure 2. Tensile strength C1, C2 and C3.

3.2. *Flexural test result*

Figure 3 shows the flexural strength results of the three composite samples. From figure (3), it can be seen that the flexural strength of sample C2 is higher than the flexural strength of sample C1 by 51 %. The enhancement achieved due to adding of 1wt % of MWCNT$_S$ into CF/epoxy resin can be due to a stronger resin matrix caused by adding MWCNTS and enhanced matrix fiber interfacial interactions based on CNT$_S$ incorporation. The results show also that the flexural strength of C3 is higher than C2 by 19%.This is because adding of MWCNT$_S$ through two routes at the same time (via epoxy resin and CF mat) increase the possibility for good mixing and good dispersion of CNT$_S$ into the CF/epoxy resin composite system and this lead to stronger of resin matrix and improve matrix-fiber interfacial interactions so improve the tensile strength.

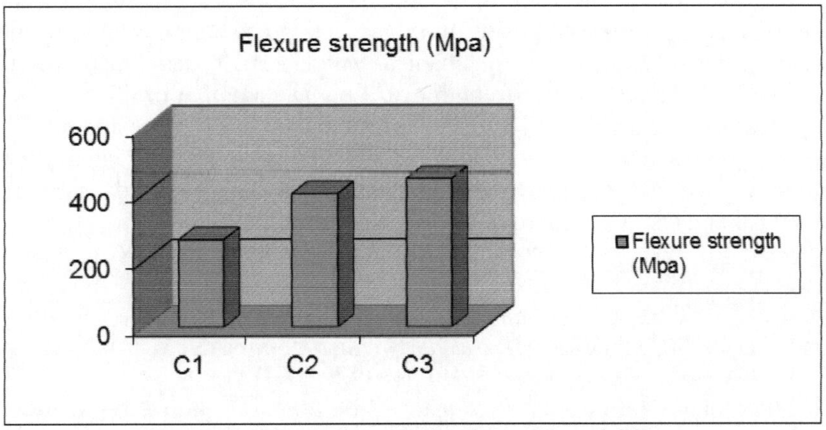

Figure 3. Flexural strength of C1, C2 and C3.

3.3. Interlaminer shear strength test results

Figure 4 shows the interlaminer shear strength of the prepared composite samples. The results show that Interlaminer shear strength of sample C2 is higher than the Interlaminer shear strength of sample C1 by 33 %. The enhancement achieved due to adding of 1wt % of $MWCNT_S$ into CF/epoxy resin can be due to a powerful resin matrix caused by adding $MWCNT_S$ and enhanced matrix fiber interfacial interactions based on CNT_S incorporation.

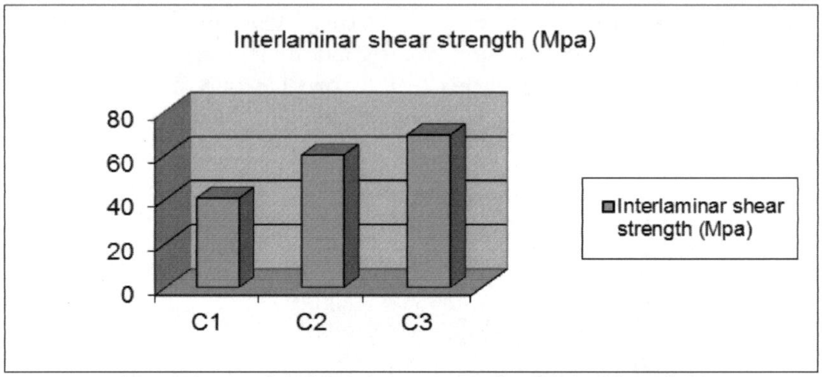

Figure 4. Interlaminer shear strength of C1, C2 and C3.

The results show also that C3 is higher than C2 by 21 %.This is because the addition of $MWCNT_S$ through two routes at the same time (via epoxy resin and CF mat) increase the possibility for good mixing and good dispersion of CNT_S into the CF/epoxy resin composite system and this lead to stronger of resin matrix and improve matrix-fiber interfacial interactions so improve the Interlaminer shear strength.

4. Conclusion

The introduction of CNT_S to the CF/epoxy resin matrix is one of the most effective methods for improving and enhancing the interface between CF and epoxy resin and so improving the structural properties as mechanical characteristics of the CF/epoxy resin composite. It was found that the method of adding CNT_S into the CF composite system play important rule for increasing the interface bonding between CF an epoxy polymer which leads to increase the mechanical properties of the prepared composite. Three composite samples based on CF and epoxy resin were prepared C1[CF/epoxy resin composite],C2[$MWCNT_S$/CF/Epoxy resin composite] in which $MWCNT_S$ added via epoxy resin and C3[CF/$MWCNT_S$/Epoxy resin composite] in which $MWCNT_S$ added via CF and epoxy resin. The results show that the mechanical properties of C2 as flexural strength, tensile strength and interlaminer shear strength higher than C1 by 51%, 9 % and 33 % respectively. This is because adding of $MWCNT_S$ to the CF/epoxy resin composite improves the epoxy bonding and the interface bonding between CF and epoxy so improve the mechanical properties. It was found that tensile strength; flexural strength and interlaminer shear strength of C3 higher than that of C2 by 10%, 19% and 21% respectively. This is because adding of $MWCNT_S$ via two routes (epoxy resin and CF mat) improves the probability of good and homogenous mixing with epoxy resin and also good dispersion of $MWCNT_S$ on the surface of CF mat which leads to stronger the epoxy adhesion and improves the interface between CF and epoxy so improve the mechanical properties.

References

[1] Schwartz M M 1992 *Composite Materials Hand Book* McGraw-Hill.
[2] Whitcomb J D 1988 *Composite materials testing and design* ASTM.
[3] Peters ST 1998 *Hand book of composites* CHAPMAN&HALL.
[4] Soutis C 2005 Fiber reinforced composites in aircraft construction *Prog Aerospace* Sci **41** 143–51.

[5] Pavia J M F, Santos A N, Rezende M C 2009 Mechanical and morphological characterizations of carbon fiber fabric reinforced epoxy composites used in aeronautical field *Mater Res* **12(3)** 367-374.

[6] Mallick P K 2008 *Fiber reinforced composites: material, manufacturing and design* **3rd Ed**. New York: CRC Press.

[7] Edie D D, McHugh J J 1999 *High performance carbon fibers* In: Burchell TD, editor. Carbon materials for advanced technologies, Pergamum p. 119–38.

[8] Liu Y, Kumar S 2012 Recent progress in fabrication, structure, and properties of carbon fibers *Polym Rev* **52(3–4)** 234–58.

[9] Sharma M, Gao S, Mäder E, Sharma H, Wei L Y, Bijwe J 2014 Carbon fiber surfaces and composite interphases *Composites Science and Technology* **102** 35–50.

[10] Pavia M C, Bernardo C A, Nardin M 2000 Mechanical, surface and interfacial characterization of pitch and PAN-based carbon fibers *Carbon* **38(9)** 1323-37.

[11] Takeichi, Furukawa N 2015 Epoxy Resins and Phenol-Formaldehyde Resins *Reference Module in Materials Science and Materials Engineering Polymer Science: A Comprehensive Reference* **5** 723-751.

[12] Chung D D L 1994 *Carbon Fiber Composites* Butterworth-Heinemann.

[13] Gipson R F *Principles of composite materials mechanics* McGraw-Hill.

[14] Zhang H, Zhang Z, Breidt C 2004 Comparison of short carbon fiber surface treatments on epoxy composites Institute for Composite Materials, University of Kaiserslautern, and Erwin Schrodinger strasse **58**.

[15] Zhang J, Ju S, Jiang D, Peng H X 2013 Reducing dispersity of mechanical properties of carbon fiber/epoxy composites by introducing multi-walled carbon nanotubes *Composites Part B: Engineering* **54** 371–376.

[16] Magdy W 2004 Production technology of composite materials based on polymeric matrices Master, MTC, Egypt.

[17] Li X, Han X, Tan Y, and Xu P 2007 Preparation and microwave absorption properties of Ni-B alloy- coated Fe3O4 particles *Journal of alloys and compounds* 240-245.

[18] Li X , Yang H, Fu W, Liu S, Zhu H, and Pang X 2006 Preparation of low-density super paramagnetic microspheres by coating glass micro balloons with magnetic nanoparticles *Materials science and Engineering: B* **135** 38-43.

[19] Rawal S and Brantley J 2013 Development of Carbon Nanotube-based Composite for Spacecraft Components In *Recent Advances in Space Technologies (RAST), 2013 6th International Conference on. IEEE* 13-19.

[20] Zhao Z, Teng K, Li N, Li X, Xu Z, Chen L, Niu J, Fu H, Zhao L, Liu Y 2017 Mechanical, thermal and interfacial performances of carbon fiber reinforced composites flavored by carbon nanotube in matrix/interface *Composite Structures* **159** 761–772.

[21] Fouda H, Guo L and Elsharkawy K 2017 Preparation and Characterizations of Composite Material Based on Carbon Fiber and Two Thermoset Resins *MATEC Web of Conferences* **88** 02002.

A new application of SAM in the non-destructive inspection for SIM card

Bin Cai[1,2], Yandong He[3], Yang Zhao[1,2], Yanning Chen[1,2], Haifeng Zhang[1,2] and Dongyan Zhao[1,2]

[1]State Grid Key Laboratory of Power Industrial Chip Design and Analysis Technology, Beijing Smart-chip Microelectronics Technology Co., Ltd. Beijing 100192, China
[2]Beijing Engineering Research Center of High-reliability IC with Power Industrial Grade, Beijing Smart-chip Microelectronics Technology Co., Ltd. Beijing 100192, China
[3]School of Electronics Engineering and Computer Science, Peking University, Beijing 100871, China

Abstract. Scanning Acoustic Microscopy (SAM) is a typical inspection method in the semiconductor IC manufacturing industry. Because the die thickness is a key parameter for SIM card, a new method to measure the internal die thickness of SIM card is proposed with SAM's reflective scanning mode. Using this method the internal die thickness of SIM card can be accurately measured without introducing any damages to SIM card. The thickness model and methodology based on the SAM signals have been established. The model was properly verified and calibrated by two real test cases.

1. Introduction

Scanning Acoustic Microscopy (SAM) is a typical instrument used to examine the adhesion between two different materials in a non-destructive way [1, 2]. Frequencies in the range of 5-300MHz are typically used for semiconductor applications. Ultrasonic frequency is related with penetration depth and resolution. The resolution is higher with increasing ultrasonic frequency, while the penetration range is shorter [3].

Based on the features of short time, high efficiency and non-destructiveness, SAM has become a significant inspection tool as non-destructive way and widely applies in the Semiconductor encapsulation industry. However, to my knowledge, SAM rarely appears as the inspection tool of SIM card, especially the measurement of the thickness of the inner die.

The die thickness is one of the key parameters for SIM card. It is highly necessary to monitor this parameter in the process. The conventional measurement of die thickness in the SIM card has to get the die out from the SIM card. It will unavoidably damage the SIM card. On the other hand, the conventional method will typically take about 1 hour, and even more, to complete the die thickness measurement due to the preparation processes. Therefore, the inspection sample size is limited due to the inspection cost and time.

In this paper, a new non-destructive method was proposed to measure the thickness of the internal die in SIM card. The inspection function of SAM has been expanded further.

2. Principle and methodology

2.1. Principle of SAM

The ultrasonic, a pulse signal of high frequency (5MHz - 300MHz), has good directivity and constant speed in the uniform material.

Generally SAM is used in two scanning mode, which is reflection and transmission [4]. The reflective scan mode takes advantage of the reflective signal from the interface and defect to analyze the location and size of failure spot. The ultrasonic which is generated by the transducer will penetrate the couplant into the sample. When the acoustic pulse hits the interface between two materials, parts of the signal get reflected. The degree of reflection depends on the acoustic impedance of the two materials, and the value of acoustic impedance is related to the characteristic of two materials [5]. Analyzing phase and amplitude of the reflected signal can be used to characterize the examined interface.

2.2. Method for thickness measurement

The reflective ultrasonic between the incoming and outgoing interface can be detected by the transducer, as shown in Figure 1. The peak of reflective signal in Figure 2 presents the location of the incoming or outgoing interface. As shown in Figure 2, the transmission time, from the incoming to outgoing interface of the material, is able to be determined exactly. As well known, the value of acoustic velocity is material related. Some typical values are listed in Table.1. Therefore, the thickness of the material can be expressed by equation (1),

$$T = V_{material} * t/2 \tag{1}$$

Where T is the thickness of the internal die, $V_{material}$ is acoustic velocity of the test material, and t is the transmission time from the incoming to outgoing interface of the material.

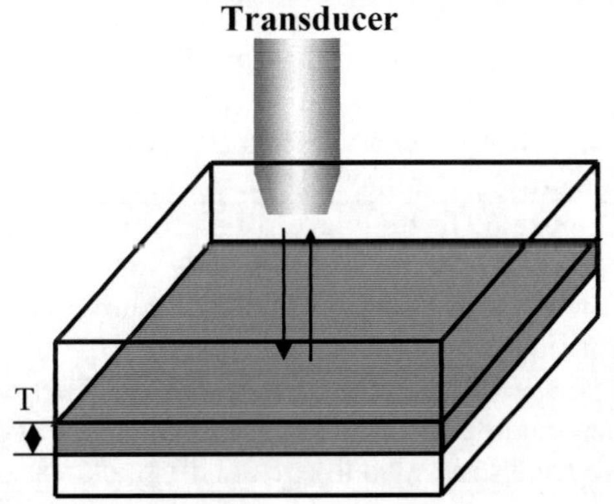

Figure 1. The reflective signal between the incoming and outgoing interface is detected by the transducer.

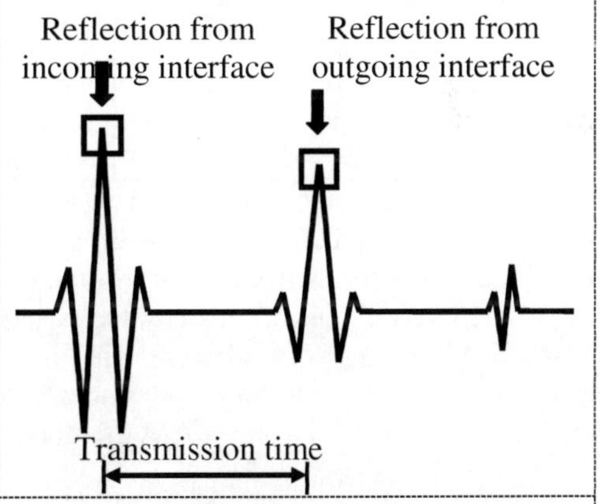

Figure 2. The peak of the reflection presents the location of the incoming or outgoing interface. The transmission time from the incoming to outgoing interface is counted.

Table 1. The value of acoustic velocity

Material	Acoustic velocity (mm/μs)
Aluminium	6.350
Steel, common	5.920
Steel, stainless	5.740
Brass	4.399
Copper	4.720

Iron	5.930
Silver	3.607
Gold	3.251
Zinc	4.170
Tin	2.960
Silicon	8.400

3. Results and discussion

3.1. SAM front-side scanning mode

The front-side scanning mode is employed to the SIM card's module as shown in Figure 3. The SAM image and waveform are generated by reflection scanning mode of SAM as in Figure 4. The SAM image by the front-side scanning is shown in Figure 5, including internal die.

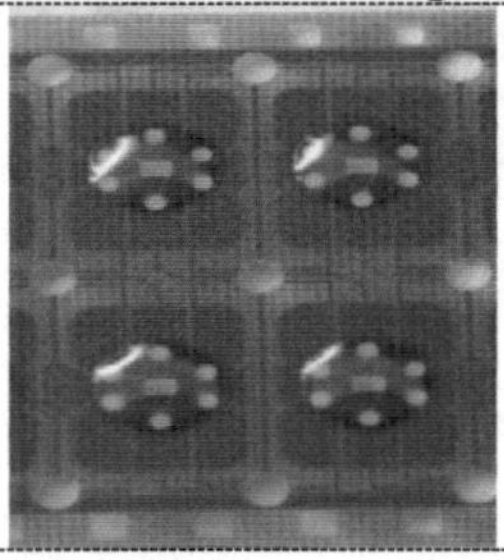

Figure 3. The top image of SIM card used in measurement.

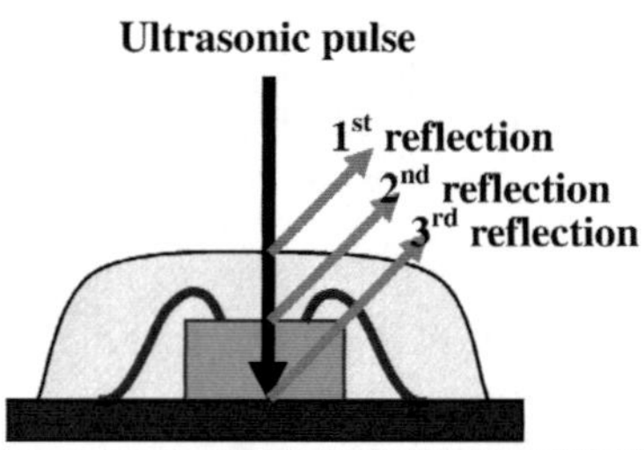

Figure 4. The schematic diagram of SIM card's drawing. The different reflective signals are generated from several layers when the ultrasonic pulse penetrate SIM card's module.

In this case, the original SAM signal is shown in Figure 6. In the waveform, the peak of the 2nd reflection presents the incoming interface, and the peak of the 3rd reflection stands for the outgoing interface. We can extract the transmission time t from the SAM reflection signal in Figure 6. Then the thickness of the internal die can be obtained based on the equation. (1). In order to verify the thickness model, two SIM cards with different die thicknesses are tested. The actual die thicknesses are measured by conventional method also. The die thicknesses are summarized in Table 2. It can be seen that the die thickness measured by SAM method is quite consistent with the actual die thickness.

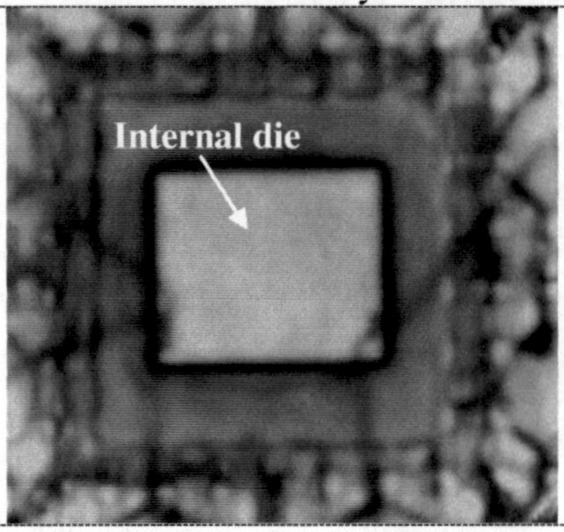

Figure 5. The SAM image of front-side scanning mode.

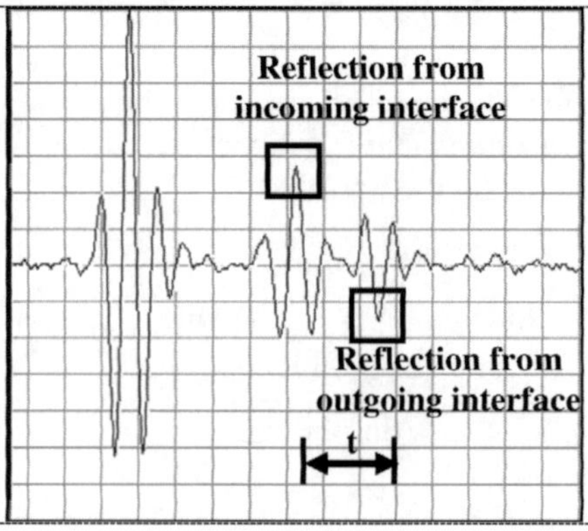

Figure 6. The reflective waveform and transmission time from the incoming to outgoing interface.

Table 2. Spec and measured internal die thickness of SIM card module

Type	Spec Thickness	SAM test Thickness		Real Thickness
		Transmission time	Thickness	
Card-A	150±10um	0.039us	163.8um	159.2um
Card-B	200±10um	0.049us	205.8um	204.3um

3.2. SAM backside scanning mode

Unlike the previous SIM card module measurement, due to the existing air among the interfaces in the SIM card, the thickness of the internal die can't be measured using front-side scanning mode. To inspect the reflective acoustic, the exposed backside die is necessary. To open a window, the center metal contact of SIM card needs to be removed by manual uncovering as presented in Figure 7 and Figure 8. The test method is described following.

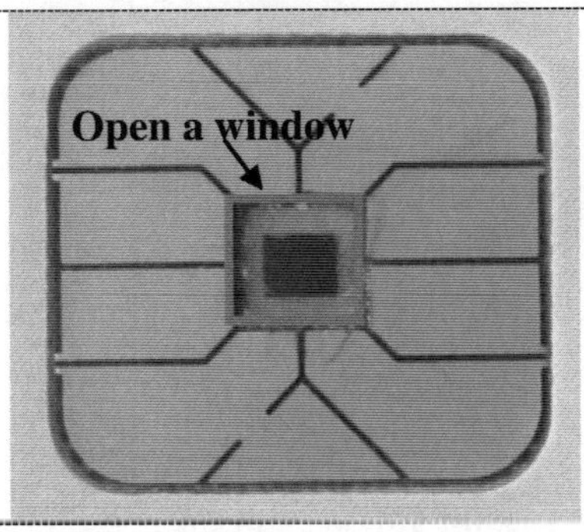

Figure 7. An open window at the center contactor.

Figure 8. The SAM image of backside scanning mode.

The SAM image and waveform are generated by reflection scanning mode of SAM as shown in Figure 9. In the waveform shown in Figure 10, the peak of the 1st reflection presents the incoming interface, and the peak of the 2nd reflection stands for the outgoing interface. From the SAM reflection signal in Figure 9, the internal die thickness can be obtained with same data processing as previously described front-side scanning mode.

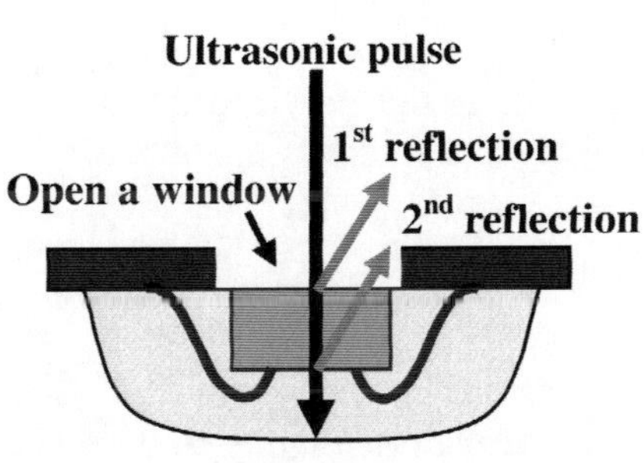

Figure 9. The schematic diagram of SIM card's drawing.

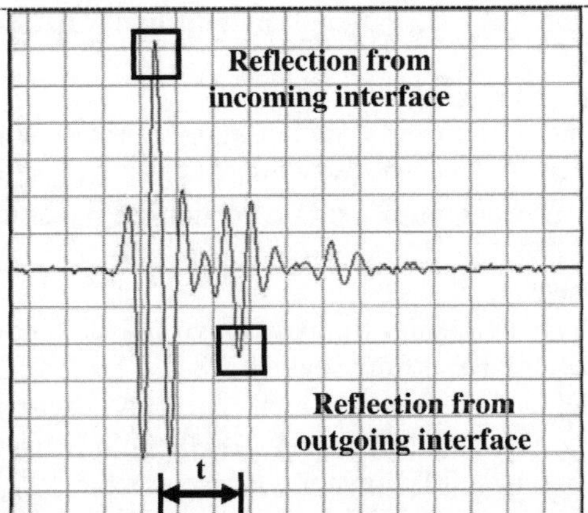

Figure 10. Image shows the reflective waveform and transmission time of incoming and outgoing interface.

The thickness summary is shown in Table 3. The SAM test results are calibrated and verified in a conventional and destructive way. The accurate SEM results are shown in Figure 11 and Figure 12 for two cases.

Table 3. SAM measures the thickness of the internal die of SIM card's module, and compares to spec thickness and real thickness

Type	Spec Thickness	SAM test Thickness		Real Thickness
		Transmission time	Thickness	
Card-A	150±10um	0.038us	159.6um	159.2um
Card-B	200±10um	0.048us	201.6um	204.3um

Figure 11. In the conventional and destructive way, the real thickness of Card-A is shown by SEM.

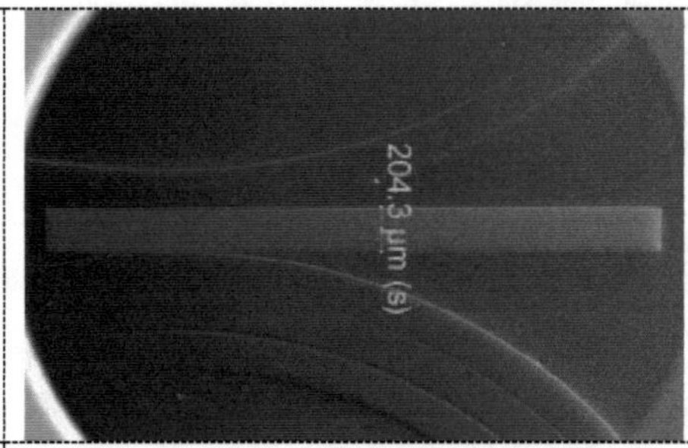

Figure 12. In the conventional and destructive way, the real thickness of Card-B is shown by SEM.

From the results we can conclude that the SAM backside scanning mode can measure the thickness accurately.

4. Conclusion

With the development of SIM card and smart card, it's necessary to create an innovative method for measuring the thickness of the die in SIM card. A new application is proposed with SAM's reflective scanning mode from front-side and backside. With non-destructive SAM measurement, we are able to demonstrate its advantage over conventional and destructive method. Preparation and measurement time by SAM can be shortened into a reasonable range. Therefore the cost is lower than the conventional method. The acceptable accuracy of SAM method was verified by physical measurement of SEM. SAM turns out to be an effective way of die thickness measurement in SIM card.

Acknowledgments

I would like to express my gratitude professor Yandong He who has offered me valuable suggestions in the paper. She has spent much time reading through each draft and provided me with inspiring advice. Without her patient instruction the paper can't be finished at a short time.

References

[1] Callister W D, Rethwisch Jr D G 2012 *Fundamentals of Materials Science and Engineering* John Wiley & Sons, Inc. 845-848.
[2] Khaled A, Brand S, et al. 2016 Investigating stress measurement capabilities of GHz Scanning Acoustic Microscopy for 3D failure analysis *Microelectronics Reliability* **64** 336–340.
[3] Poschgan M, Maynollo J, Inselsbacher M 2012 *Microelectronics Reliability* **52** 2115-2119.
[4] Wang S 2015 Fabrication of a novel device by release the stress in a reliable way *IEEE IPFA* 404-406.
[5] Chizhik D 1991 Surface Characterization Using the Restricted Aperture Acoustic Microscope(PhD Dissertation).

IOP Conf. Series: Materials Science and Engineering **220** (2017) 012023 doi:10.1088/1757-899X/220/1/012023

Parameters affecting morphologies and efficiencies of mesoporous perovskite solars cells

M M Abdoul-latif[1], J Xu[1], J Yao[1] and S Dai[1]

[1]State Key Laboratory of Alternate Electrical Power System with Renewable Energy Sources, North China Electric Power University, Beijing 102206, China

Abstract. The mesoporous perovskite morphologies and optical and electronic properties are influenced by many parameters with which the device performance could be affected. Firstly, the working temperature can be mentioned. Secondly the substrate temperature which combined with spin coated speed could permit to load perovskite inside the mesoporous independent to working temperature. And other factor as the methyl ammonium iodide (CH_3NH_3I) solution concentration could be temperature dependent on the lead iodide, means the reaction is most probable done when the concentration increase with the increasing temperature, whereas the reaction between lead iodide and methyl ammonium can be decrease with lead iodide thickness. It is most probable that the solution concentration also has a significant effect on the morphology of the perovskite that is confined in the pores of the mesoporous TiO_2. It is most probable also the spin coating speed has added at this decreasing, meaning the large size of crystal decreased the transformation lead iodide to Perovskite. Andhere it will try to explain thickness of capping layer can improve the short-circuit photocurrent (J_{sc}) but this enhancement decrease with high working temperature.

1. Introduction

Several perovskite morphologies are been reported to date solvent or deposition dependent [1, 2, 3] and the optical and electronic properties of perovskite materials have been also studied thoroughly in recent decades as the band gap [4, 5], the absorption coefficient [2], and the carrier mobility [6]. The reaction between PbI_2 and CH_3NH_3I can be increased the optical absorption due to the formation of the perovskite. The increase can be observed at different wavelengths. In general, the hybrid perovskites have a very strong tendency to form and crystallize, even at room temperature. Due to the high energy of formation, the hybrid perovskites form even in a solid-state reaction, when the two precursors are brought into contact. Whilst, Poglitsch et al. reported that the position of the $CH_3NH_3^+$cation in $CH_3NH_3PbI_3$perovskite is only fixed in the orthorhombic phase at low temperature and could not be fixed at room temperature due to the cubic symmetry requiring eight identical positions for the cations [7]. As a result, the tetrahedral-coordinated C and N atoms exhibit a disordered state inside the eight tetrahedral of the cuboctahedron around the normal A positions (1/2, 1/2, 1/2) of the ABX_3 perovskite. Therefore, the $CH_3NH_3PbI_3$ perovskite has tetragonal phase, with a slightly distorted PbI_6 octahedra around the c axis at room temperature [7, 8]. Recently, the crystal chemistry of $CH_3NH_3PbI_3$ was studied in detail [9]. Using a differential scanning calorimeter, a reversible phase transition was found from the tetragonal to cubic transition occurring at 57.3°C upon heating and a cubic to tetragonal transition at 56°C upon cooling. A low-temperature stabilized orthorhombic phase was detected at 100 K, where its crystal structure was investigated. A room-temperature stabilized tetragonal phase was calculated to have a band gap of 1.51eV. Otherwise at which temperature limit we could see a real variation of J_{sc} of perovskite solar cell. Also the rate of conversion could be depending on the concentration of the CH_3NH_3I solution, so a lower solution concentration leads to slower reaction kinetics, it could have that the reaction kinetics were temperature dependent.

This is not a real study leading to proving the decrease the J_{sc} with the temperature but the observations after many making the devices perovskite solar cells. This investigation is leaded by using AFM, XRD and I-V test. There are many parameters which could decrease the J_{sc}, among them, hole transporting material (HTM) which has played an important role in the decreasing of J_{sc}. But high working temperature decreased the J_{sc} with thickness of perovskite in depositing twice lead iodide. And this article will try to explain this phenomenon in basis on optical and electronic properties.

2. Experimental

2.1. Fabrication of perovskite solar cells

Solar cells were fabricated on fluorine-doped tin oxide glass substrates with a sheet resistance of 15Ω/square (Pilkington). FTO substrate was patterned by chemical etching with Zinc powder and HCl aqueous solution. The conductive substrates were cleaned with detergent solution, de-ionized water and ethanol. A thin dense compact TiO_2 layer (c-TiO_2) was deposited on the etched FTO substrate by spin coating (3000 rpm, 30s) using a titanium diisopropoxidebis (acetylacetonate) solution in ethanol and HCl at a volumetric ratio of 0.35:5:1 followed by calcinations at 500°C. Porous TiO_2 films got by commercially available TiO_2 paste (18-NR-T, Dyesol, average particle size ~ 20 nm) diluted in ethanol (1:3.5 weight ratio) were directly deposited on top of the c-TiO_2 by spin coating at 5000 rpm for 30s and annealed at 500°C. Thus, the prepared mesoporous TiO_2 films were infiltrated with PbI_2 by spin-coating at 5500 rpm or 6500 rpm for 30s and dried at 70°C. PbI_2 solution is diluted in DMF (dimethylformamide) (462mg/ml) and was kept at 70°C. After drying, the films were spin coated at 5500rpm or 6500rpm at 30s after loaded 80μl for each substrate a solution of CH_3NH_3I in 2-propanol (18mg/ml) for 30s. After drying, the hole-transporting material (HTM) was deposited by spin-coating a solution of spiro-MeOTAD, tert-butylpyridine, lithium bis(trifluoromethylsulphonyl)imide and tris(2-(1H-pyrazol-1-yl)-4-tert-butylpyridine)cobalt(III) bis(tri-fluoromethylsulphonyl)imide in chlorobenzene. Silver (66nm) was thermally evaporated on top of the device to form the back contact.

2.2. Characterization

The morphology of films was characterized by scanning electron microscopy (SEM, SU8010), we have also used atomic force microscopy (AFM) for getting more information about perovskite morphologies. The crystal structure was examined using X-ray Diffractometer (XRD, D8 ADVANCE, Cu Kα radiation, λ=1.54Å, tube voltage = 40kV, tube current = 30mA). The photocurrent density – voltage characteristics of the perovskite solar cells were evaluated by a source meter (Keithley 2400) under a solar simulator (Oriel 92250A-1000, Newport, $100mWcm^{-2}$). The absorption of different films was measured using a UV-visible Spectrophotometer (UV-2450, Shimadzu). Incident photon-to-current conversion efficiency (IPCE) spectra were recorded using the QE-R (Enlitech) quantum efficiency measurement system. A calibrated silicon solar cell was used as reference.

3. Results and discussion

3.1. Top-view SEM images and cross-sectional

Firstly in Figure 1 is shown the different morphologies with variation loading content, PbI_2 film becomes dense with increasing loading content. Secondly, in figure 1(d) and 1(e) are shown the variation of perovskite morphologies using same CH_3NH_3I solution concentration but different PbI_2 loading content. Perovskite crystal size increases also with increasing PbI_2 loading content.

Figure 1. (a), (b) and (c) PbI_2 deposited into TiO_2 at different loading content, (d) and (e) perovskite on TiO_2 with different volume loading of PbI_2.

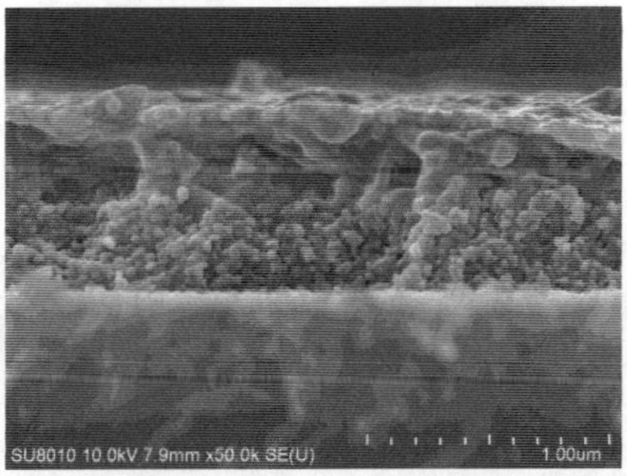

Figure 2. Cross-sectional SEM of a complete photovoltaic device FTO/c-TiO_2/m-TiO_2/ $CH_3NH_3PbI_3$/spiro/Ag

Figure 2 shows cross-sectional SEM of a photovoltaic device. The pore filling of TiO_2 with perovskite increase with TiO_2 thickness but Power Conversion Efficiency (PCE) is best with the thinner TiO_2 layer about 260 nm still have large capping layer of perovskite[10]. It's also confirmed that the pore filling of TiO_2 depend with temperature of substrate when the substrate temperature is above 100°C before PbI_2 deposition, the J_{sc} and PCE become strongly better [11, 12].

3.2. Atomic Force Microscopic (AFM)

The investigation with AFM is allowed to show that the capping layer decrease with decreasing the content of PbI_2 as shown in figure 3. (a) and (d) and also with increasing of spin coating speed for this reason there are different views with different speeds deposition as shown in figure 4. (a) and (b).The morphology of perovskite has changed the form in the same substrate at different scales as shown in figure 3. (b) and (c).

Large capping layer of perovskite with thinner layer of TiO_2 is the best architecture device but at room temperature this capping layer of perovskite couldn't be over than 140 nm due to in the triiodide films are limited by about 100 nm diffusion length, so that photogenerated charge in thicker films cannot be efficiently extracted before recombining [13, 14, 15]. Other studies have got results optimum devices performances are obtained for perovskite layer thickness around 210 nm [16] and

300nm [17]. In fact the diffusion length decreases at higher temperature [18], and generally increases charge-carrier recombination rate.

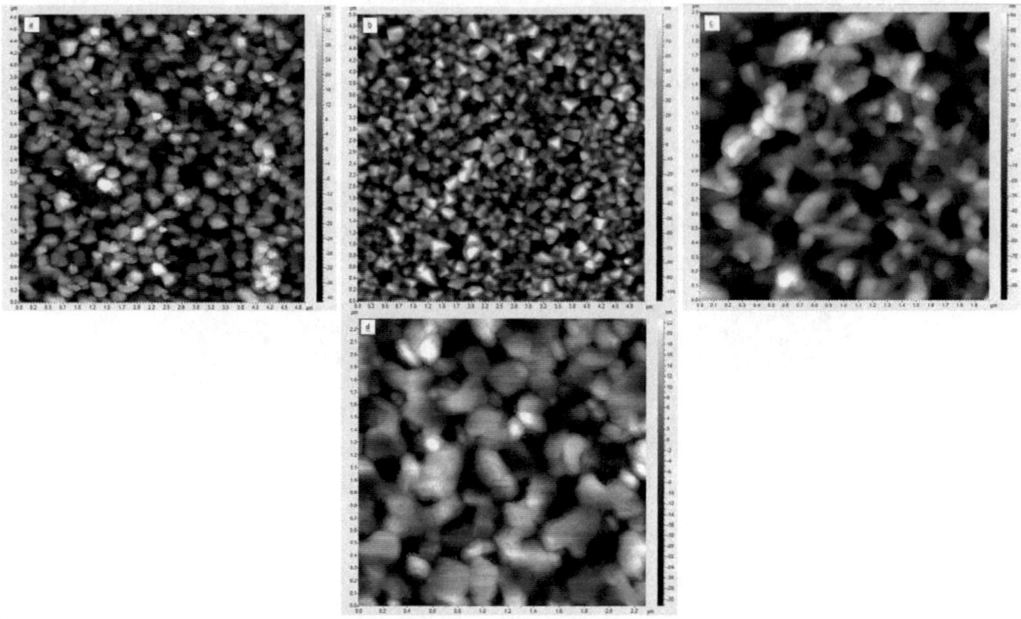

Figure 3.(a) PbI_2 on TiO_2 without capping layer (15µl PbI_2 loading on TiO_2) deposited at speed 6500rpm. (b) and (c) perovskite on TiO_2. (d) PbI_2 deposited twice at 6500 rpm.

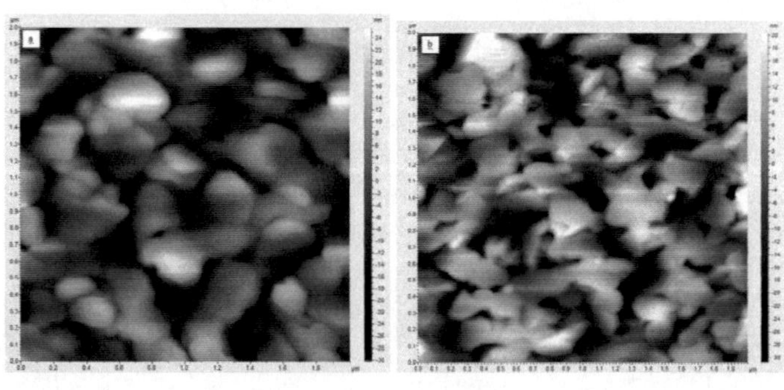

Figure 4.PbI_2 deposited on TiO_2 once (30µl) with different spin coating speed, (a) with 5500 rpm and (b) with 6500 rpm.

3.3. X-Ray Diffraction (XRD)

The curves in Fig.5a show X-ray diffraction spectra measured before the TiO_2/PbI_2 film is brought into contact with the CH_3NH_3I solution with PbI_2 deposited into TiO_2 at different loading volume. Crystal grow in a preferential orientation along the c axis [1, 19, 20], then the main diffraction peak corresponding to the (001) lattice plane. The diffraction peak at $2\theta = 38°$ can be attributed to the diffraction from (110) lattice plane of 2H polytype, suggesting that the anatase scaffold induces a different orientation for the PbI_2 crystal growth, but the diffraction peak attributed to the (111) lattice plane doesn't appear in the curve at different loading volume. The curves in Fig.5b show X-ray diffraction spectra measured after the TiO_2/PbI_2 film is brought into contact with CH_3NH_3I solution. The conversion of PbI_2 nanocrystals in the mesoporous TiO_2 film should be practically complete on a timescale of seconds, as is evident from the immediate disappearance of its most intense diffraction peak corresponding to the (001) lattice plane, and the concomitant appearance of the XRD reflections for the tetragonal perovskite. The peaks at 14.17, 28.47 and 43.27° are respectively assigned to the (110), (220) and (330) planes. The presence of these peaks indicates that the conversion to perovskite, while incomplete, is done. The conversion of PbI_2 to perovskite decreases with film thickness at high working temperature as is shown in Fig.5b.

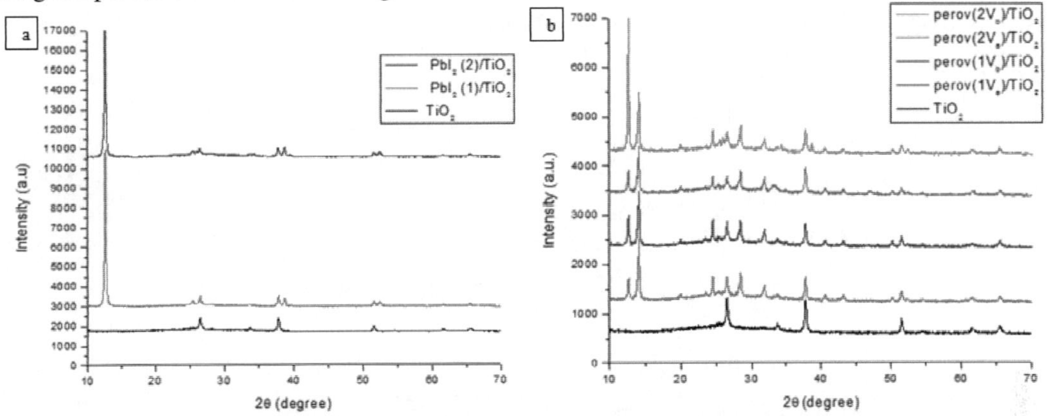

Figure 5. (a) Spectra of PbI_2 deposited on TiO_2 at once and twice (20µl). (b) Spectra of perovskite with different loading content on TiO_2.

3.4. Current density-voltage (I-V) characteristics

The current density (J)-voltage (V) characteristics of the solar cells is measured under simulated air mass 1.5 global (AM1.5G) solar irradiation as is shown in Fig.6. The short-circuit current increases at low working temperature but the decreasing of the fill factor (FF) parameter is not only due to high temperature, while open circuit voltage remains same at different working temperature.

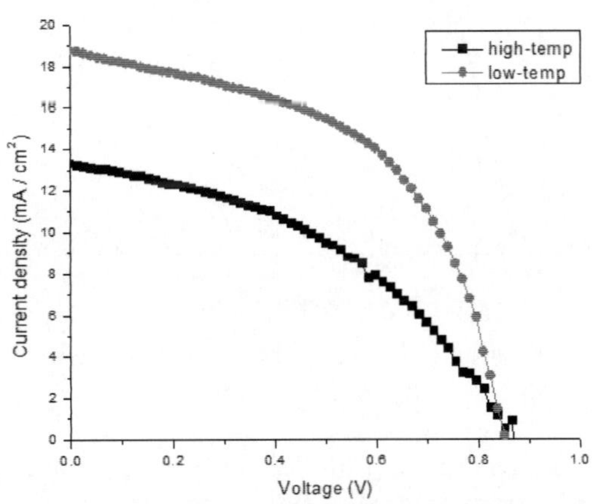

Figure 6. J-V curves for photovoltaic devices with same volume of PbI_2 loading at different working temperature.

3.5. Absorption

The optical properties of $CH_3NH_3PbI_3$ have been investigated, where the exciton binding energy and exciton Bohr radius have been determined to be 50 meV and 22 Å. Since the Bohr radii is relatively large and the exciton binding energy small, the excitons from $CH_3NH_3PbI_3$ will be Wannier-type [7]. And a material with a low exciton binding energy, the exciton is rapidly quenched with increasing temperature. Apart from photoluminescence fact, the formation of perovskite could be incomplete and decrease the absorption spectrum. The band gap of PbI_2 is about 2.5 eV corresponding to an absorption onset of about 500 nm. Any wavelength between 500 nm and the absorption onset of the perovskite absorber (800nm) should therefore be suitable to monitor the formation of $CH_3NH_3PbI_3$. In figure 7 is shown the curves of absorbance spectra which are almost same with a small decreasing between 600 and 800 nm when the thickness of perovskite increases. This means that the transformation of PbI_2 to $CH_3NH_3PbI_3$ isn't complete [11]. It's also most probable that the twice deposition of lead iodide doesn't improve the absorption.

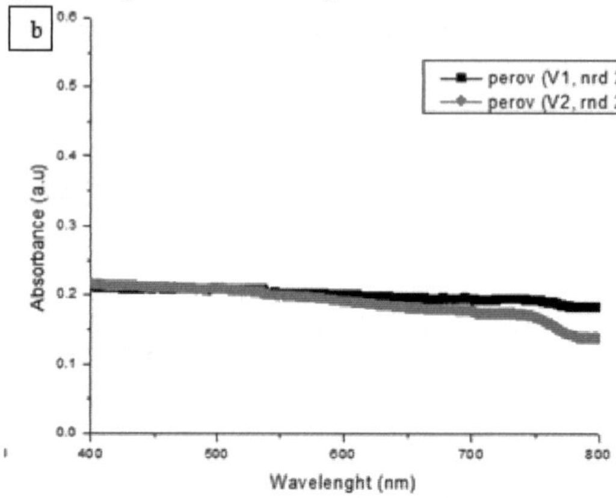

Figure7. Spectra of perovskite absorbance with different volume of PbI_2 deposited on TiO_2.

3.6. IPCE

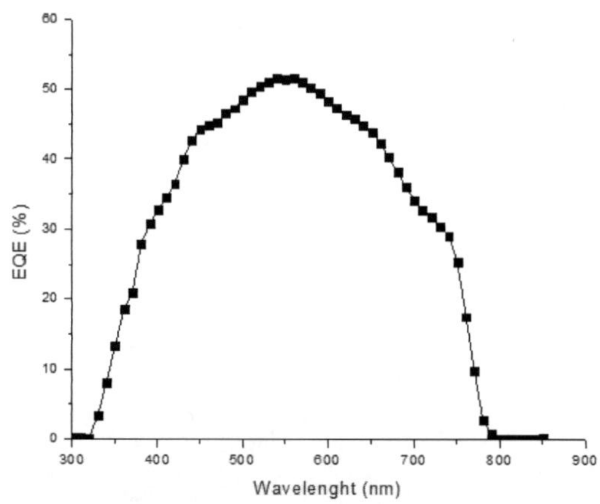

Figure8. IPCE spectrum of photovoltaic device without capping layer at high-working temperature (J_{sc}= 10.3 mA/cm^2).

The Equation (1) of IPCE (incident photon-to-electron conversion efficiency) or EQE (external quantum efficiency) corresponds to the internal efficiencies, meaning the electronic processes within the devices, where LHE is the light harvesting efficiency, and η_{inj} and η_{coll} are the electron injection and charge collection efficiencies respectively.

$$IPCE(\lambda) = LHE(\lambda) \times \eta_{inj}(\lambda) \times \eta_{coll}(\lambda) \qquad (1)$$

In figure 8 is shown IPCE spectrum as a curve with a maximum around 550 nm, it means that the diffusion length is short due to a high working temperature.

4. Conclusion

XRD spectra have been allowed to see that the conversion PbI_2 to perovskite is incomplete at high working temperature furthermore became insignificant with increase thickness. Capping layer with twice deposition of lead iodide is the best way for improving the J_{sc} but high working temperature is an inconvenient factor for this enhancement by decreasing the diffusion length. The precise mechanism which working temperature is responsible for reducing the J_{sc} is not fully understood and the parameter such as decreasing the speed coating could decrease the J_{sc}. And the absorption decreases with the conversion PbI_2 to $CH_3NH_3PbI_3$ which becomes incomplete with the increase perovskite thickness.

References

[1] Burschka J, Pellet N, Moon S-J, Humphrey-Baker R, Gao P, Nazeeruddin M K and Grätzel M 2013 *Nature* **499** 316-319.
[2] Im J-H, Lee C-R, Lee J-W, Park S-W and Park N-G2011 *Nanoscale* **3** 40088.
[3] Etgar L, Gao P, Xue Z, Peng Q, Chandiran A K, Liu B, Nazeerudin M K and Grätzel M 2012 *J. Am. Chem. Soc.* DOI: 10.1021/ja307789s.
[4] Kim H K, Lee C R, Im J H, Lee K B, Moehl T, Marchioro A, S. J. Moon, Humphry-Baker R J, Yum H, Moser J E, Grätzel M and Park N G 2012 *Scientific Reports* **2** 591 DOI: 10/1038/srep00591.
[5] Quarti C, Mosconi E, Ball J M, D'Innocenzo V, Tao C, Pathak S, Snaith H J, Petrozza A and De Angelis F 2016 *Energy & Environmental Science* **9** 155DOI: 10.1039/c5ee02925b.
[6] Brenner T M, Egger D A, Rappe A M, Kronik L, Hodes G and Cahen D 2015 *Journal of Physical Chemistry Letters* **6** 4754-4757 DOI: 10.1021/acs.jpclett.5b02390.
[7] Kim H S, Im S H and Park N G 2014 *Journal of Physical Chemistry C* **118** 5615-5625.
[8] Motta C, El-Mellouhi F, Kais S, Tabet N, Alharbi F and Sanvito S 2015 *Nature Communications* **6** 7026DOI: 10.1038/ncomm8026.
[9] Baikie T, Fang Y, Kadro J M, Schreyer M, Wei F, Mhaisalkar S, Graetzel M and White T J 2013 *Royal Society of Chemistry*doi: 10.1039/c3ta10518k.
[10] Leijtens T, Lauber B, Eperon G E, Stranks S D and Snaith H J 2014 *Physical Chemistry Letters* **5** 1096-1102.
[11] Lee Y H, Luo J, Humphry-Baker R, Gao P, Grätzel M and Nazeeruddin M K 2015*Adv. Funct. Mater* DOI: 10.1002/adfm.201501024.
[12] Ito S, Tanaka S, Vahman H, Nishino H, Manabe K and Lund P 2014*CHEMPHYSCHEM* **15** 1194-1200DOI: 10.1002/cphc.201301047.
[13] Xing G, Mathews N, Sun S, Lim S S, Lam Y M, Grätzel M, Mhaisalkar S and Sum T C 2013 *Science* **342** 344 DOI: 10.1126/science.1243167.
[14] Stranks S D, Eperon G E, Grancini G, Menelaou C, Alcocer M J P, Leijtens T, Herz L M, Petrozza A and Snaith H J 2013 *Science* **342** 341DOI: 10.1126/science.1243982.
[15] Marchioro A, Teuscher J, Friedrich J D, Kunst M, Krol R, Moehl T, Gratzel M and Moser J E *Nature Photonics* **8** 374 DOI: 10.1038/NPHOTON.2013.374.
[16] Chen L-C, Chen J-C, Chen C-C and Wu C-G 2015 *Nanoscale Research Letters* **10** 312 DOI: 10.1186/s11671-015-1020-2.
[17] Momblona C, Malinkiewicz O, Roldán-Carmona C, Soriano A, Gil-Escrig L, Bandiello E, Scheepers M, Edri E and Bolink H J 2014 *APL Mater.* **2** 081504 DOI: 10.1063/1.4890056.
[18] Millot R L, Eperon G E, Snaith H J, Johnston M B and Herz L M 2015*Adv. Funct. Mater.* **25** 6218-6227.
[19] Song Z, Watthage S C, Phillips A B, Tompkins B L, Ellingson R J and Heben M J 2015 *Chem. Mater.* **27** 4612-4616.
[20] Berdiyorov G R, Madjet M E, El-Mellouhi F and Peeters F M 2016 *Solar Energy Materials & Solar Cells* **148** 60-66.

IOP Conf. Series: Materials Science and Engineering **220** (2017) 012024 doi:10.1088/1757-899X/220/1/012024

Establishment of magnetic sensors in above ground marking system

L Y Sun[1], Y B Li[2], Y T Wu[1], W P Li[1], P Q Wu[1]

[1]School of Control and Mechanical Engineering, Tianjin Chengjian University, Tianjin, 300384China

[2]State Key Lab. of Precision Measurement Technology and Instrument Tianjin University Tianjin, 300072, China

Abstract. Magnetic sensors used in Above Ground Markers (AGM) to measure the magnetic field signals are one of the main parts of AGM. By collecting the change of magnetic field parameter using magnetic sensors, AGM can record the precise time when MFL_PIG is passing by the AGM. In this paper, the magnetic sensors, such as induced coil and magneto-resistive sensor, are investigated in characteristics and performance of measurement. With advantages of low cost, high resolution and wide range of measurement, the induced coil is adopted in a novel AGM. The traction test, seal ability test and temperature test show that the AGM with induced coil as magnetic sensor works well under extreme condition and reacts to change the magnetic field with high sensibility.

1. Introduction

Pipeline transportation plays significant role in modern industry and national economy. However, pipeline accidents occur frequently due to aging, corrosion, construction defects, man-made damages which result from third party engineering construction or theft, and other reasons. Pipeline accidents may cause serious pollution of human living environment and great economic losses. Therefore, pipeline should be measured regularly to avoid these problems [1].

Among the methods of pipeline defect detection, leakage flux detection method is the most widely used one at present. The magnetic flux leakage detection system (MFL_PIG) mainly consists of three parts: the pipeline detector, the ground marking system and the data analysis software [2]. The ground marking system (also called above ground marker or AGM [3]) is the outer part of the magnetic flux leakage detection system, usually put above the pipeline and under the ground.

AGM detects the exact time when MFL_PIG passes through it, as well as marks the pipeline position points. When MFL_PIG completes the whole length of the pipeline detection, the large amount information which contains the position and corrosion characteristics of the pipeline, would be processed with the reference of position data given by above ground marker. The precise position of AGM is of extreme importance. This paper mainly discusses the magnetic sensors used in above ground marking system.

2. Detection principle of pipe leakage flux detection (MFL_PIG)

The theory of magnetic flux leakage detection is based on the fact that the ferromagnetic material is magnetized under the induction of the external magnetic field [4]. If the material is not defective, the distribution of the magnetic field lines is regular, as it is shown in figure1 (a). At the defect of the material, the magnetic field is bent due to the increase of the magnetic resistance. A portion of the material inside the magnetic field leaks out of the material surface to form a magnetic flux leakage, see figure 1 (b).

In the process of MFL_PIG detection, the leakage of magnetic field is detected by sensors which produce corresponding induced signals. These signals are filtered, amplified, processed, and recorded in the mass memory of the detector. With the reference of the position data supplied by above ground marking system, the geometry size and severity of defects will be evaluated.

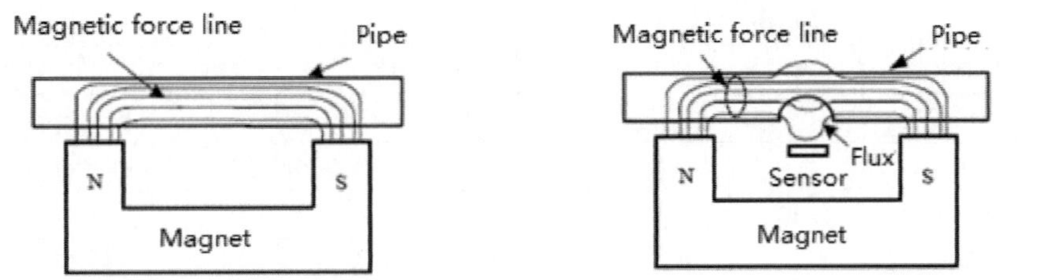

(a) Detection of normal pipe(without defect) （b） Detection of pipe with defects
Figure1. Principle of leakage flux detection.

3. Magnetic sensors using in above ground markers

Magnetic sensor which is used in ground marking system to detect magnetic signals and to transfer to electric signals is the crucial part of the whole system. The magnetic sensors should have high resolution and wide detection range in order to improve the performance of the tracking and positioning. At present, the main magnetic sensors used in ground marking system are coil sensor, flux-gate, magneto-resistive sensor, hall sensor, nuclearprecession, optically pumped sensor and quantum interference devices, etc [5]. Table 1 shows the detection range and resolution of magnetic field intensity of these sensors [6]. The detection range of induced coil and Quantum interference devices-SQUID is the widest. It can be seen the resolution of the SQUID and the coil is the highest.

When pipeline leakage flux detector works inside the pipeline, the above ground markers (AGM) are buried along the pipeline. The distance between each AGM is 1 to 2 km. The length of long-range oil or gas transmission pipeline is usually dozens or hundreds of kilometers, at least dozens of AGMs are needed for one test. During the test, the AGM work in the field using the energy supplied by lithium battery they carry because the detection test usually takes several days. Therefore, low power consumption is also required for the magnetic sensors used in AGM.

Table 1. Magnetic sensors and detection range of magnetic induction

Magnetic Sensors	Approximate detection range	Resolution
	1nT 1μT 1mT 1T	$1fT/\sqrt{Hz}$ $1pT/\sqrt{Hz}$ $1nT/\sqrt{Hz}$
Induced Coil		
Flux-gate		
Magneto-resistive sensor		
Hall-effect		
Optically pumped		
Quantum interference devices -SQUID		

In the following part, the main characteristics and performance of coil sensor and magneto-resistive sensor will be investigated in detail.

4. Coil sensors using in above ground markers

4.1. Working principle

The use of induction coil to measure the magnetic field has been widely used for a long time. The working principle of induction coil measuring magnetic field is based on Faraday's law of electromagnetic induction. The alternating magnetic field induces a voltage in the coil, and the intensity of the alternating magnetic field can be obtained by measuring the voltage.

Table 2 compares the characteristics and performance of coil, flux gate sensor, and magneto-resistive sensors, which could be used in above ground marking system.

Table 2. Characteristics of magnetic sensors

Characteristics	Coil	Flux-gate	Magneto-resistive sensor
Resolution	High	High	Relatively High
Energy Consuming	Low	Relatively low	Relatiely high
Volume	Big	Small	Small
Cost	Low	Relatively low	High
Detection Range of magnetic field intensity	10^{-10}~10^{6}Oe	10^{-10}~10^{-6}Oe	10^{-10}~10^{-6}Oe

Induction coil is classified into two kinds: hollow coil and magnetic core coil. Hollow coil usually has large volume, having better linearity, but lower sensitivity comparing with magnetic core coil. Core coil adds a magnetic core in the hollow coil. The linearity of core coil is not as good as hollow core, but the sensitivity is better. In MFL_PIG detection, the magnetic core induction coil is used. The typical core coil includes the core, coil, shielding, and shell composition, ect.

The equation of the induced electromotive force on the core coil is:

$$e(t) = -N\frac{d\varphi}{dt} = -NS\frac{dB}{dt} = -\mu_c NS\frac{dH}{dt} \tag{1}$$

Where: N is the number of turns of the coil; S is the cross-sectional area of the coil;μ_c is the effective permeability of the core, which is related to the material and size of the core.

4.2. Characteristics and performance

Induced coil, as a traditional passive magnetic sensor, has advantages of high resolution, low cost, low power consumption, wide measurement range, high accuracy, and easy to manufacture, which enables it to be widely used in measurement of magnetic field. However, the volume of induced coil is relatively big comparing with other magnetic sensors, which limits its application.

For the application in magnetic field detection of above ground marker, the volume of induced coil is acceptable if it is carefully designed in shape.

After considering all the advantages and disadvantages of the magneto-resistive sensor and induced coil sensor, induced coil is adopted as the magnetic sensor in design of above ground markers.

5. Experiment and discussion

In order to test whether the AGM and the main clock can meet the technical specifications and the stability of the work in various environments, a series of reliability tests have been designed and completed. It verified that the above ground marker using induced coil sensor meet the requirements of leakage flux detection for pipeline. Some of the tests, which was carried out to test the performance of above ground markers with induced coil sensors are listed below.

5.1. Traction Experiment

The AGMs are placed parallel to each other and perpendicular to the travel path of the test magnet which is placed on a small cart moving back and force to stimulate MFL_PIG. The poles of the MFL_PIG were simulated by using a NdFeB magnet with a magnetic pole pitch of 40 cm. The experiment results of traction experiment are shown in Table 3.

IOP Conf. Series: Materials Science and Engineering **220** (2017) 012024 doi:10.1088/1757-899X/220/1/012024

5.2. Seal ability test

Due to the harsh environment of the site, the above ground markers are most likely to withstand rain or soak, so the marker should have good sealing performance to protect the components it contains. During the seal ability test, the marker is set to work mode and placed in sealed compartment after tightening the coil cover and bottom cover. The pressure in the compartment is 1MPa, which is equivalent to the pressure at the depth of 100m. After one hour, open the sealed compartment and observe whether the seams are leaked and check if the marker is working properly.

Table 3. Result of traction experiment

AGM	Item	Hour	Minute	MC value	Traction times	Time Interval/s	Layout position
8	003	11	51	00042504	1	0	0.5m, Right below the pipe
	002	12	15	00182730	2	1402.26	Facing, 3.6m
	000	13	26	00613596	3	4308.86	Facing, 4m
	006	13	53	00771675	4	1580.79	Facing, 4m
9	006	18	22	00041059	1	0	0.5m, Right below the pipe
	000	19	57	00612333	3	3995.81	Horizontal along the axis, 1m
2	001	12	11	00044781	1	0	0.5m, Right below the pipe
	005	12	39	00214939	2	1701.58	1.4m
	004	13	46	00616271	3	4013.32	Perpendicular, 1m
1	001	13	31	00024224	1	0	Facing, 2.7m
	000	14	38	00425560	2	4013.36	Facing, 4m
	004	15	04	00583619	3	1580.59	Facing, 4m
5	006	12	12	00057370	1	0	0.5m, Right below the pipe
	000	13	14	00428957	3	3715.87	Perpendicular, 1m

5.3. Temperature test

Since the working environment of the above ground marker is in the wild, or even the desert and ice sheets, it is required to work in a wide temperature range. To verify the temperature performance of the instrument, the above ground marker were put into temperature thermostat test chamber with temperature range can be set from -40 to 130 ℃. The temperature stability in the chamber is 0.1 ℃.The experiment results of temperature experiment are shown in Table 4.

Table4. Result of temperature experiment

Temperature/ ℃	AGM	Start Time	Passing Time	MC (Passing)	MC (difference)	Note
70	A	9:30	10:00	330987	300742	Highest
	B	9:30	10:00	331002	300757	
50	A	10:15	10:45	601481	571236	
	B	10:15	10:45	601494	571249	
20	A	11:02	11:30	872023	841778	Normal
	B	11:02	11:30	872039	841794	
0	A	11:40	12:15	1142641	1112396	
	B	11:40	12:15	1142653	1112408	
-10	A	12:31	13:00	1411781	1381536	
	B	12:31	13:00	1411790	1381545	
-20	A	13:20	13:50	1712792	1682547	Lowest
	B	13:20	13:50	1712808	1682563	

Note: 1. The synchronization time of theAGMs is 9:10, and value of MC is 30245.
2.MC (difference) is the difference between the synchronization moment and when the stimulating magnet passing by.

During the temperature experiment, the main clock (MC) is synchronized with the two above ground markers - A and B at room temperature, and the PIG magnetic pole is simulated with a small magnet passing through the ends of the coils of ground markers. First, the induced electromotive force waveform was recorded at room temperature. Then, put the markers into thermostat test chamber with the end of the coil towards the observation window, and adjust the temperature in accordance with the order from high temperature to low temperature. After adjusting the temperature of the thermostat chamber, make the markers work for half an hour, and record the waveform when the small magnet passing by.

5.4. Experimental results and analysis

Above ground markers using induced coil as magnetic sensor to collect the magnetic signal when MFL_PIG passing by the AGMs works well during the traction test. The newly designed AGM also passed the extreme temperature test and seal ability test with ideal result.

It is to be mentioned that the traction test carried out in this paper is based on that the above ground markers are placed perpendicularly to the axis of the pipeline. If the AGMs are laid deflected to the pipeline because of the subsidence of the AGMs in the soil,errors may occur. And the error is proportional to the deflected angle. The detailed information of the experiment with different layout of the above ground marker are investigated in other articles, see reference [7].

6. Conclusion

As a magnetic sensor, induced coil has the advantage of low cost, simple structure, wide measurement range and high resolution. After being carefully designed in structure, it can be put inside a cup-shaped AGM to measure the change of magnetic induction of the field. It fulfils the requirements of magnetic sensors used in AGM in traction test, temperature test and seal ability test.

Acknowledgment

The authors would like to thank foundation of Demonstration Project of Tianjin Marine Economic Innovation and Development Region (Grant No. 2015120024000473), and Applied Basic and Frontier Research Project of Tianjin (Tianjin Natural Science Foundation) (Grant No. 17JCYBJC19300).

References

[1] Zhang Y 2011 Development of the above ground mark system for pipelines inspection using a magnetic flux leakage detector[PhD dissertation], Tianjin, Tianjin University.

[2] ParkG S, Park E S 2002 Improvement of the sensor system in magnetic flux leakage-type non-destructive testing (NDT)*IEEE Transactions on Magnetics*38(2)1277-1280.

[3] Pei Y, Yu N, Liu Q, Liu J 2004 Theory and application of anisotropic magneto-resistive sensor*Instrument Technique and Sensor*826-29.

[4] WuG, Li Y, Jin S, Hu X 2004 Application of Magneto-resistive Sensor on above GroundMaker of MFL-PIG*Journal of Instrument & Instrumentation*25 129-133.

[5] LiuZ W, DeanA P 2004 Precision Positioning AGM System United States Patent US 6,816,110 B1.

[6] Song X C, Wu X J, Kang YH 2014 An inspection robot for boiler tube using magnetic flux leakage and ultrasonic methods*Insight*46(5)275~277.

[7] Yang L, Geng H,Gao S 2016 Magnetic flux leakage internal detection technology of the long distance oil pipeline*Chinese Journal of Scientific Instrument* 37(8) 1737-1746.

IOP Conf. Series: Materials Science and Engineering **220** (2017) 012025 doi:10.1088/1757-899X/220/1/012025

Cause Analysis of Pressure Plate Breakage of Valve Limit Switch in Turbine Bypass System of Nuclear Power Plant

Li Nan[1], Wang Ming chang[1], Guan Jian jun[1], Li Guo dong[1]

[1]Suzhou Nuclear Power Rsearch Institute Co. Ltd., China

Abstract. The limit switch plates of nuclear power plant unit 1 in the turbine bypass system valve are finding multiple fractures. On the basis of metallographic analysis and vibration analysis, the stress state of the pressure plate is simulated and calculated. The results show that there are some creases in the original plate of the limit switch and the installation error of the pressure plate is the main reason for the break.

1. Introduction

Turbine Bypass System (GCT) consists of the Condenser Steam System (GCT-c) and the Atmospheric Steam System (GCT-a).When the reactor power is not consistent with the turbine load, the steam turbine bypass system (short for GCT) discharges excess steam to the condenser, deaerator, the atmosphere, according to the actual situation. So that ensure unit safety by avoiding the nuclear steam supply system temperature and pressure exceeds the protection threshold. According to the different power levels of the unit, the steam turbine bypass line condenser control valve is used to maintain the power balance of the primary and secondary circuits during the turbine load rejection stage, manual control power plant operation stage or special steady-state operation phase[1]. GCT-c emission control valve has three groups; the opening order is GCT121VV, GCT117VV, and GCT113VV. As the operating characteristic of the GCT121VV is priority to open, so its action is relatively frequent.

GCT121VV valve limit switch pressure plate material is the 316 stainless steel, it triggers limit switch when running up and down. The failure of the pressure plate will cause the logic signal of fully open / fully closed cannot produce when the valve closing, these may affect the water level setting and the final power setting of the steam generator (short for SG) generate. And that will affect water level control of the SG, and it also influence the power, temperature, and water level of regulators automatically adjustment. By metallological analysis, vibration influence analysis and failure simulation analysis, this paper analysis manufacturing defects, operation problems and installation problems. Based on the analysis of the vibration, the failure model of the plate of the steam turbine bypass system is established. The root causes of GCT-c valve's limit switch plate failure reason are explored in the nuclear power plant.

2. Metal performance analysis

The breakage pressure plate of the GCT121VV valve has CP1, OP and CP 3 limit switches. As shown in Fig.1, the OP side is used to display the fully open position of the valve, the CP side is used to show the valve clearance position, and participate in the final power setting settings functions, and the CP1 side is used to the feedback valve is open. On both sides of the stem connector, the limit switch has a pressure plate. There are two fracture plates on the site, see Fig2.

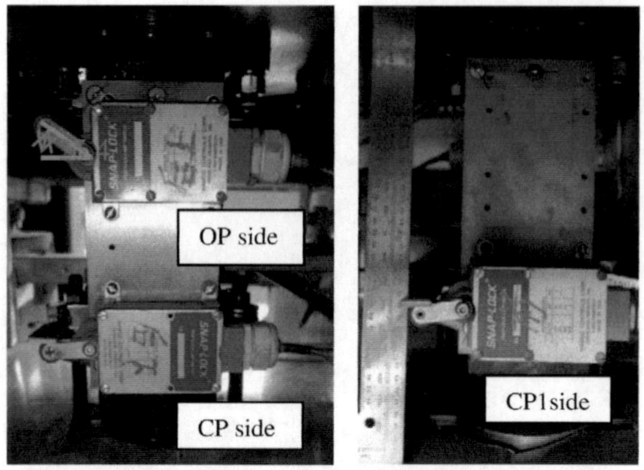

Figure 1. The limit switch of GCT121VV valve.

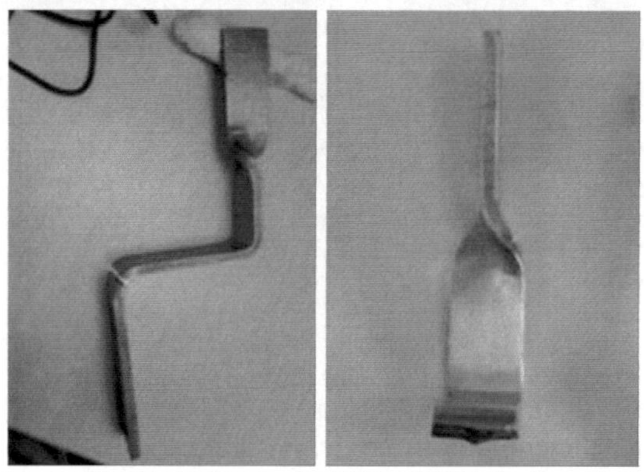

Figure 2. The fracture morphology of GCT121VV limit switch pressure plate.

By inspecting, there are significant processing creases at the corner of the pressure plate. The cross section of the pressure plate is located at the crease. There was no significant plastic deformation in the rest of the area except for the slight plastic deformation of the outer side. The fracture surface can see fatigue beige line, see Fig.3.

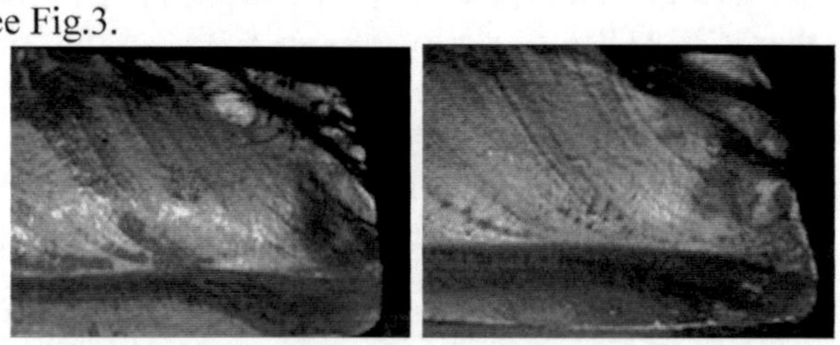

Figure 3. The analysis photo of pressure plate.

2.1. Electron microscopic observation of the pressure plate

Scanning electron microscopy (SEM) was used to observe the microstructures of the fracture plates; the scanning electron microscope is TESCAN VEGA TS5136XM. Macroscopic observation and scanning electron microscope observation of the crease at the corners are shown in Fig.4.

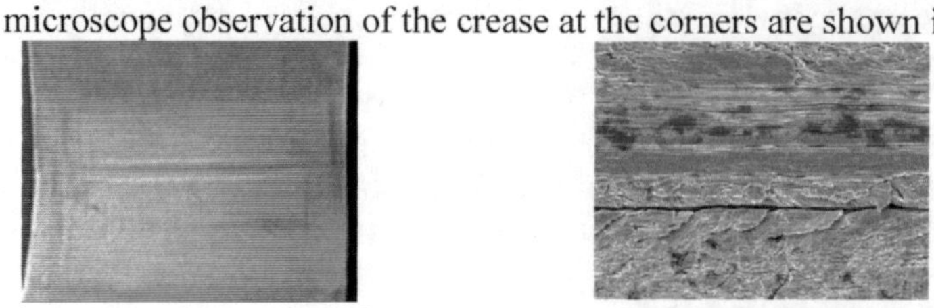

a) the macro picture CP side; b) the crease topography at 1 000

times of the CP side

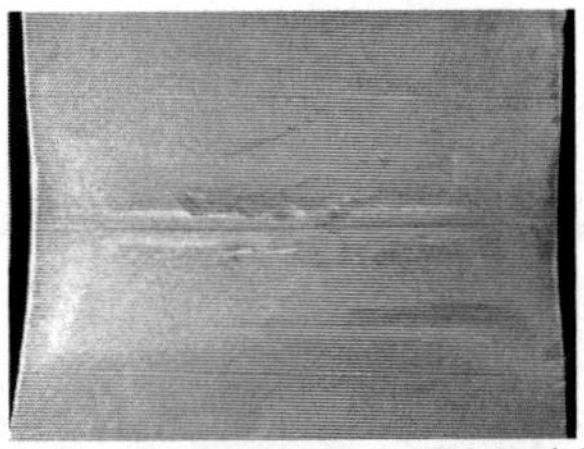

c) the macro picture CP1 side

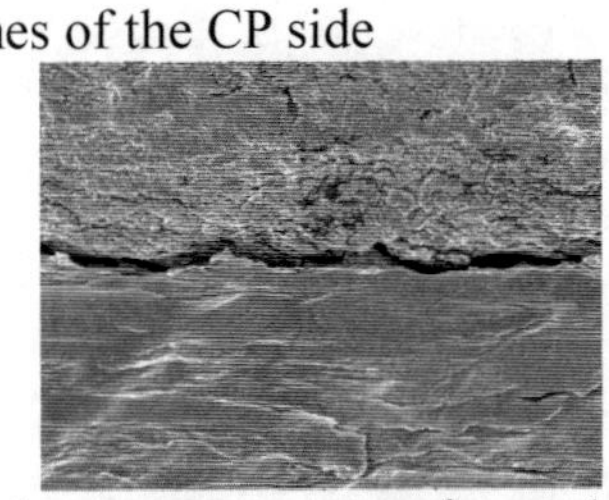

d) the crease topography at 1 000 times of the CP1 side

Figure 4. Microstructure of crease at the corner of the pressure plate.

Microscopic observation of the crease at the corners show that there are obvious micro cracks on the edge of the CP side and the CP1 side, and the micro cracks of surface are the original manufacturing defects [2].

2.2. Fracture surface analysis

Scanning electron microscopy (SEM) was used to observe the microstructures of the fracture plates; the scanning electron microscope is TESCAN VEGA TS5136XM. The extension area of fracture surface of CP side and CP1 side are shown in Fig.5 and Fig.6.

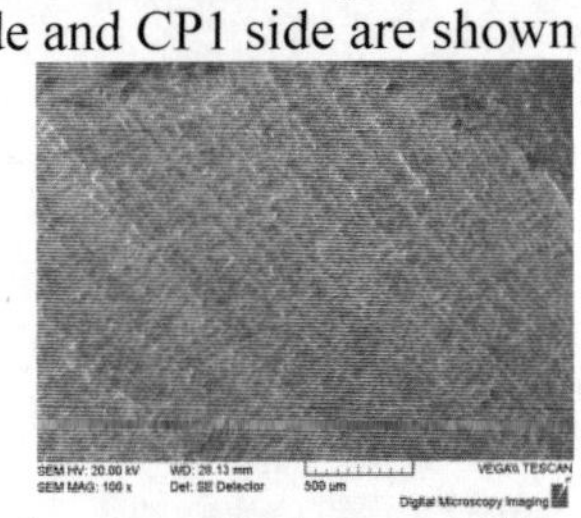

a) the Morphology of Fracture area under 100 times;

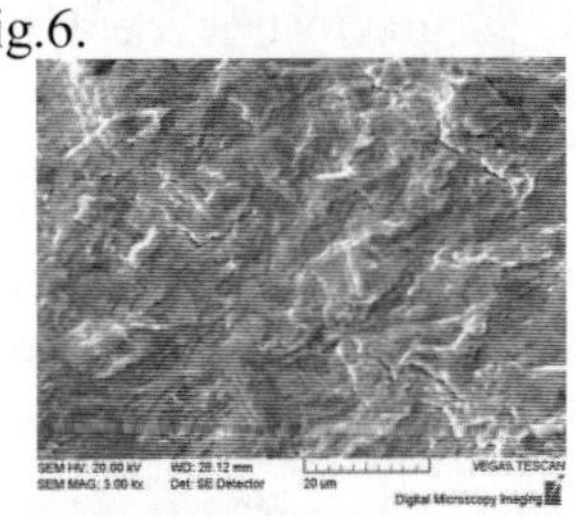

b) the Morphology of Fracture area under 3000 times;

Figure 5. The Fracture surface of CP side (extension area).

a) the Morphology of Fracture area under 100 times;

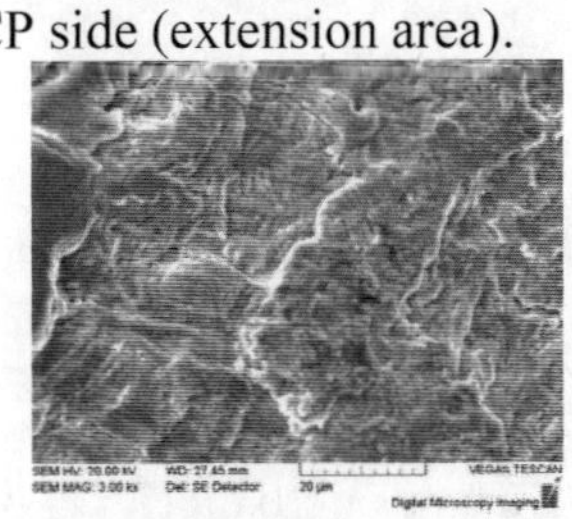

b) the Morphology of Fracture area under 3000 times;

Figure 6. The Fracture surface of CP1 side (extension area).

It is found that the fracture of CP side, CP1side fracture plates are cracked by the inner surface of the inner arc and expanded at the angle of the outer arc side, and finally at the corner of the outer arc. Fracture extension area macroscopic visible white stripe lines, microscopic visible obvious fatigue characteristics.

2.3. Hardness analysis

The micro hardness test is carried out on each part of the press plate. The test equipment is 402MVD digital micro-Vickers hardness tester. The test load is 1kg and the holding time is 10s. According to the characteristics of the fracture, the micro hardness of the curved section, middle layer, outer arc section, straight section inner layer, middle layer and outer layer is tested. The results of hardness test on the CP side and the CP1 side plate is shown in Table 1.

IOP Conf. Series: Materials Science and Engineering **220** (2017) 012025 doi:10.1088/1757-899X/220/1/012025

The results of hardness test on the CP side and the CP1 side plate show that the hardness of the inner arc and the outer arc are higher than that of the middle layer. There is no obvious abnormality in the hardness of the press plate.

Table 1. The results of hardness test on the CP side and the CP1 side plate

NO.	position	hardness (HV1)				average value
CP side	the curved surface of the bend	329	320	329	338	329
	the outer arc of the bend	318	309	311	306	311
CP1 side	the curved surface of the bend	323	327	327	330	327
	the outer arc of the bend	293	296	294	293	294

2.4. Metallographic analysis

Metallographic specimens are prepared according to ASTM E3 "Standard Method for Fabrication of Metallographic Samples". Metallographic tests are performed on the CP side and CP1 side press plates of the fracture plates. The microstructure of the pressure plate is shown in Fig.7.

The result of metallographic examination shows that the structure of CP side press plate is austenite and ferrite. There is a deformation flow line and a slight dimple defect in the curved part. The inner arc and outer arc surface are not found cracks, and other defects. The organization of the pressure plate of CP1 side is also the austenite and ferrite, and on the plate, there are deformation flow lines. On the bending parts, the inner arc surface and the outer arc surface of straight parts are not found cracks, fine crystal layer defects.

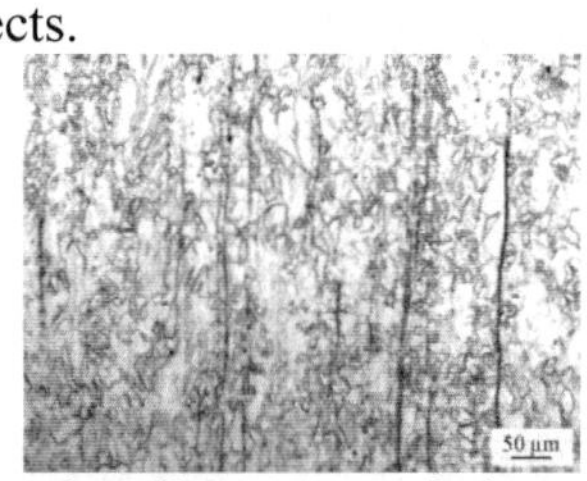

a) the bending part of the middle layer of CP Side;

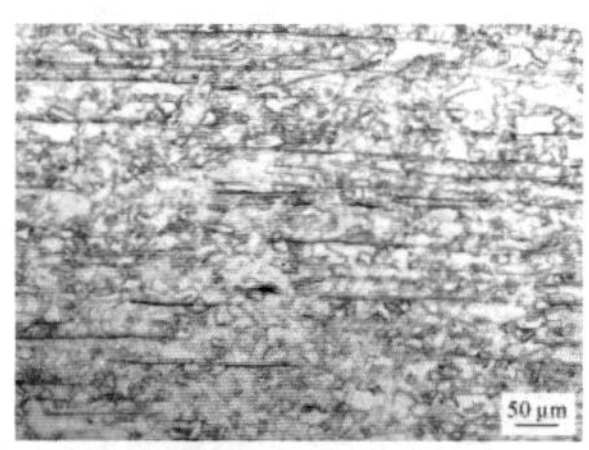

b) the straight part of the middle layer of CP1 Side

Figure 7. The microstructure of the pressure plate.

The results of scanning electron microscopy, metallographic analysis and hardness analysis show that, there are many surface micro cracks in the creases of the fracture plates; the grain structure of the inner surface of the pressure plate is not abnormal; there was no obvious abnormality in the hardness of the parts; the fracture surface of the arc surface side is the multi-source crack, expansion to the outward arc side[3]; Fracture microscopic observation shows typical fatigue flaky features on the fracture[4].

3. Vibration analysis

The vibration of GCT121VV valve is analysed, vibration measurement position is shown in Fig.8, The vibration values of GCT121VV valve fully closed, 100% opening, 35% of the opening are measured, shown in Table 2. The vibration results show that the vibration of the GCT121VV valve is low frequency vibration. Comparing the results of vibration measurement with the vibration data of the same position of other nuclear power plants, the vibration level of GCT-c valve in this nuclear power plant is not obvious abnormality.

IOP Conf. Series: Materials Science and Engineering **220** (2017) 012025 doi:10.1088/1757-899X/220/1/012025

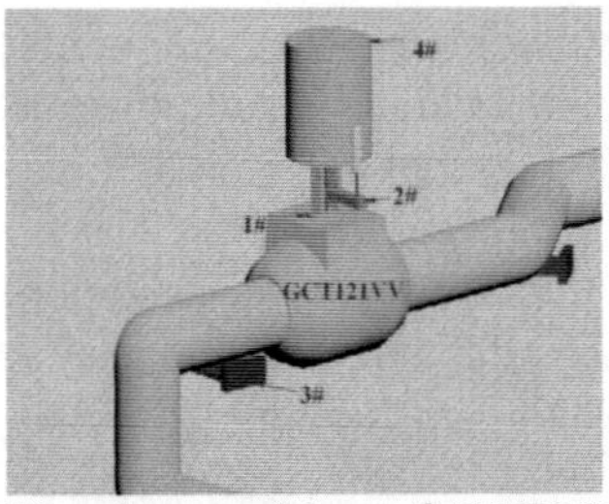

Figure 8. Vibration position of GCT121VV valve.

Table 2. The valve vibration value of GCT121VV (mm/s)

opening degree	1H	1V	1A	2H	2V	3H	3V	4H	4V	4A
0%	1.60	4.87	2.24	2.42	4.90	2.50	2.68	10.30	13.00	3.16
100%	9.82	9.13	10.30	14.70	13.30	11.80	13.10	15.70	13.90	10.32
35%	10.60	10.00	14.70	10.60	9.97	12.20	11.10	9.43	11.30	9.73

H—Horizontal vibration; V—Vertical vibration; A—Axial vibration

4. Analysis of operating conditions

Measuring the stroke of the GCT121VV valve when it is in fully open and fully closed state, the whole stroke is 167mm. The stroke of the valve is consistent with the Equipment Operation and Maintenance Manual (short for EOMM); the process of positioning the feedback plate and other equipment did not scratch the situation.

Through the action test, it is found that there is no obvious impact phenomenon, and the pressure plate is not obvious deformation when the valve in the fast opening and closing process.

5. Failure simulations

According to the size of the pressure plate and the installation operation, the stress state of the pressure plate is simulated. The limit switch action torque is 23in • lbs (about 2.599Nm), the swing arm length is 38.4mm. The force is about 67.4N when the limit switch just action (action stroke 10 °) by calculating, and the force is about 170N when the limit switch act to pressure to the limit position (Full stroke: 37 °). The force situation of limit switch is shown in Fig.9.

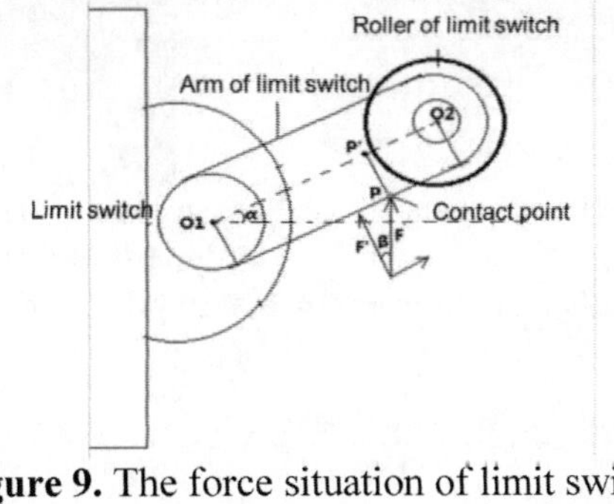

Figure 9. The force situation of limit switch.

When the lower limit switch of GCT121VV valve is operating, if the pressure plate cannot be pressed to the roller directly during the action, the force arm will be shortened. Inspecting GCT121VV valve installation, finding that the upper limit switch arm and the pressure plate has contact marks. When the wheel directly contacts the arm it will cause the force increasing.

According to the size of the pressure plate and the installation and operation of the field, the finite element simulation analysis of the stress state of the pressure plate is carried out. The geometrical dimensions used in the modeling process are as follows: the actual thickness of the press plate is 5.5

mm, the crease depth is about 0.2 mm, the crease and the width are 0.2 mm. The inner surface of the bolt hole is set to a fixed confinement boundary condition, and a concentrated force F which is vertically downward in the plate width direction is applied at the C of the right side of the platen. The simulation model is shown in Fig.10.

According to the angle of the scene, the position of the contact point and the analysis of the action stroke data, the pressure check force is 67.4N and 170N respectively.

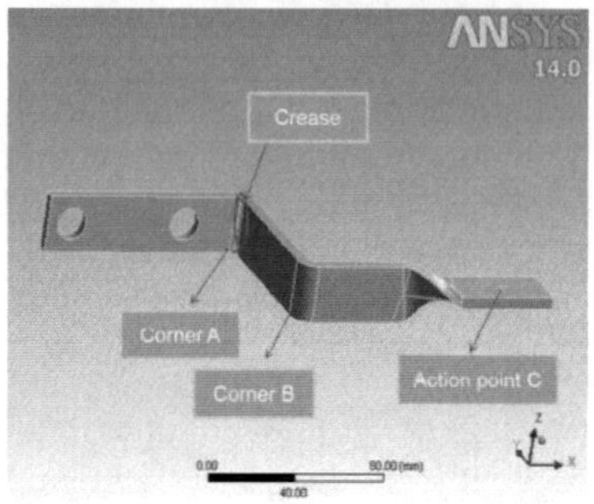

Figure 10. The Simulation model of pressure plate.

The stress nephogram of pressure plate under different force are shown in Fig.11, and the stress calculation results are shown in Table 3.

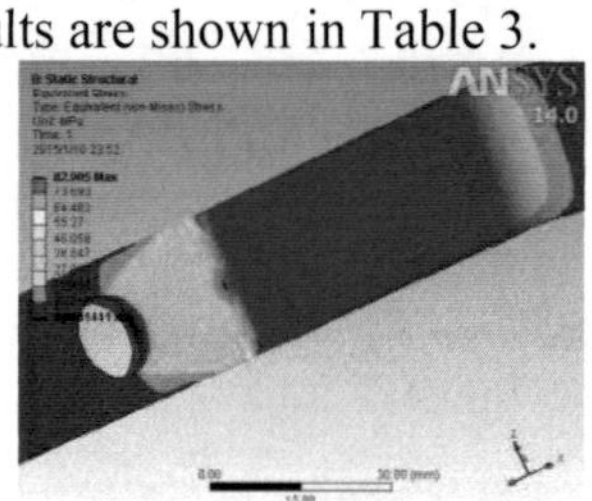

(a) Stress distribution without crease under 67.4N force

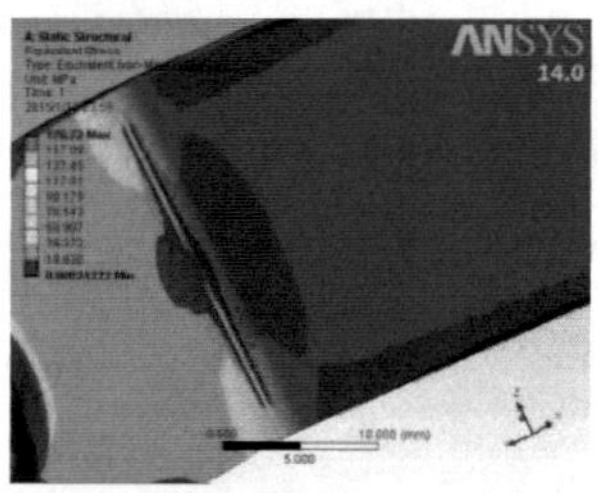

(b) Stress distribution with crease under 67.4N force

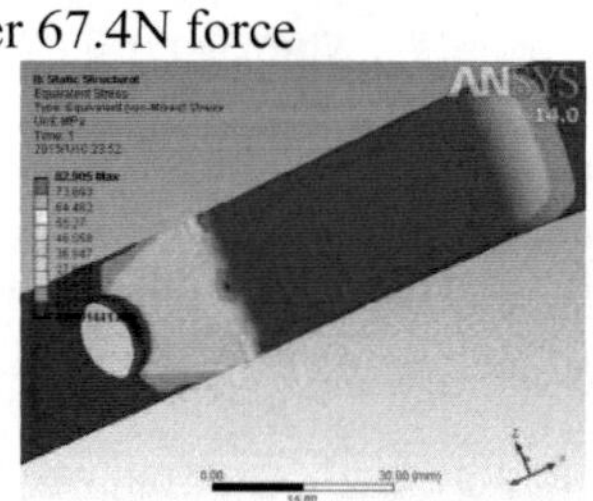

(c) Stress distribution without crease under 170N force

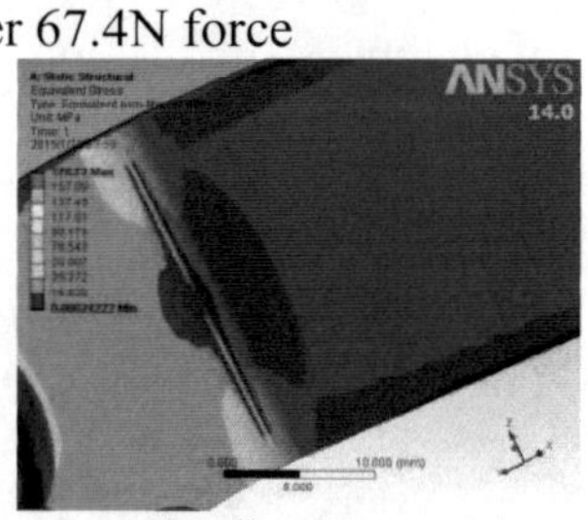

(d)Stress distribution with crease under 170N force

Figure 11. The stress nephogram of pressure plate under different force.

Table 3. Stress calculation results (Mpa)

maximum equivalent stress	the roller touches the press plate	the roller is not touch the press plate
The corner of plate without creases	82.9	209.0
The corner of plate with creases	176.0	445.0

The stress distribution of the two kinds of press plates under different loads is simulated by finite element simulation, which is with and without processing creases. When the limit switch act, the roller is not contact with the pressure plate, and the force is 67.4N; when roller contacts plate, the force is

170N. The simulation results are shown in Table 3. The results show that the maximum equivalent stress of the corners is 445.7MPa, and the 316 stainless steel generally allowable stress of 480MPa.

The maximum stress of the crease plate at 170.0N (i.e. the plate pressures on the roller and close to the limit switch) is 445.0MPa, closing to the allowable stress of the material 480.0MPa. Compared with the non-crease plate, the stress of the crease part of the crease plate is obviously increased, and the stress concentration at the crease is obvious which is consistent with the failure of the failure plate, and it greatly reduces the strength of the pressure plate[5]. Inspecting the valve and finding that the valve is improper install. the upper limit switch on the valve arm and the pressure plate has contact marks, resulting in full stroke force of valve is more than 170.0N, the pressure plate crease stress exceeds the allowable stress.

6. Conclusions

The pressure plate of the nuclear power plant GCT121VV valve limit switch are not found abnormal in the microstructure, operating conditions and other aspects. However, there are original defects which are processing creases in the pressure switch plate. And in the crease, there is micro-crack, which results in stress concentration in the crease of the press plate during operating. The maximum stress of the creased plate at 170.0 N is close to the allowable stress of the material by stress check calculation, when the plate pressing on the roller and approaching the full stroke.

Under normal operating conditions, due to the problem of valve installation, resulting in the pressure plate cannot always press the roller and direct contact with the arm during the operation. So the valve full stroke action force is more than 170.0N, making the crease of the pressure plate beyond the material allowable stress. Limit switch plate at the corner of the crease part is the crack source, fatigue to the corner of the arc outside the expansion, and eventually break. Indicating that the original plate processing has problems, the pressure plate manufacturing quality is poor.

This limit switch type is different with other domestic nuclear power plant limit switch. It uses a bent steel plate to trigger the limit switch, if the limit switch is installed properly; it is easy to cause pressure plate cannot press on the roller and easily overpressure. At the same time, due to the actual installation of the difficulties, this type of GCT121VV limit switch cannot accurately adjust the limit switch position, which could avoid the upper limit switch arm contact with pressure plate

7. Improvement measures

Through the above analysis, the creases and micro cracks exist in the bending angle is the main reason of GCT121VV valve limit switch plate fracture. To prevent the occurrence of similar incidents, the following precautions should be taken:

1) Replace the same type of valve limit switch plate for GCT-c system of the nuclear power plant to ensure that the replacement plate has no quality defects, including no creases and micro cracks at the corner. And to check the blanks spare parts, requiring the manufacturer to provide quality qualified pressure plate.
2) To verify the installation of the nuclear power plant limit switch, adjust the limit switch which does not meet the requirements.

References

[1] Guangdong Nuclear Power Training Center 2007 *Devices and Systems of 900MW PWR* Beijing: Atomic Energy Press.
[2] Lv S, Gong J, Lu Q 2011 Fracture Analysis of Hoisting Short Joints *Mechanical Engineering Materials* **2** 90-92.
[3] Liu H 2011 *Material mechanics* Beijing: Higher Education Press.
[4] Fang H 2004 *Practical manual for the design and manufacture of commonly used data and standards* Beijing: Contemporary Chinese Audio and Video Publishing House.
[5] Zhao D 2004 Fracture Analysis of 20CrMnTi Press Plate *Aerospace Manufacturing Technology* **5** 25 - 27.

IOP Conf. Series: Materials Science and Engineering **220** (2017) 012026 doi:10.1088/1757-899X/220/1/012026

Leakage Current Analysis of Bushing with Water-cut Sheds in Heavy Rain Test

Chai Zhen[1], Yu Xinzhe[1]

[1]China Electric Power Research Institute, Haidian District Beijing, 100192 China

Abstract: In this article, heavy rain tests under two rainfall intensities are introduced. Three kinds of bushings with different water-cut-sheds were used in these tests. Then leakage currents monitored during tests are analysed and contrasted to obtain characteristics and tendency. The results show that maximum value of leakage current is affected by voltage applied, profile of water-cut sheds, and the rainfall intensity. Higher AC voltage applied on bushing leads to higher peak of leakage current. More and larger water-cut sheds makes bushing able to withstand higher leakage current. Heavier rain may cause leakage current rise totally. These results may be helpful in online monitoring of insulation.

1. Introduction

The property of pollution and wet is always taken into account on the external insulation selection of the outdoor substation equipment. The main method of the external insulation design is on the basis of the pollution flashover property. Previous research [1] [2] showed that: when in heavy rain condition, there are rain curtains formed by water with high conductivity beside the surface. Flashover occurs along the rain curtains, not on the surface. So mechanisms and characteristics of pollution flashover and heavy rain flashover are different. The outdoor equipment whose external insulation is selected through the method above may flashover in heavy rain condition. And research on the method of protecting equipment from heavy rain flashover is needed.

Water-cut shed is generally accepted as an effective protective method in heavy rain condition. It can cut the rain curtains off, decreasing the probability of heavy rain flashover. Water-cut sheds are usually made from silicone rubber, and bonded with the bushing sheds.[2] But there are some problems on bonding strength and mechanical property. Porcelain bushings manufactured with several water-cut sheds on it may avoid these problems.

Leakage current analysis is an auxiliary method to monitor the status of the insulated surface while high voltage is applied on it. Researchers in STRI carried out rain test of hydrophilic porcelain post insulators, and calculated the resistive component and capacitive component of the total leakage current. They observed the change of currents under different rain density. [3] Indonesian researchers presented the characteristics of leakage current at different shed of epoxy resin insulator under rain contaminants, and compared characteristics of leakage current under rain and dry test. [4] But they only gave single values, ignoring the tendency of leakage current during the period from voltage applying to flashover or withstand.

In recent years, there were several heavy rain flashover accidents on bushings occurred in EHV substations in China. To improve the ability of withstanding heavy rain flashover, effective and economical protection measures should be raised according to EHV bushing rain tests.

In this article, heavy rain flashover tests were carried out on porcelain bushings with different water-cut sheds. Leakage currents monitored during the flashover tests are contrasted to find the change rule of it.

2. Specimens

Three types of 500 kV porcelain bushing with different water-cut shed configurations were used in this test. Their insulation height are all 4.7 m. One of them is common bushing with no water-cut shed, the others are with four and eight water-cut sheds, respectively. Table 1 gives parameters of those tested porcelain bushings, and figure 1 shows the pictures of three porcelain bushings.

Table 1. Parameters of tested specimens

Number	Leakage Distance (mm)	Number of Water-cut Sheds	Spacing of Water-cut Sheds (mm)	Overhang of Water-cut Sheds (mm)
1	18775	0	-	-
2	18775	4	1165	50
3	19200	8	580	95

Figure 1. Profiles of tested specimens.

Figure 2. Set-up of artificial rain test.

3. Test facilities and procedures

3.1. Test facilities

Artificial heavy rain flashover tests were carried out at the Pollution and Environment Laboratory in State Grid UHV DC Test Base, which is located in Changping, Beijing. Tested bushings were installed vertically in the artificial environmental chamber which has a test space of 20 m in diameter and 25 m in height. It is capable for environmental tests such as icing, raining and fogging. Figure 2 shows the set-up of artificial rain test in the chamber.

The main circuit of AC power supply of the artificial heavy rain flashover test is shown in figure 3. It meets the requirement of IEC 60507 [5]. T1 is 4800 kVA regulator. T2 is the AC test transformer of 800 kV, 4800 kVA. R1 is the protecting resistance, and V.D is the AC voltage divider. T.O is the specimen and R2 measures the leakage current during the test. The leakage current measurement system is designed by Graduate School at Shenzhen, Tsinghua University. The sampling rate of this system is 5 kHz. figure 4 shows the signal processing module and leakage current monitoring module.

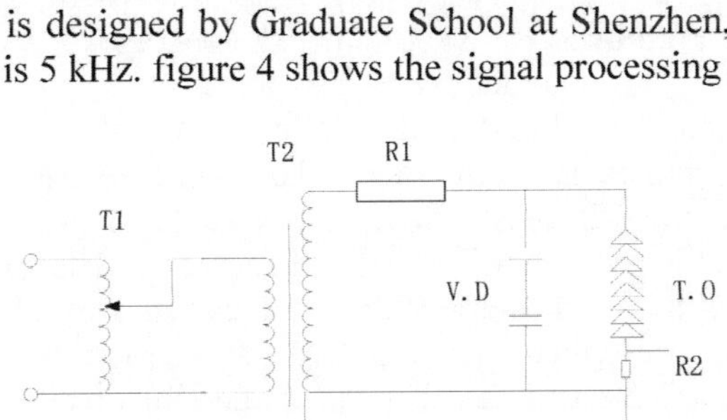

Figure 3. Profiles of tested specimens.

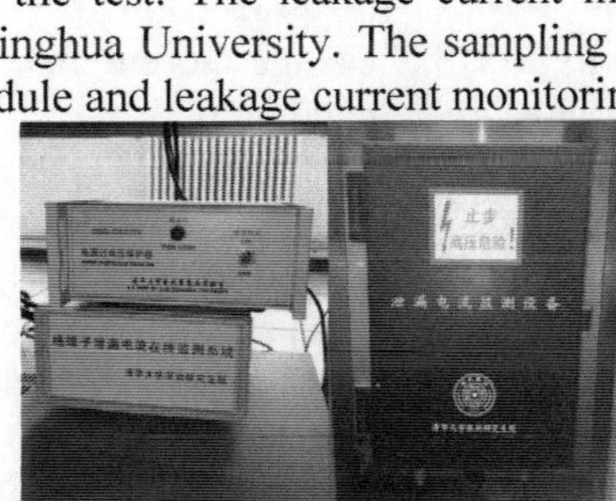

Figure 4. Set-up of artificial rain test.

The rain simulator is consist of water container and shower device. The shower device is installed on a platform which could move up and down freely. There is a 4×12 nozzle matrix on the shower device. Each nozzle can change the shooting direction of the water. Through adjusting nozzles, we make sure the artificial rain dropping on the bushing slantingly. The angle between rain direction and horizon is about 45°, which means horizontal component and vertical component of the rain are almost the same. The 45° raining is regarded as the worst raining condition because rain may wet the upper and bottom of sheds evenly.

The measurement of rainfall intensity used special device. It has a long stick. There are two metal container on the top of the stick. One has a horizontal opening, collecting the vertical component of the rain. The other has a vertical opening, collecting the horizontal component of the rain. Rainfall intensity was calculated according to the water in two containers after collecting rain for 1 min. Rainfall intensity of 5 mm/min and 10 mm/min were required in this test for continuous rain curtains with high conductivity.

3.2. Test procedures

The pollution layer was sprayed on the surface of specimen according to IEC 60507. The insoluble contamination was kaolin. To stop the pollution layer being washed away too quickly, gelatin was added into the suspension. The value ratio between sodium chloride and kaolin is 1:6, and the value ratio between gelatin and kaolin is 1:10. After spraying, wash test should be done according to IEC 60815 [6] to make sure the ESDD is 0.1 mg/cm^2 within the error range of ±10%.

After dried completely in shady place, the tested bushings were installed in chamber. Firstly, a preset voltage was applied on the specimen, and then the rain started. Rain duration was at least 5 min if no flashover occurred.

4. Characteristics of leakage current

4.1. Leakage currents of one bushing under different voltages

Firstly, the leakage current under different voltages applied on bushing with four water-cut sheds are compared. Because leakage waveforms change on the 50 Hz frequency, we use upper envelope instead to find out how amplitude changes. Figure 5 shows the upper envelopes of each leakage current wave under U_1 、 U_2 and U_3 after 5mm/min rain starting. The order of three voltage is $U_1 < U_2 < U_3$.

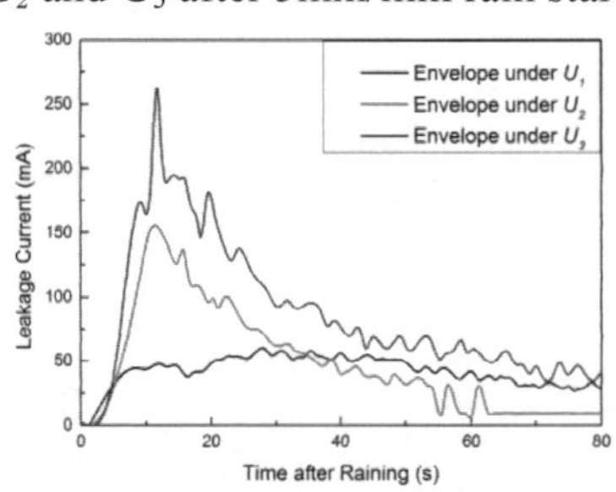

Figure 5. Leakage current envelope under Different Voltage.

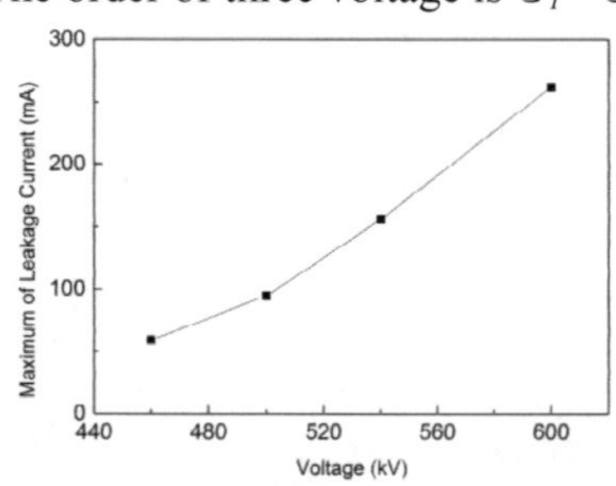

Figure 6. Maximum of leakage current under different voltage.

As is displayed in figure 5, when U_1 is applied and rain starts, the leakage current increases to about 50 mA, then waves around. When U_2 is applied and rain starts, the leakage current firstly increases and reaches the maximum of 150 mA, then decreases to a very low value. When U_3 is applied and rain starts, the leakage current increases to more than 250mA and decreases. The curves of leakage current aren't smooth, and little peaks on them indicate that there are partial arcs appearing and disappearing on the surface. Those processes described above are all happened in 80 s after artificial rain showering on the surface of bushing. We can see that the maximum of current has positive correlation with the applied voltage from direct comparison in figure 6. Trend of curves are the same if voltage is high enough and flashover doesn't occur.

4.2. Leakage currents of three bushings under identical rainfall intensity

There is significant diversity between 50% flashover voltage of three bushings. It's impossible to contrast currents of bushings under identical voltage. Therefore, leakage currents under the highest withstand voltage of each bushing are contrasted. Figure 7, figure 8 and figure 9 are the upper envelopes of three bushings' leakage currents. Tests were carried out at 5 mm/min rainfall intensity.

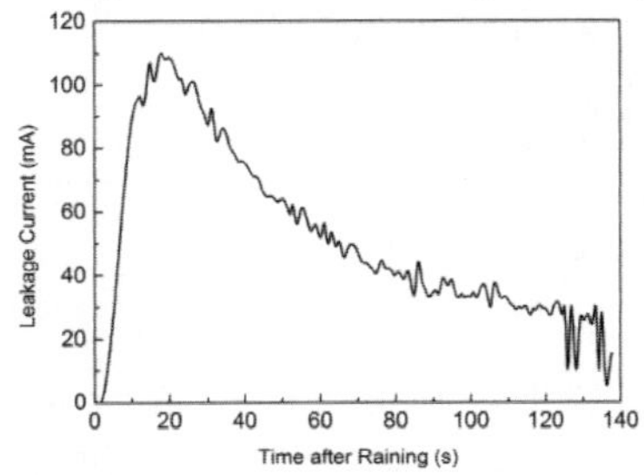

Figure 7. Leakage current envelope of No.1 bushing.

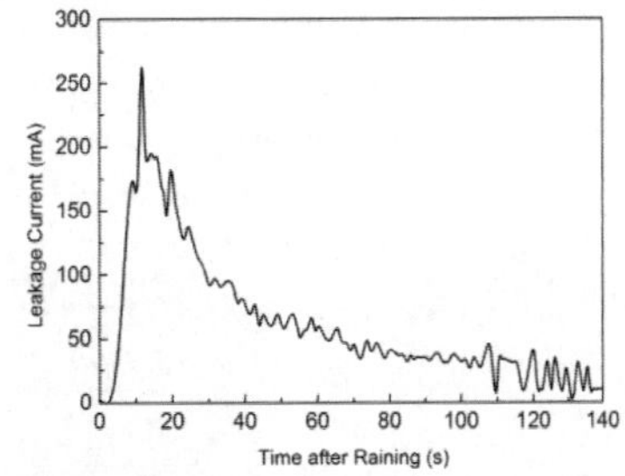

Figure 8. Leakage current envelope of No.2 bushing.

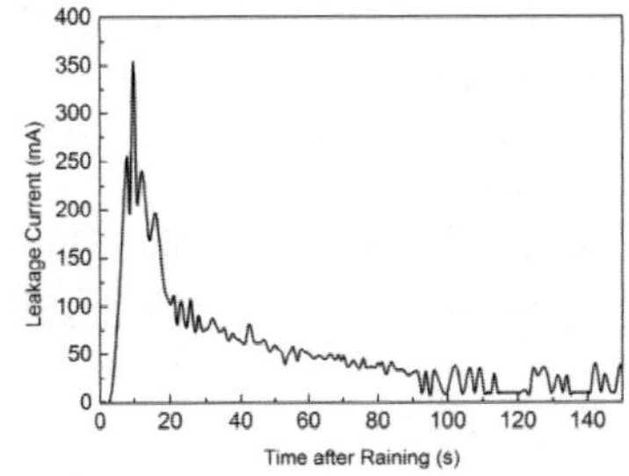

Figure 9. Leakage current envelope of No.3 bushing.

As shown in figure 7, figure 8 and figure 9, the trends of curves are all firstly rising quickly to the maximum, then decreasing to less than 50 mA, finally fluctuating in a limited range. It matches three stages of bushing surface. Firstly, rain washes away the contamination on the surface. Resolving salt in contamination, high conductivity rain curtain makes leakage current rise rapidly. Possibility of flashover is highest in this stage. Then if flashover doesn't occur, leakage current decreases as salt is resolved less in the rain curtain. Finally leakage current fluctuates because of the appearance and disappearance of partial arcs.

The maximum value of leakage current is about 110 mA on the surface of bushing with no water-cut shed, about 260 mA on the surface of bushing with four water-cut sheds and about 350mA on the surface of bushing with eight water-cut sheds. So water-sheds protect bushing from flashover when higher leakage current flows on the surface.

Harmonics of leakage currents on three bushings are also analysed. Ten cycles of waveform around the maximum value are chosen. The fundamental component of leakage current is calculated through FFT (fast Fourier transform), as well as 3rd, 5th, and 7th harmonics. Since the maximum values of three bushings' leakage currents are different, direct compare of harmonics is meaningless. Instead, relative values of harmonics are shown in figure 10, supposing relative value of fundamental component is 100%. Obviously, 3rd harmonic is the main harmonic in leakage current. The relative value of 3rd harmonic rises if bushing has more and lager water-cut sheds. 5th and 7th harmonics can be ignored because their values are less than 5% of fundamental component.

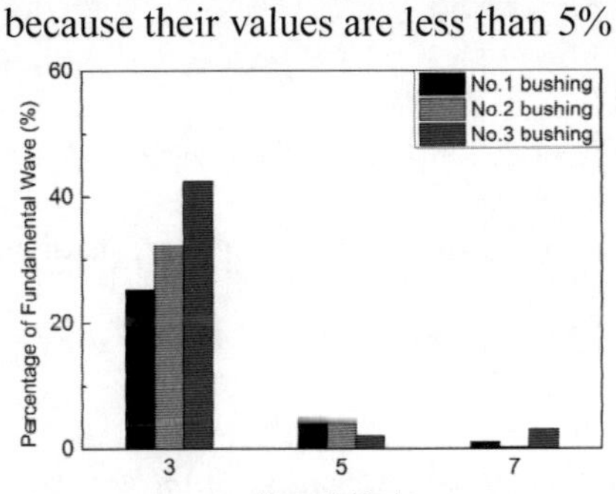

Figure 10. Relative values of harmonics.

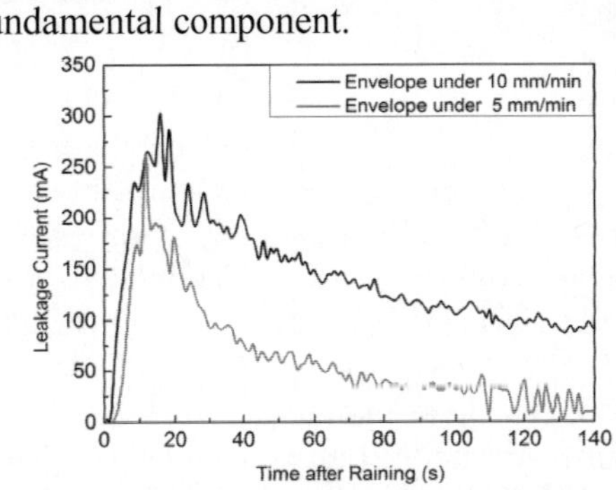

Figure 11. Leakage current envelope of No.2 bushing under different rainfall.

4.3. Leakage currents of one bushing under different rainfall intensities

Bushing with four water-cut sheds (No.2) was tested under 5mm/min and 10mm/min rainfall intensity. The leakage current under these two conditions are compared in figure 11. As rainfall intensity increases from 5 mm/min to 10 mm/min, the maximum value of leakage current increases from 260 mA to 310 mA approximately. After raining for enough long time, the leakage current value maintains during 50 mA and 100 mA under 10 mm/min rainfall intensity. Contrastingly, the leakage current decreases lower than 50 mA under 5 mm/min.

The diversity between leakage currents under different rainfall intensity has relationship to the rain curtains. Figure 12 shows the distribution of curtains on the bottom of bushing under 5 mm/min (a) and 10 mm/min (b) rainfall intensity. Significantly, rain curtains under 5 mm/min is less than that under 10 mm/min. Less rain curtains means leakage current flowing longer distance on the contamination layer. Since the layer's resistance is larger than rain curtain, leakage current's maximum value and stable value are larger as the rainfall intensity increases.

(a) (b)

Figure 12. Distribution of rain curtains on bottom of No.2 bushing.

5. Conclusion

From the analysis above, we can conclude that leakage current in heavy rain test is affected by voltage applied, profile of water-cut sheds, and the rainfall intensity. Following is the conclusions in detail.
a) The maximum value of leakage current has positive correlation with the applied withstand voltage.
b) The maximum of leakage current is higher with more and larger water-cut sheds under the same rainfall intensity and withstand voltage.
c) The maximum and stable values of leakage current on the same bushing are higher when the rain is heavier.

Through analysing the of rain test under the highest withstand voltage, the typical tendency and characteristics of leakage current are obtained. Those are helpful in online monitoring of insulation in substation. And the status of equipment insulation can be diagnosed easily through leakage current monitoring.

References

[1] Farouk R 1975 Mechanism of Insulator Flashover under Artificial Rain *Proc. of the Insulation of Electrical Engineers* **122** pp. 449-454.
[2] Youwei L 2001 Investigation of rain contamination flashover and application of aided sheds *High Voltage Apparatus* **37** pp. 6-10.
[3] Ralf H 1994 The AC-behavior of a hydrophilic and hydrophobic post insulator during rain *IEEE Trans. Power Delivery* **9** pp. 1584-1592.
[4] Abdul S and Hermawan 2014 Leakage current characteristics at different shed of epoxy resin insulator under rain contaminants *1st Int. Conf. on Information Technology, Computer and Electrical Engineering* pp. 413-418.
[5] IEC 60507-2013: *Artificial Pollution Tests on High Voltage Insulators to Be Used on a.c. Systems.*
[6] IEC/TS 60815-1:2008: *Selection and Dimensioning of High Voltage Insulators Intended for Use in Polluted Conditions—Part I: Definitions, Information and General Principles.*

IOP Conf. Series: Materials Science and Engineering **220** (2017) 012027 doi:10.1088/1757-899X/220/1/012027

Ultra-High Voltage DC Convertor Station Equipment Condition Data Access Technology Based on multi-Source Heterogeneous Fusion

Wang Feng[1], Zhang Bo-wen[1], Han Shuai[1], Ren Wei[2], Xu Hai-jun[2], Fu Long-ming[3]

[1]China Electric Power Research Institute, Beijing, 100192
[2]Operation corporation of SGCC, Beijing, 100052
[3]Beijing Join Bright Digital Power Technology Co. Ltd, Beijing, 100085

Abstract. With the large-scale construction of special high-voltage project, as well as power supply reliability, security, economic and other increasingly demanding, state monitoring equipment involved in more and more monitoring projects and more and more monitoring data, because these data exist in multiple isolated systems in the Ultra-High Voltage(UHV) AC-DC substation, there is no data sharing mechanism, so a holistic analysis, application and sharing approach for the data set will need a deep consideration. In this paper, the equipment condition monitoring system frame of the UHV converter station and the scheme of the equipment state data access of UHV converter station based on the multi-source and heterogeneous data fusion are presented. Then, data exchange technology of UHV equipment state early warning center was introduced, and a data access and conversion device in the Zhongzhou converter station was deployed to solve the timeliness and functionality difficult of the existing system to meet the requirements of UHV operation and maintenance support.

1. Introduction

AC-DC transformer equipment is very important in UHV power transmission project, and its reliability and operation directly determine the stability and security of the whole system. In order to ensure the safety of the large power grid, according to the relevant documents and the requirements of State Grid Corporation of China (SGCC), the UHV AC-DC substation belongs to first-class substation, directly managed by SGCC Inspection Department. China Electric Power Research Institute (CEPRI) established SGCC UHV equipment status warning center in order for professional equipment management and the whole process of technical supervision, including monitoring and early warning, state detection and diagnosis, analysis and evaluation for UHV AC-DC equipment. At present, SGCC has built the transmission and distribution equipment monitoring system with the provincial company and SGCC headquarters two deployments, but because the data for the state monitoring system were huge, the master station of the provincial company bore the traffic pressure of power supply companies of the whole province and SGCC headquarters, the information channel restricted the SGCC UHV equipment status warning center to obtain equipment monitoring information timely. What's more, the existing condition monitoring system could not satisfy the demand of service support of the UHV equipment condition warning center, as in [1].

For the characteristics of the UHV DC converter station equipment state information, such as wide sources, Information heterogeneity, Huge quantity and various attributes, the paper put forward a

UHV DC converter station equipment monitoring system framework which met the timeliness requirements and a UHV DC converter station equipment condition data access scheme based on multi-source heterogeneous data fusion; It introduced the data exchange technology between the station and SGCC UHV equipment status warning center, and proposed multi-source cross platform data access interfaces and data access methods. The data access and conversion device has been deployed in the Zhongzhou converter station pilot to implement related data cross platform interactive sharing, such as charge detection, on-line monitoring and robot inspection. And it solved the problems which the existing system could not meet the requirements of UHV operation support in terms of timeliness and function.

2. UHV DC converter station equipment monitoring system framework

The project established a unified UHV equipment condition monitoring system which included the two level deployment, namely UHV converter station and SGCC headquarters, and three level application, namely SGCC headquarters, provincial company and provincial maintenance company; Also, the project established a standard open information technology framework. And it established two data sources, including the UHVDC converter station and the SGCC UHV equipment status warning center. The UHV equipment state warning center could obtain data directly from the UHV station, and did not need to go through the Production Management System (PMS) on the side of the net or SGCC headquarters. The UHV converter equipment condition monitoring system frame is shown in Figure1. The data access and conversion device is deployed in the internal of convertor station. The device accesses to online monitoring data through the Condition Access Controller (CAC), and it should follow Q/GDW 740-2011 Transformer Equipment On-line Monitoring I2 Interface Specification when communicating with the CAC, as in [2]. What's more, all data within access station, such as valve hall infrared temperature measurement data, fault data and meteorological data, should follow UHVDC Converter Station Equipment Condition Data Access Specification; After the data is collected, the device uses the Web Service interface service to push the data to the UHV equipment state warning center through the internal information network, as in [3] and [4]. The PMS on the side of the net provides data, such as Equipment, test, defects and adverse conditions, for UHV equipment status warning center by setting the database read-only access, as in [5].

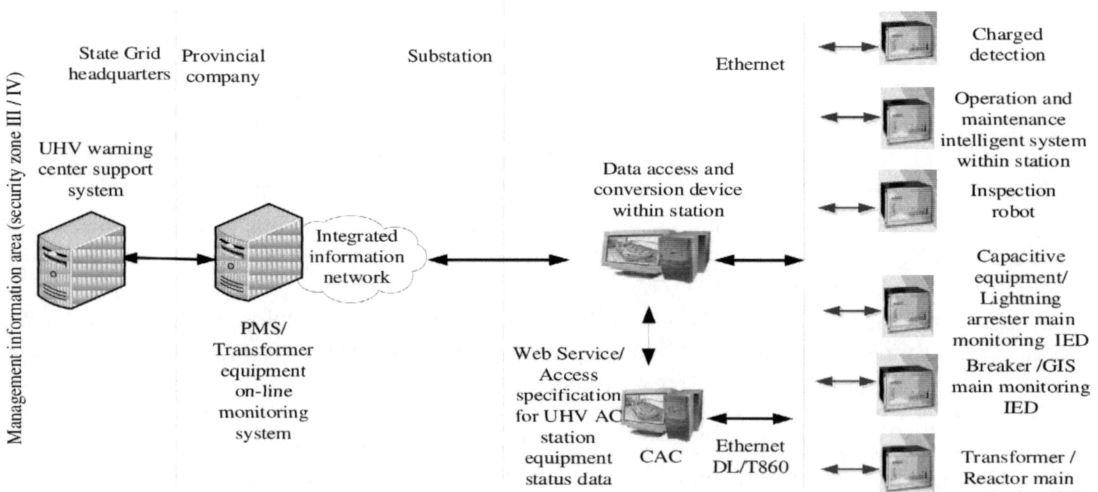

Figure 1. Equip condition monitoring system framework of UHV DC converter station.

3. UHV DC converter station transformer state data access scheme

3.1. The data access framework

The UHV DC converter station equipment condition data access framework is shown in Figure 2:

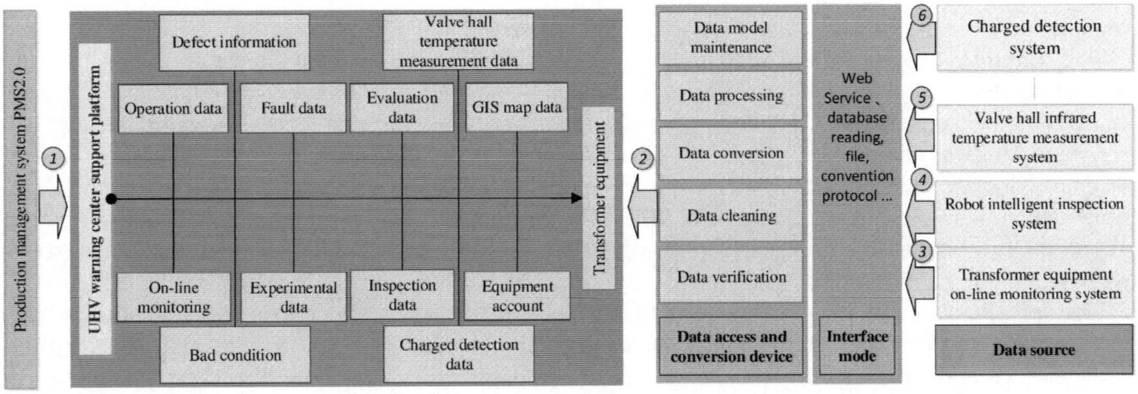

Figure 2. Framework of equipment condition data access for UHV DC converter station.

PMS2.0 syncs basic data, substation account, primary device account within station, defect information, fault information, adverse condition information, hidden information and test data to the UHV warning center support system;

Data access and conversion device within station syncs online monitoring data, robot inspection data, the valve hall infrared temperature measurement data and live test data to the UHV warning center support system;

Online monitoring system syncs data to the data access and conversion device within station, such as basic data and monitoring device account, dissolved gases in oil, top-oil temperature, top-oil temperature and SF6 gas water, as in [6];

Robot intelligent inspection system syncs inspection data to the data access and conversion device within station;

The valve hall infrared temperature measurement system syncs data, for example, infrared temperature measurement data, to data access and conversion device within station;

Live test system syncs data, such as oil chromatogram, partial discharge, infrared and ultraviolet to data access and conversion device within station.

3.2. Methods of obtaining data in the station

The main data sources of the UHV converter station mainly included online monitoring system, electric testing system and the inspection robot system. The specific data sources and data types are shown in Table 1. In UHV DC converter station equipment condition data access and conversion device, access station used three ways to get data, namely the database direct reading, file import and Enterprise Service Bus (ESB) , as in [7].

Table 1. Data types within station

Data source	Data type
On-line monitoring	transformer equipment condition monitoring parameter, hydraulic oil temperature, oil chromatogram, partial discharge, grounding current, casing capacitor dielectric loss, OLTC condition, vibration spectrum, etc.; GIS equipment condition monitoring parameters, mainly including partial discharge, gas density and water content, gas leakage, gas composition, circuit breaker operating mechanism and the schedule, etc.
Charged detection	infrared temperature measurement, lightning arrester charged testing, ultrasonic/UHF partial discharge testing, SF6 micro water and decomposition product testing, etc.

Inspection robot	visible light image and infrared heat maps, infrared measuring temperature, ultraviolet spectrum, etc.

4. The function design of UHV DC converter station equipment condition data access and conversion device

The UHV DC converter station equipment condition data access and conversion device was an integrated hardware device integrating the function of related software. The input of the device was the data associated with device status, such as on-line monitoring, live testing, robot inspection and valve hall temperature measurement within the UHV station. Through a series of data processing, data become believable and could be displayed, while meeting the requirements of data storage and data analysis, as in [8].

The specific functions of the device included data extracting, data transformation, data loading, data output, etc. For internal information network in the business system data (such as online monitoring data, the inspection data, charged detection data, etc.), the system directly got data pushed by business system. For business system and data (such as weather information, lightning monitoring data, etc.) deployed in the external information network, it carried out data transmission under the aegis of the isolating device, as in [9] and [10]. Functional process framework is shown in Figure 3:

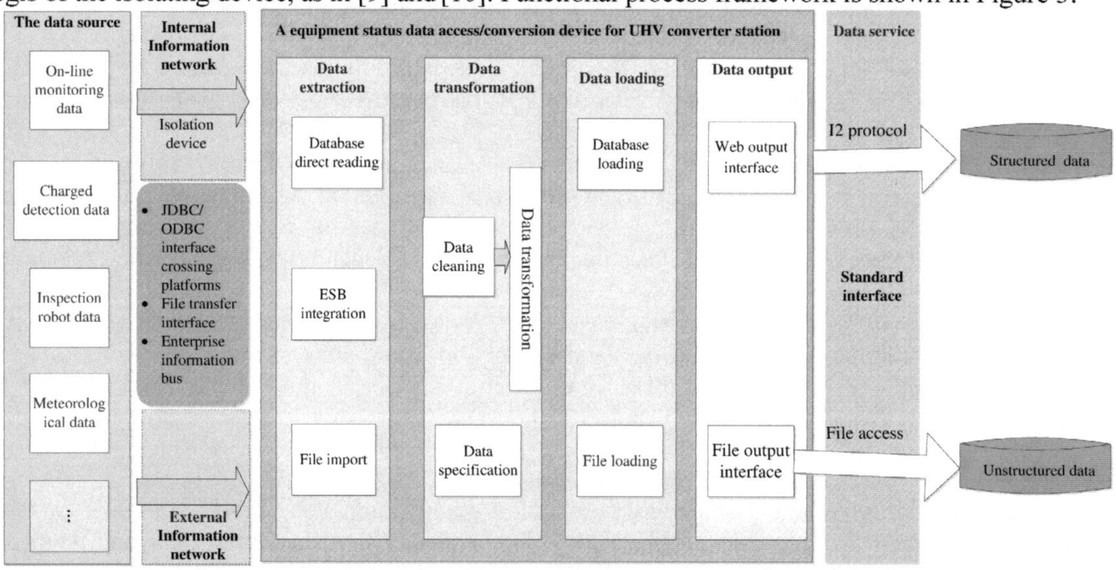

Figure 3. Functional flow diagram of data access and conversion device.

5. Conclusion

Because UHV equipment status information was scattered, and the existing system was difficult to meet the requirements of UHV operation support system in terms of timeliness and function, this paper carried out technical research on UHV transformer equipment state monitoring system architecture and equipment status management, control and early warning.

This paper established UHV transformer equipment warning center support platform based on multi-source heterogeneous data. Finally, this paper implemented the functions of equipment condition data access and processing, information interaction and control in the information island within UHV station, and achieved the functions of UHV substation equipment state centralized monitoring, condition evaluation and analysis and diagnosis, and forecasting and early warning to ensure the safe and stable operation of UHV.

The paper carried out in-depth research and beneficial exploration on the UHV equipment condition monitoring system architecture and UHV DC converter station equipment condition data access and conversion technology.

- The paper put forward a UHV DC converter station equipment monitoring system framework which met the timeliness requirements, and a UHV DC converter station equipment condition data access scheme based on multi-source heterogeneous data fusion.
- The paper carried out the data exchange technology research between the UHV DC converter station and the SGCC UHV equipment status warning center; proposed multi-source cross platform data access interfaces and data access methods; designed the UHV DC converter station equipment condition data access and conversion device function frame. And the device has been deployed in the Zhongzhou converter station pilot to implement related data cross platform interactive sharing. And it solved the problems which the existing system could not meet the requirements of UHV operation support in terms of timeliness and function.

Acknowledgement
This work is in part supported by the National High-tech R&D Program of China (863 Program) (No.2015AA050204) and Science and Technology Program of State Grid Corporation of China (No.GY71-15-045).

Reference

[1] Dong X 2004 Key Problems and Solution of Distributed Data Integrating and Sharing [PhD Dissertation] Jilin University.
[2] Q/GDW 740-2011 Transformer Equipment On-line Monitoring I2 Interface Specification Beijing: State Grid Corporation of China 2012.
[3] Xu S 2015 Research of Substation Information Integration on Based on SOA [PhD Dissertation] Kunming University of Science and Technology.
[4] Zhou S, Jin W, Li W 2014 Study on Data Access Method of New Real-time/historical Database Platform *Zhejiang Electric Power* **02** 55-58.
[5] Xie S, Yang Q, Wang B, Zhou G 2014 Design and implementation of open information platform for transmission and transformation equipment condition monitoring *Power System Protection and Control* **23** 125-130.
[6] Chen X, Hu G, Shen X, Chen W 2015 Realization of Monitoring Innovation of Auxiliary Equipment Integration of ±800 kV EHV Converter Station *Power System and Clean Energy* **05** 54-57.
[7] Zheng Y, Zhou Y, Xiao B, et al. 2015 Research on the Online Monitoring Data Access Technology for Intelligent Substation *Qinghai Electric Power* **02** 47-50.
[8] He F, Zheng H, Zhang Y 2013 Design of Intelligent Substation Online Monitoring System Based on Internet of Things Technology *Instrument Technique and Sensor* **6** 64-66.
[9] Ren X, Tang X, Sun W, et al. 2012 Research and Application of Substation On-line Monitoring and Auxiliary Equipment Monitoring System *Jiangsu Electrical Engineering* **28** **(6)** 31-34.
[10] Zhou S, Tao M 2012 Study of Universal Access Method in Real-time/historical Database Platform *Zhejiang Electric Power* **12** 94-98.

Response characteristic of high-speed on/off valve with double voltage driving circuit

P X Li[1], M Su[2], D B Zhang[1]

[1]School of Mechanical Engineering, Guizhou University, Guiyang Guizhou, China
[2]School of Big Data and Computer Science, Guizhou Normal University, Guiyang Guizhou, China

Abstract. High-speed on/off valve, an important part of turbocharging system, its quick response has a direct impact on the turbocharger pressure cycle. The methods of improving the response characteristic of high speed on/off valve include increasing the magnetic force of armature and the voltage, decreasing the mass and current of coil. The less coil number of turns, the solenoid force is smaller. The special armature structure and the magnetic material will raise cost. In this paper a new scheme of double voltage driving circuit is investigated, in which the original driving circuit of high-speed on/off valve is replaced by double voltage driving circuit. The detailed theoretical analysis and simulations were carried out on the double voltage driving circuit, it showed that the switching time and delay time of the valve respectively are 3.3ms, 5.3ms, 1.9ms and 1.8ms. When it is driven by the double voltage driving circuit, the switching time and delay time of this valve are reduced, optimizing its response characteristic. By the comparison related factors (such as duty cycle or working frequency) about influences on response characteristic, the superior of double voltage driving circuit has been further confirmed.

1. Introduction

High-speed on/off valve, an important part of turbocharging system, Due to the high speed on/off valve could influence the pressure ratio of the turbocharger, the actuator links to the control port of valve. Times delay in response of the valve affect the leakage of the exhaust gas, and then affecting the stability of the pressure ratio. In order to improve the response characteristic, a number of studies have been done. In Ref.[1], the soft magnetic material of high specific resistance. Fe-Cr alloy, is used to replace original low carbon steel and a significant improvement on control response was achieved. In Ref.[2], presenting an optimal design method on the HSV's magnetic field by making a full consideration of the effects of various soft magnetic material's properties and geometries on the HSV's electronic performance. In Ref.[3], a method for investigating the armature mass importance on the EFI performance is proposed. It was found a range of values in which the response characteristic can be improved. In Refs.[4,5,6],the giant material is used to make the electromechanical converter and the switching time of the high-speed on/off valve is shortened. In Ref.[7], a parallel coil is presented to improve the dynamic performance of high-speed on/off valve. In Ref.[8], the influence of the mass of valve core, spring stiffness, spring preload and coil number of turns on solenoid valve responses are analyzed. In Ref.[9], simulation research on high-speed solenoid Valve, the authors found out that the factors that affect the closing and opening response characteristics of solenoid valve are drive current and the pretightening force of solenoid valve spring. These moves can improve the response characteristic but there are still many limitations. For instance, the less coil number of turns, the solenoid force is smaller. The special armature structure and the magnetic material will raise cost. The pretightening force increasing, the closing time of solenoid valve becomes longer while the opening time is shortened.

In this paper, a new and effective method of shorting the times delay in response of high-speed on/off valve is proposed. The original driving circuit of this valve is replaced by double voltage driving circuit. To begin with, the mathematical model and simulation model of a high-speed on/off valve are established; next simulation and analysis between the different driving circuits are carried out to verify the improvements of response time by the double voltage driving circuit; finally, by studying the influences of related factors on response characteristic, the superior of the double voltage driving circuit is further confirmed.

2. Structure and working principle of the high-speed on/off valve

Fig 1 shows the structure of the high-speed on/off valve which is used in turbocharging system. In order to achieve the switch "on" and "off", by using PWM (Pulse-Width Modulation) controls the current of coil, the electromagnetic force is changed. The resultant of the electromagnetic force and the spring force then changes the position of the armature. Pushed by the armature, the motions of the valve core is then regulates the opening of this valve.

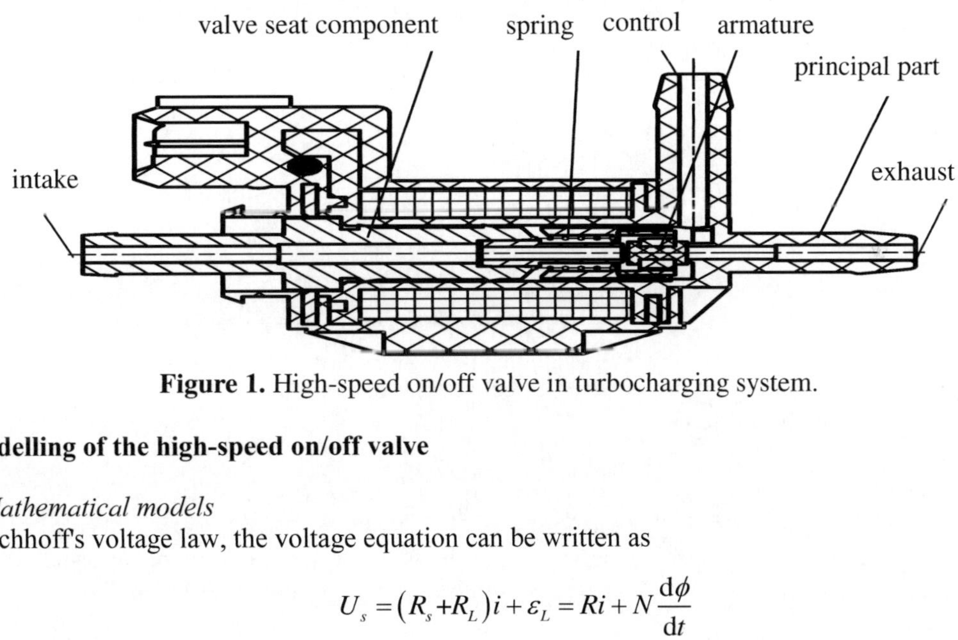

Figure 1. High-speed on/off valve in turbocharging system.

3. Modelling of the high-speed on/off valve

3.1. Mathematical models

By Kirchhoff's voltage law, the voltage equation can be written as

$$U_s = (R_s + R_L)i + \varepsilon_L = Ri + N\frac{\mathrm{d}\phi}{\mathrm{d}t} \tag{1}$$

Where i is current of the coil; ϕ is total magnetic flux; N is the number of coil turns; R is the resistance in coil circuit.

By magnetic circuit Kirchhoff's law

$$iN = \phi R_m \tag{2}$$

Where R_{m} is the total magnetic resistance, it consists of the reluctance of ferromagnetic material in magnetic circuit, solenoid valve work air gap reluctance and non-working air gap reluctance. For solenoid electromagnet, due to the magnetic field strength is mainly focused on the solenoid inside, so the magnetic circuit of magnetic potential are mainly concentrated in the working air gap, the non-working air gap and magnetic resistance of the yoke can be ignored [10]. Air gap magnetic resistance of magnetic circuit is a very important parameter, due to the magnetic resistance rate of the air is far outweigh than the magnetic resistance rate of ferromagnetic materials, so the air gap magnetic resistance calculation is accurate or not has highly affect to the accuracy of the calculation results of magnetic system.

The total magnetic resistance computation formula is as follows

$$R_m = \frac{\delta - x}{\mu_0 S} \tag{3}$$

Where δ is the air gap width of valve core and the core; x is the movement displacement of valve core. In particular, $\mu_0 = 4\pi \times 10^{-7}$ Vs / Am.

Substituting equation (3) into equation (2), It can be get

$$iN = \phi \frac{\delta - x}{\mu_0 S} \tag{4}$$

By Maxwell's electromagnetic suction formula, the electromagnetic force generated as a result of magnetic flux. And Tsinghua University's Jia-Li Qian, she obtained the expression of electromagnetic attraction by a large amount of calculation and analysis:

$$F_{mag} = \frac{1}{2} i \frac{\mathrm{d}\varphi}{\mathrm{d}\delta} \tag{5}$$

She pointed out that this electromagnetic force equation (5) is applicable for any occasion of the electromagnet.

In general:

$$\varphi = N\phi \tag{6}$$

φ is electromagnet flux.

Put the equation (4) into the equation (5), it can be get

$$F_{mag} = \frac{1}{2} i \frac{\mathrm{d}\varphi}{\mathrm{d}\delta} = \frac{iN}{2} \frac{\mathrm{d}\phi}{\mathrm{d}\delta} = \frac{\mu_0}{2} \frac{(iN)^2}{\delta^2} \tag{7}$$

Referring to Fig. 1 and applying Newton's second law of dynamic to the moving elements, the valve core equation can be written as

$$m_v \frac{\mathrm{d}^2 x_v}{\mathrm{d}t^2} + C_s \frac{\mathrm{d}x_v}{\mathrm{d}t} = F_{mag} - F_q - kx_v \tag{8}$$

Where F_{mag} is electromagnetic force, m_v is mass of the valve coil, including armature and pole; C_s is coefficient of viscous friction; F_q is aerodynamic force; k is spring coefficient; x_V is valve core displacement.

From the equation (7) and equation (8) can see the conclusion, when the structural parameter of the solenoid remain unchanged, if the current changes rapidly to a higher value, it can help to gain a larger force. The greater the changing rate of coil current is, the shorter the time for current value to reach a certain degree is, the shorter the time for electromagnet to provide a greater electromagnetic force is. Meanwhile, the higher the coil current increases, the greater the electromagnetic force is, the shorter the time for solenoid valve to close and the longer time for it to open.

In general, the factor of drive current affects the response characteristics of the high-speed on/off valve.

3.2. Dynamic model and simulation analysis

According to the structure and mathematical equations of the high-speed on/off valve, the dynamic model based on AMESim is built as showed in Fig 2 The electromagnetic force is changed by current of coil, and affects the responses of high-speed on/off valve. In the circuit system, the electric varies directly with the voltage. Voltage may thus be said to produce an effect to the electromagnetic force, affecting the response characteristic of this valve. In order to get the principle of response, the voltage is set by 13.5V and 25V.

Table1 shows that the main definitions and values of notations used in the model.

Based on the emulate analysis, the response characteristics of high-speed on/off valve in different driving voltages are obtained. In the parameter mode of the AMESim software, parameter settings are provided with reference to table 1. In simulation mode, the related graph curves are shown as shown in Fig 3, it shows the influences of different voltages on the valve response.

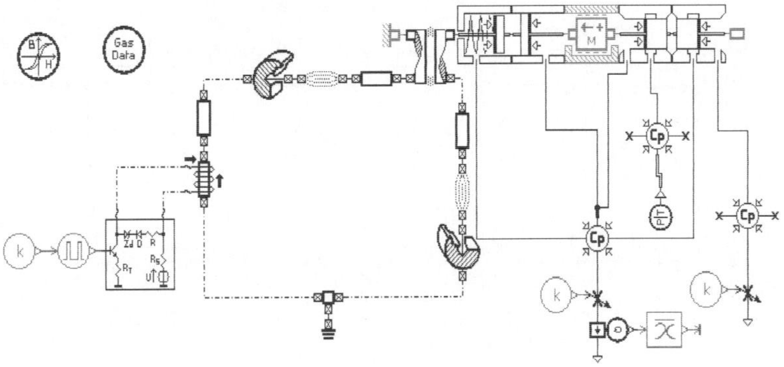

Figure 2. The simulation model of high-speed on/off valve.

Table 1. Simulation parameters

Parameter	number
voltage driving[V]	13.5（25）
gas pressure [kPa）	1.813
gas constant [J•kg/K]	287.1
gas temperature[K）	293.15
mass of valve coil[kg]	0.00175
coil turns[tr]	1460
frequency[Hz]	30
initial air gap[mm]	1.16
pole diameter[mm]	8
Spring preload [N]	0.34
spring coefficient[N/mm]	0.1446
chamber length at zero[mm]	2.35
electrical resistance[Ω]	23
duty ration	0.5

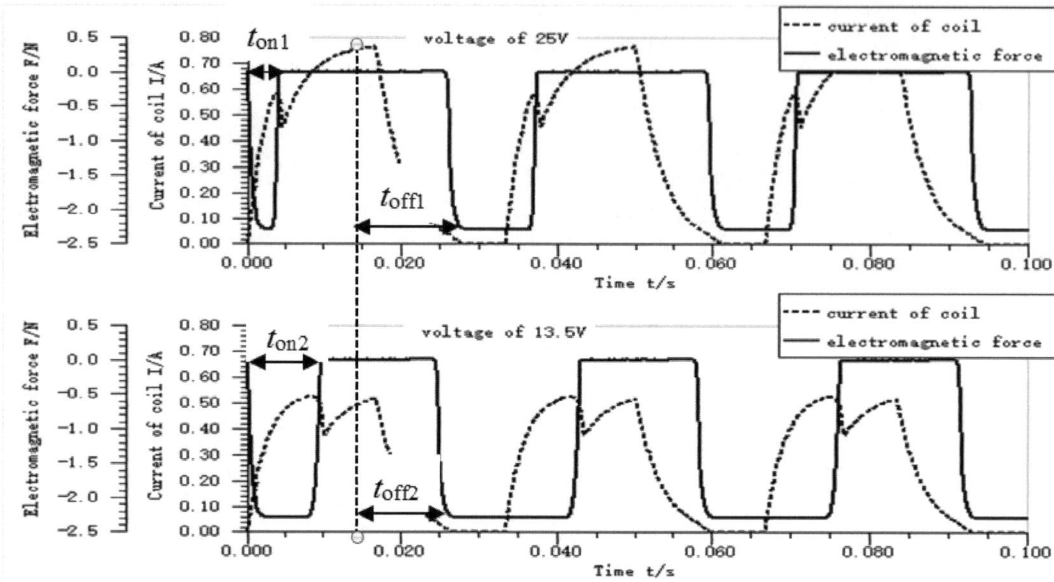

Figure 3. The influences of different voltages on the valve response.

Nomenclature:

t_{on}—opening time;

t_{off}—closing time.

From Fig 3, it is understood that the opening time of the high-speed on/off valve is reduced but the closing time is increased when the voltage is increased. This phenomenon occurs because of the variation of electromagnetic force and the maximum current. The facts as follow: during the process of power, when the voltage is set to 25V, the maximum electromagnetic force accomplished in shorter time than the voltage is 13.5V, then load force is overcome and the valve core is opened. But in the process of power failure, when the voltage is set to 25V, the current is greater than the voltage is 13.5V, and high current comes down to zero in more time. In order to improve the response characteristic, the voltage cannot simply be increased or reduced.

4. Driving circuit optimize

On the one hand, a high driving voltage is used in high-speed on/off valve which makes the acceleration of the high-speed on/off valve closing time and the reduced of opening time. On the other hand, using a low voltage has the opposite effect. In order to improve the response times, the driving circuit needs to be designed to have a characteristic of variable voltage output.

4.1. Double voltage driving circuit

Double voltage driving circuit is respectively using two constant working voltage of different sizes in different stages of the working process, in order to provide large driving current and small maintaining current to adjust the voltage modulation current.

In this paper, the scheme of double voltage driving circuit is applied to this high-speed on/off valve. Fig 4 shows the principle of double voltage driving circuit.

In this circuit diagram, the driving circuit consists of two switches, two diodes and a DWZ regulator. Two switches are identical IRFP150, two diodes are D1 and D2 respectively. PWM_1 and PWM_2 are control signals, respectively controlling two electronic switches 1 and 2; this measure is aimed to control the high voltage source and the on/off valve. When the PWM_1 signal is high, the PWM_2 signal is low, the switch 1 is closed, the switch 2 is disconnected, and the PWM driving voltage is driving voltage is U_1 with U_2, after through R_1, R_2, and then getting the voltage of the is higher. But when the PWM_1 signal is low, the PWM_2 signal is high, the switch 1 is disconnected, the switch 2 is closed, and only the low voltage is operated at this time, so the driving voltage is only U_2. D2, DWZ and RZ constitute the high speed switch valve release circuit.

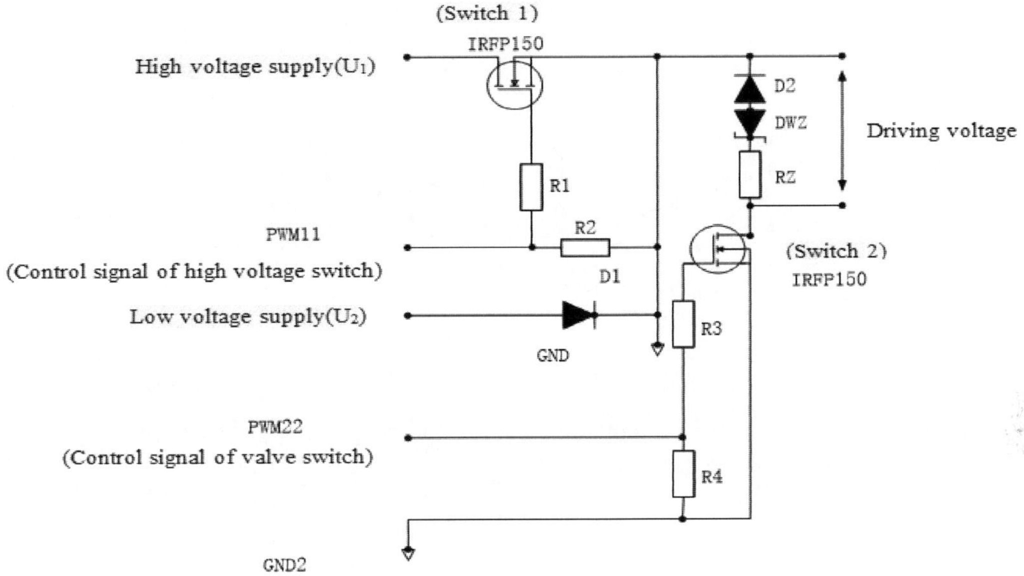

Figure 4. Principle of double voltage driving circuit.

4.2. Simulation analysis
Two sets of simulations were carried out for the original driving circuit scheme and the double voltage driving circuit scheme. Fig 5 shows the simulation model of high-speed on/off valve on the double voltage driving, the driving voltage is set to 25V and 5V, high voltage is a driving in short time, set the duty ration of PWM_1 to 10% and the duty ration of PWM_2 to 50%(adjustable). Fig7 shows the voltage and current of coil in the double voltage driving circuit. It can be learned from Fig 6 that high-voltage load, current of coil is increased rapidly to 0.636A within 2.6ms; next, when the high-voltage closed but low-voltage load, the current of coil is reduced to 0.22A and keeps a steady; finally, when the low-voltage closed, the current is rapidly reduced to zero.

The original circuit is set to 25V or 13.5V. The simulation results of the influences of different driver circuits on the valve response are shows in Fig 7 Table 2 shows the comparison of the different response times under different driver circuits of high-speed on/off valve.

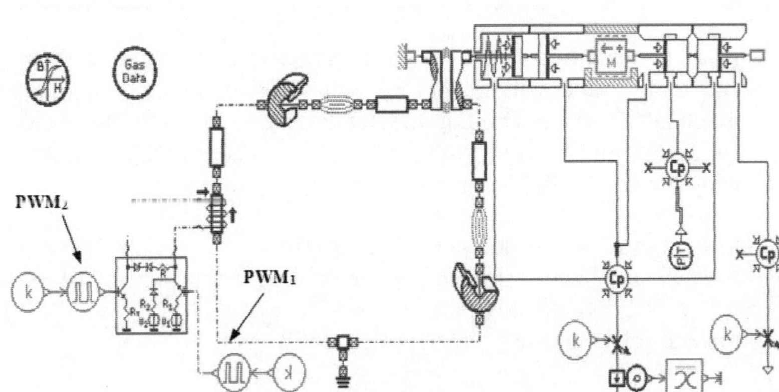

Figure 5. The simulation model of high-speed on/off valve on the double voltage driving.

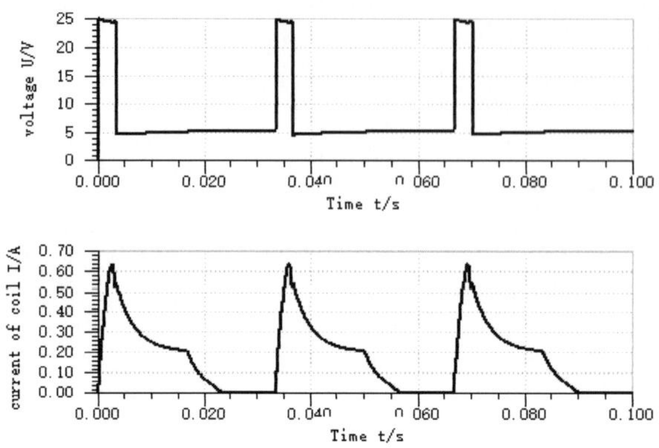

Figure 6. Double voltage driving circuit.

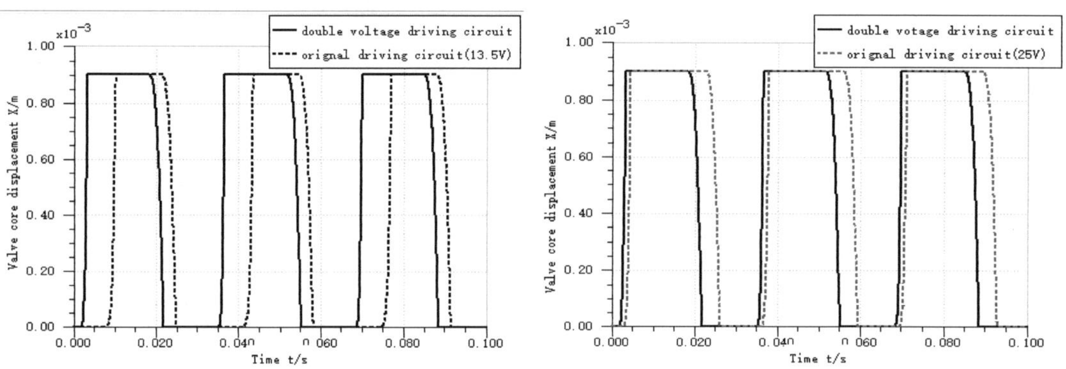

Figure 7. The influences of different driver circuits on the valve response.

Table 2. Comparison of the response times

parameter	Switching time(ms)		Delay time(ms)	
	Opening	Closing	Opening	Closing
Double voltage driving circuit	3.3	5.3	1.9	1.8
Original driving circuit(13.5 V)	10	8.2	7.7	4.6
Original driving circuit(25 V)	4.6	9.5	3	5.9

From Table 2, it can be seen that the delay times and switching times of the high-speed on/off valve is reduced by using the double voltage driving circuit. This is mainly due to the high voltage can make the current to rise faster at the beginning, the better initial condition of the closing of valve core will be get by using a low maintaining voltage after the valve totally opened. It can avoid the drawbacks of only using the high voltage or the low voltage driving.

5. Influences of related factors on response characteristic

In order to further study the superior of double voltage driving circuit, under the existing condition of the high-speed on/off valve, by the comparison related factors (such as duty cycle or working frequency) about influences on response characteristic.

5.1. Influence of duty ratio on response characteristic

The variation of duty cycle restricts the movement of valve core in some extent, Fig 8 shows the influence of different driver circuits on the valve response characteristic when changes the duty cycle.

Both of them are driven by a working frequency of 30HZ. Different driving circuits are the original driving circuit of 13.5V and the double voltage driving circuit, duty cycles respectively are 100%, 80%, 60%, 50%, 30% and 10%.

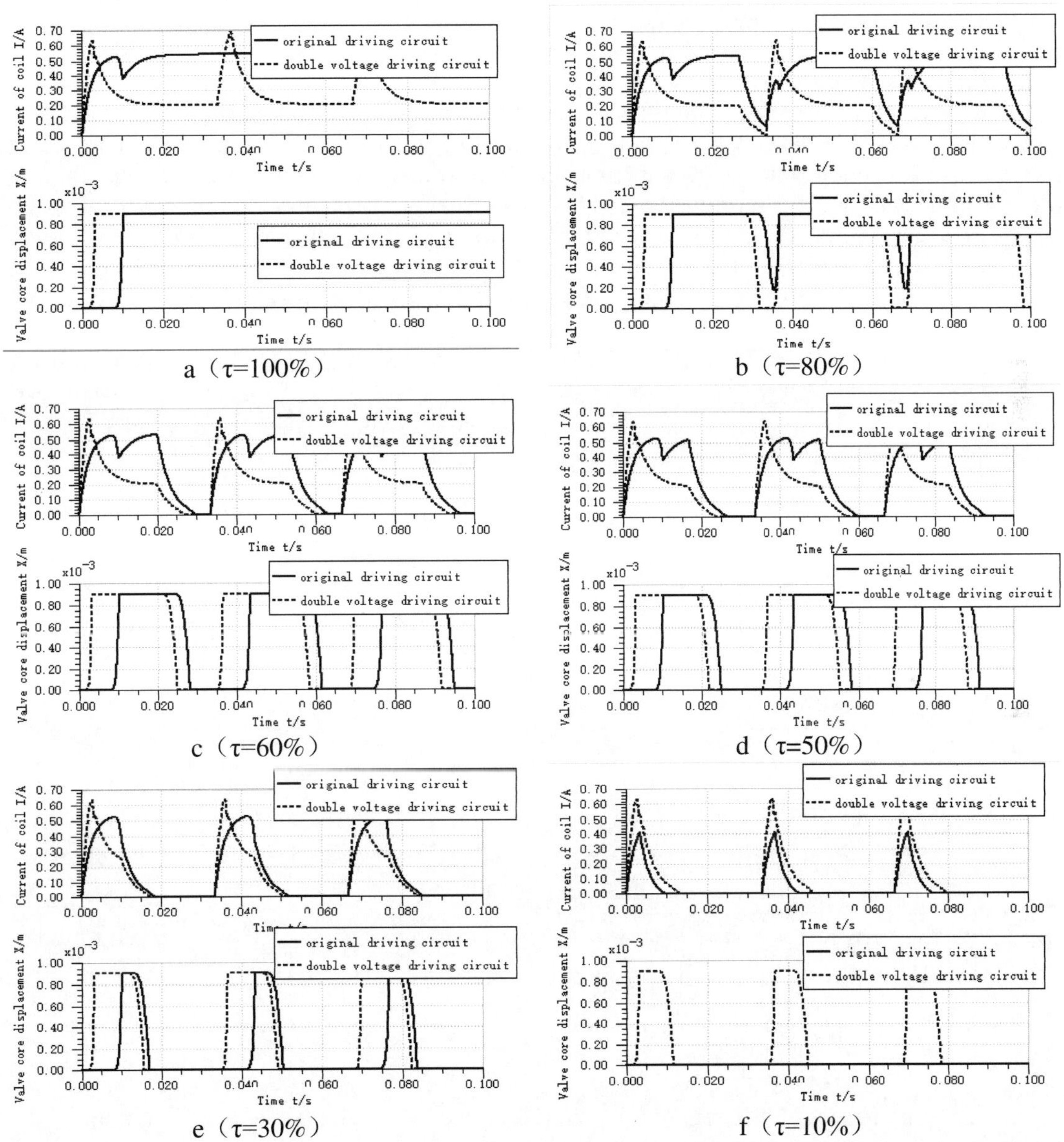

Figure 8. The influences of different driving circuits on the valve response when changes the duty cycle.

From Fig 8, the switching times and the movement process of the high-speed on/off valve are changed by the duty ration. The following results are obtained: (1) At the same duty ration, the high-speed on/off valve closing time and opening time are reduced when the valve is driven by the double voltage driving circuit, the response performance of the high-speed on-off valve is more superior. (2) At the small duty ratio (such as τ=10%), when the valve is driven by the original circuit, the electromagnetic force cannot overcome the load force, and the spool cannot be turned on. But the valve has the normal process of opening and closing because the valve is driven by the high voltage during the double voltage driving circuit. (3) At the big duty ratio (such as τ=80%), when the valve is driven by the original circuit, the valve has not the process of completely closing because the PWM pulse rising edge has come but the current of coil has not dropped to zero. But the valve has the normal process of opening and closing because the valve is driven by the low maintaining voltage during the double voltage driving circuit.

5.2. Influence of frequency on response characteristic

The PWM signal of the high-speed on/off valve has better function of reproduction. Due to a delay of response time of high-speed on/off valve, the working frequency of PWM signal cannot be selected very high. Using a double voltage driving circuit can affect the response time to a certain extent, Fig 9 shows the influences of different driver circuits on the valve response characteristic when changes the working frequency. Both of them are driven by the 50%'s PWM_2 duty cycle and the 10%'s PWM_1 duty cycle. Different driver circuits are the original driving circuit of 13.5V and the double voltage driving circuit, they are driven by a working frequency of 20HZ, 30HZ, 50HZ and 70Hz, respectively.

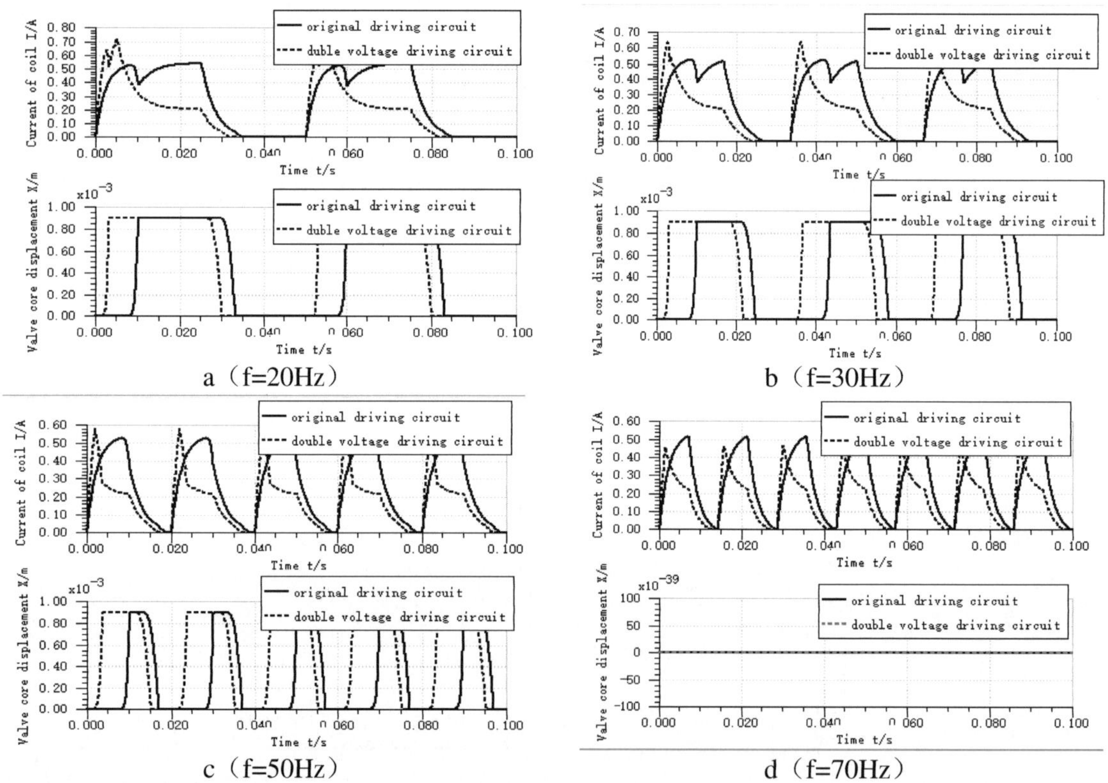

Figure 9. The influences of different driving circuits on the valve response when changes the working frequency.

From Fig 9, the switching times, delay times and the movement process of the high-speed on/off valve are changed by the different working frequency. The following results are obtained: (1) when the working frequency is increased, the dynamic characteristics of the electromagnetic force need to be improved, that makes the response characteristics of the high-speed on/off valve become poor. The main reason is that the working frequency is too much, that causing the current rise cannot make the magnetic field into the saturated state. Controlling the working frequency in a certain range is necessary. (2) At the same working frequency, the high-speed on/off valve closing time and opening time are reduced when the valve is driven by the double voltage driving circuit, and the response performance of the high-speed on-off valve is more superior. (3) Double voltage driving circuit improves the response characteristics of this valve, the higher the working frequency, the greater the effect (such as f=50Hz). (4) When the high-speed on/off valve is driven by working frequency of 30 Hz, it has a better response characteristic. Therefore, 30HZ is always taken as working frequency in this valve.

6. Summary

In this paper, based on the analysis of limitations for the response characteristic of high-speed on/off valve, the double voltage driving circuit scheme is put forward and investigated by theoretical analysis and simulation. The comparison of the simulating results confirms the superior of the scheme. The following conclusions were reached.

(1) When the high speed on/off valve is driven by the double voltage driving circuit, the switching times and delay times of this valve are reduced, optimizing its response characteristics.
(2) The current of coil and the valve core displacement of the high-speed on/off valve are changed by the duty ration.
(3) At the same duty ratio, the high-speed on/off valve is driven by double voltage driving circuit that has advantage in the movement process than original driving circuit.
(4) When the working frequency is increased, the dynamic characteristics of the electromagnetic force need to be improved, that make the response characteristics of the high-speed on/off valve become poor.
(5) Double voltage driving circuit improves the response characteristics of the valve, the higher the working frequency, the greater the effect.

References

[1] Yang M, et al. 1998 Study of the injection control valve in a new electronic diesel fuel system (No. 980813) *SAE Technical Paper.*
[2] Tao G, Chen H Y, He Z B 2002 Optimal design of the magnetic field of a high-speed response solenoid valve *Journal of Materials Processing Technology* **129(1)** 555-558.
[3] Passarini L C, Nakajima P R 2003 Development of a high-speed solenoid valve: an investigation of the importance of the armature mass on the dynamic response *Journal of the Brazilian Society of Mechanical Sciences and Engineering* **25(4)** 329-335.
[4] Shi C, Liu Y, Zhang Y 2004 Design and study of a new kind of larger flow rate high-speed on-off valve *Chinese Journal of Mechanical Engineering* **4** 040.
[5] Shi Y P, Zhang Y Z, Liu C W 2004 Study and design of a high-speed on-off valve taken rare earth alloy material as actuator *Journal-Nanjing University of Science and Technology* **28** 385-389.
[6] Meng A 2007 Design of Pulsed Jet On-Off Valve Based on Giant Magnetostrictive Actuator *2007 First International Conference on Integration and Commercialization of Micro and Nano systems* American Society of Mechanical Engineers.
[7] Kong X, Li S 2014 Dynamic performance of high speed solenoid valve with parallel coils *Chinese Journal of Mechanical Engineering* **27(4)** 816-821.
[8] Li S, Guo P, Jiang W, et al. 2015 Research on response characteristics and parameters optimization of high-speed solenoid valve *Control Conference(CCC), 2015 34th Chinese. IEEE.*
[9] Zhang H, Zhang Y 2010 Simulation Research on Dynamic Response Characteristics of High-Speed Solenoid Valve *2010 International Conference on Future Information Technology and Computing.*
[10] Tao R, Zhang H, Fu D, et al. 2010 Simulation of ABS hydraulic system and optimization of solenoid valve *Transactions of the Chinese Society of Agricultural Engineering* **26(3)** 135-139.

IOP Conf. Series: Materials Science and Engineering **220** (2017) 012029 doi:10.1088/1757-899X/220/1/012029

Performance analysis of 'Perturb and Observe' and 'Incremental Conductance' MPPT algorithms for PV system

Ehtisham Lodhi[1*], Zeeshan Lodhi[2], Rana Noman Shafqat[1] and Fieda Chen[1]

[1]School of Electrical Engineering, Beihang University, Beijing 100191, China
[2]Department of Electrical Engineering, CIIT, Abbottabad, 22060, Pakistan

*Email: ehtishamlodhi1@hotmail.com

Abstract. Photovoltaic (PV) system usually employed The Maximum power point tracking (MPPT) techniques for increasing its efficiency. The performance of the PV system perhaps boosts by controlling at its apex point of power, in this way maximal power can be given to load. The proficiency of a PV system usually depends upon irradiance, temperature and array architecture. PV array shows a non-linear style for V-I curve and maximal power point on V-P curve also varies with changing environmental conditions. MPPT methods grantees that a PV module is regulated at reference voltage and to produce entire usage of the maximal output power. This paper gives analysis between two widely employed Perturb and Observe (P&O) and Incremental Conductance (INC) MPPT techniques. Their performance is evaluated and compared through theoretical analysis and digital simulation on the basis of response time and efficiency under varying irradiance and temperature condition using Matlab/Simulink.

1. Introduction

Renewable Energy (RE) is the best choice among all resources to fulfil the plea of energy [1]. The (RE) sources compromises of solar energy, wind energy, Tidal energy and etc. These sources going to be famous due to these advantages e.g. reliable, easy to available, friendly to environmental, has long life, has low maintenance cost and pollution free [2-3]. The energy from sun is a vital source of energy for generation of electricity [4]. Solar energy is assumed to be safe, favourable and long lasting source of energy. The (PV) Module PV array shows a non-linear style for V-I curve and maximal power point on V-P curve also varies with changing environmental conditions. The efficiency of PV module is in the range (10-25) % [5]. To maximize their proficiency and decrease the cost of modules, (MPPT) algorithms are connected with overall system [6-9]. PV system usually employed The (MPPT) techniques to increase its efficiency under these varying conditions. The performance of the PV system perhaps boosts by controlling at its apex point of power, in this way maximal power can be given to load [10-14]. The (MPPT) approaches are employed to maximize the power of the system. The commonly used methods are Hill climbing, Incremental Conductance (INC), Perturb and Observation (P&O), Constant Current (CI) and Constant Voltage (CV) [15-20]. This paper gives analysis between two commonly employed Perturb and Observe (P&O) and Incremental Conductance (INC) MPPT techniques. The performance of these two methods are evaluated and compared through theoretical analysis and digital simulation under varying irradiance and temperature condition using Matlab/Simulink.

2. Modelling of a solar cell

The solar cell is usually made up of crystalline silicon (Si) material, which conducts electricity when sunlight is made to fall on solar cell and converts radiations from sun to electrical energy. The proportionate model of a solar cell is presented in Figure 1. The overall equation for a single diode PV cell is given by equation (1-2) [5].

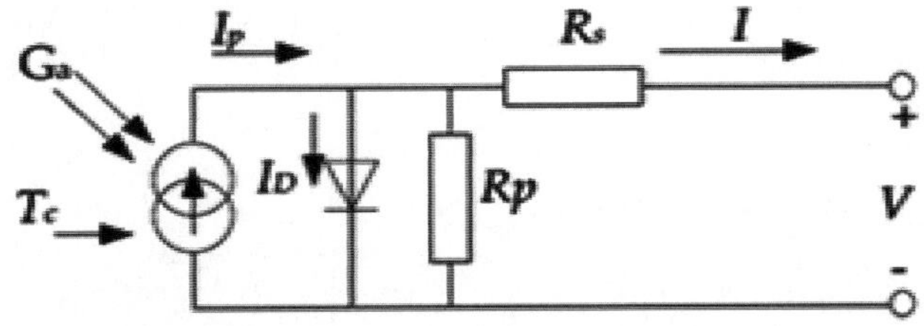

Figure 1. Single diode model of solar Cell [5].

$$I = Ip - ID[\exp(qV / aKTc) - 1] \quad (1)$$

$$I = Ip - ID[\exp(V + Rs / Vta)] - (V + RsI / Rp) \quad (2)$$

Where I_p is the current induced due to incident of light, I_D is the current due to reverse saturation, q is the Charge on electron, K is the Boltzmann constant, T_c is the temperature for p-n junction, V_t is the thermal Voltage for PV array, R_s is the Series Resistance, R_p is the Parallel Resistance. BP MSX 120 is employed in modelling of PV module. Electrical specifications of this module are shown in Table 1.

Table 1. Specification of the BP MSX 120 panel at 1000 W/m2 irradiance and 25 °C temperature

Parameters	1 Module	12 KW array
Iph	3.8713 A	38.71 A
ID	0.323 μA	3.23 μA
A	1.3977	1.3977
Rs	0.473 Ω	0.18 Ω
Rsh	1367 Ω	520

3. Characteristics of PV cell

PV array shows a non-linear style for V-I curve and maximal power point on V-P curve also varies with changing environmental conditions as shown in Figure 2. The changing of the temperature has a great effect to the array's output voltage as compared to output. The power-voltage curve which shows that as irradiance decreases the MPP of the PV module also going to be decreased.

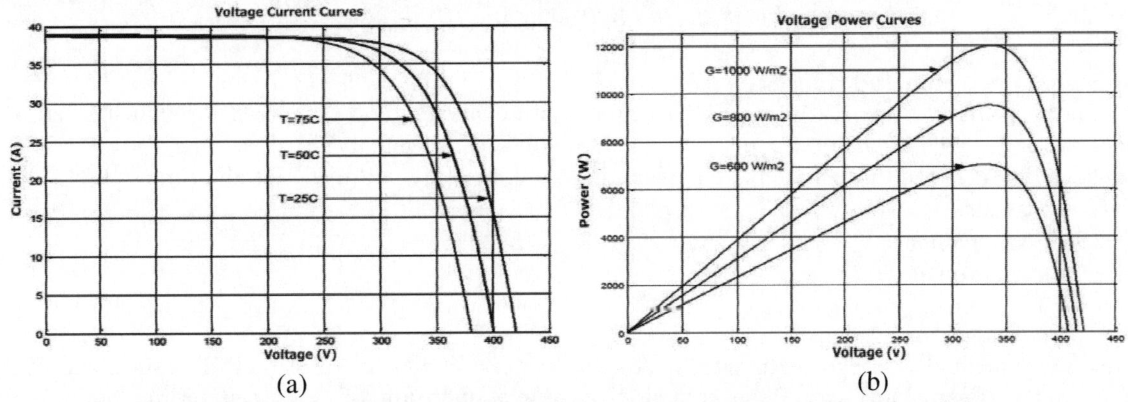

(a) (b)

Figure 2. (a) V-I features of PV module; (b) V-P features of the PV module.

4. Perturb and Observe (P&O) algorithm

P&O is frequently used (MPPT) approach for most of PV systems [6-7]. This approach regularly adds or subtracts the assigned voltage value back on checking the last power pattern value. The P&O is primitive technique that has less cost and its implementation is so much easy. Drawback is that some time its operating point is fluctuating about MPP. As a result, some power will be lost. (P&O) algorithm flow chart is described in detail in Figure 3.

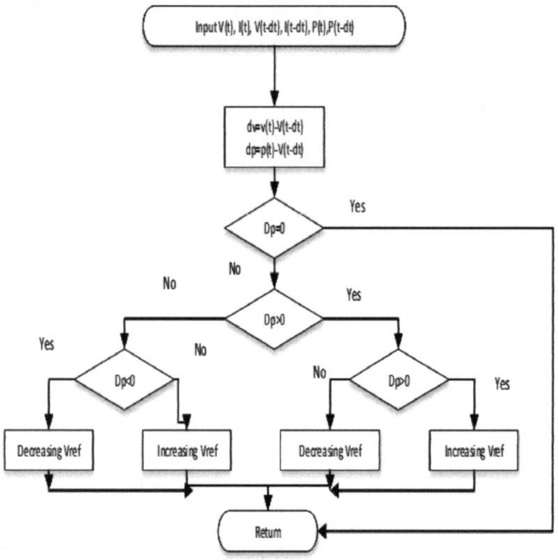

Figure 3. P&O algorithm Flow Chart.

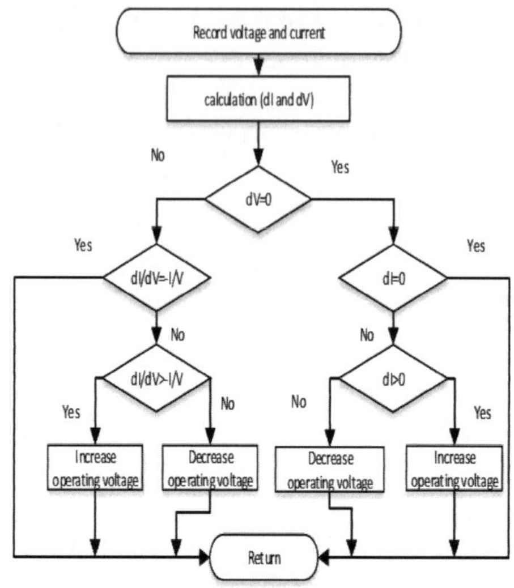

Figure 4. INC algorithm Flow Chart.

5. Incremental conductance (INC) Algorithm

INC algorithm is also very common MPPT algorithm for many PV system [4]. This approach can be extracted by differentiating two quantities i.e. power against voltage and keeping the resultant value equal to zero as shown in equation (3).

$$dP / dV = d(IV) / dV = I + V \cdot dI / dV = 0 \quad \text{on MPP} \tag{3}$$

When the controlling point is not lie on MPP and going to be changed, than that condition will be show one of these equations that are given below (3-5). When MPP arrived, INC method try to keep at that point as long as fluctuation in atmospheric condition is occurred. Figure 4 presents a flowchart of INC algorithm.

$$dI / dV = -I / V, (dP / dV = 0) \quad \text{on MPP} \tag{4}$$

$$dI / dV > -I / V, (dP / dV > 0) \quad \text{on left of MPP} \tag{5}$$

$$dI / dV < -I / V, (dP / dV < 0) \quad \text{on right of MPP} \tag{6}$$

6. Simulation result and Analysis

The performance of two (P&O and INC) methods are evaluated and compared through theoretical analysis and digital simulation under two conditions by using Matlab/Simulink. The PV module is connected to the load through converter which in turn control by two different algorithms: P&O and INC, one by one.

6.1. Steady Condition
6.2. Dynamic Condition

6.1. Steady condition

First, these methods are evaluated under steady weather conditions i.e. 1000W/m^2 irradiance and 25C temperature. Figure 5(a) shows the comparative simulation result of (P&O) algorithm for power achieved by PV array and boost converter. Figure 5(b) shows the comparative simulation result of (INC) algorithm for Power achieved by PV module and boost converter. The output of the PV module is represented by the blue line while orange line is used to represent the output of the boost converter. It can be seen from fig. 10 that INC method has good and less response time and it has nearly minor oscillations in power around MPP. On the other side, P&O method requires high response time and it

IOP Conf. Series: Materials Science and Engineering **220** (2017) 012029 doi:10.1088/1757-899X/220/1/012029

has immense oscillations in power at MPP. So, for uniform conditions INC shows best result than P&O.

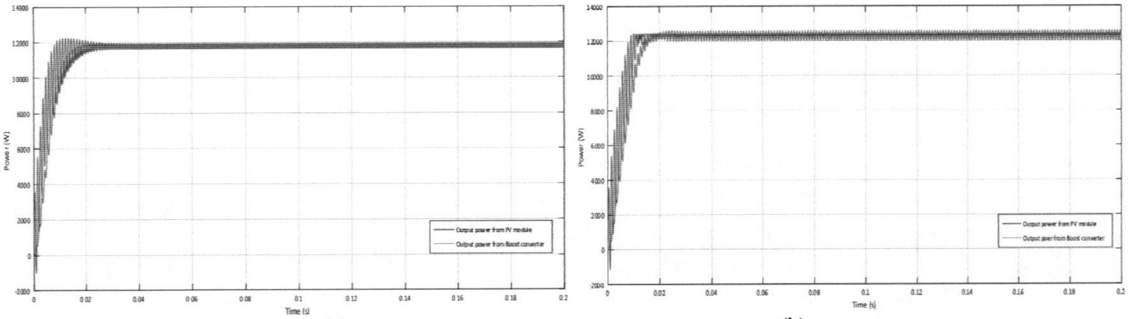

(a) (b)

Figure 5. Steady condition simulation result (a)P&O MPPT algorithm; (b) INC MPPT algorithm.

6.2. Dynamic condition

In this section of simulation, (P&O and INC) MPPT methods are analysed and tested under dynamic weather conditions. Figure 6(a) shows the comparative simulation result of (P&O) algorithm for power achieved by PV array and boost converter under dynamic condition. Figure 6(b) shows the comparative simulation result of (INC) algorithm for Power achieved by PV module and boost converter under dynamic condition. The output of the PV module is represented by the blue line while orange line is used to represent the output of the boost converter. This shows that the PV module maximum power declines, when the irradiance decreases and also current declines significantly when the irradiance decreases. So from above discussion, we can say that the PV array MPP varies with variations in shading, temperature and radiation. It has been seen that the PV system output power boosts with rise in amount of solar irradiance and drops in cell temperature. Consequently, the proficiency of the PV cell is better in winter than summer season.

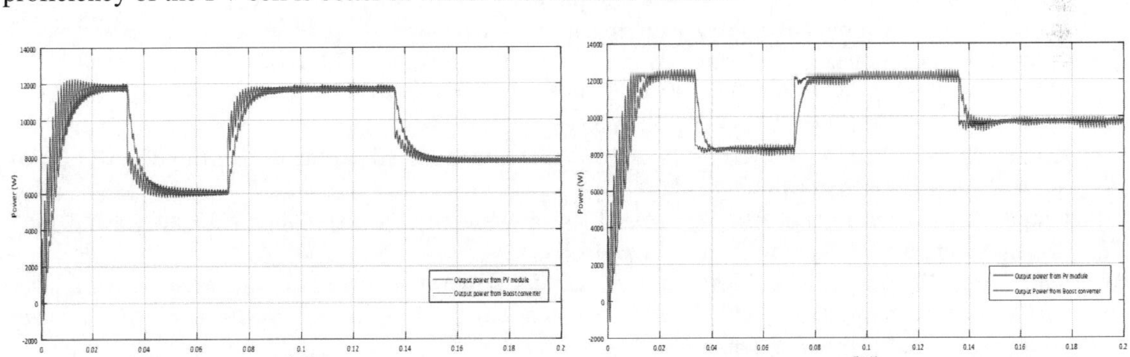

(a) (b)

Figure 6. Dynamic condition simulation result (a) P&O algorithm (b) INC algorithm.

INC algorithm has high response speed and can be controlled accurately. However, it has complex hardware design and needs precise sensors. The features of this method huge response speed, working under rapid changes in atmospheric conditions. On the other hand, P&O methodology is very simple, but its operating point is fluctuating about MPP so as a result some power will be lost. Additionally, its exhibits less accuracy and response is also slow.

7. Conclusion

This paper gives the detailed analysis about two widely employed MPPT algorithms under uniform and non-uniform conditions of temperature and irradiance in the Matlab/Simulink background. These two MPPT methods are compared on the basis of efficiency and response time. The INC shows the best proficiency among these two techniques. It exhibits the perfect efficiency and response time in steady as well as in dynamic conditions of weather. So, INC algorithm is terrific as compared to P&O.

References

[1] Khatib T, Mohamed A and Sopian K 2012 A review of solar energy modelling methods *Renew. Sustain. Energy Rev.* **16** 2864–2869.

[2] Mekhilef S, Saidur R and Safari A 2010 A review on solar energy used in industries *Renew. Sustain. Energy Rev.* **15** 1777–1790.

[3] Kapoor K, Pandey K K, Jain A K and Nandan A 2014 Evolution of solar energy in India: a review *Renew. Sustain. Energy Rev.* **40** 475–487.

[4] Solangi K H, Islam M R, Saidur R, Rahim N A and Fayaz S 2011 A review on solar energy policy *Renew. Sustain. Energy Rev.* **15** 2149–2163.

[5] Parida B, Iniyan S and Goic R 2011 A review of solar photovoltaic technologies *Renew. Sustain. Energy Rev.* **15** 1625–1636.

[6] Bialasiewicz J T 2008 Renewable energy systems with photovoltaic power generators operation and modelling *IEEE Trans. Ind. Electron.* **55** 2752–2758.

[7] Faranda R, Leva S and Maugeri V 2008 MPPT techniques for PV Systems: Energetic and cost comparison *Power and Energy Society General Meeting IEEE* 1-6.

[8] Villalva M G, Gazoli J R et al. 2009 Comprehensive approach to modelling and simulation of photovoltaic arrays *IEEE Transactions on Power Electronics* **24(5)** 1198–1208.

[9] Jain S and Agarwal V 2007 Comparison of the performance of maximum power point tracking schemes applied to single-stage grid-connected photovoltaic systems *IET Electr. Power Appl.* **1(5)** 753-762.

[10] Patel U, Sahu M D and Tirkey D 2013 Maximum power point tracking using perturb & observe algorithm and compare with another algorithm *IJDACC* **2(2)** 2319-4863.

[11] Saibabu T C C and Kumari J S 2011 Modeling and simulation of PV array and its performance enhancement using MPPT (P&O) technique *IJCSCN* **1** 9-16.

[12] Esram T and Chapman P L 2007 Comparison of photovoltaic array maximum power point tracking techniques *IEEE Transactions on energy conversion*, **22(2)**, 439-449.

[13] Mohan N, Robbin W P, and Undeland T 1995 *Power Electronics: Converters, Applications, and Design* 2nd Edition, New York, Wiley.

[14] Zainuri M M, Radzi M and Rahim N 2014 Development of ada control maximum power point dc-dc converter *Renewable* 183-194.

[15] Kinjal P, Shah K B and Patel G R 2015 Comparative Analysis of P&O and INC MPPT Algorithm for PV System *2015 (ICEESCO).*

[16] Kumar P S, Balamurali P and Ravikumar C 2015 Comparison of perturb and observe MPPT for PV system conjunction with buck-boost converters *International Journal of Engineering Research and General Science* **3(6)**.

[17] Atallah A M, Abdelaziz A Y, and Jumaah R S 2014 Implementation of perturb and observe MPPT for PV with direct control method using buck and buck-boost converter *Emerging Trends in Electrical, Electronics & Instrumentation Engineering: An international Journal (EEIEJ)* **1(1)** 31-44.

[18] Tofoli F L, Pereira D C, and Paula W 2015 Comparative Study of Maximum Power Point Tracking Techniques for Photovoltaic Systems Hindawi Publishing Corporation *International Journal of Photo Energy*.

[19] Bayod-Rjula A A, and CebolleroAbin J A 2014 A novel MPPT method for PV system with irradiance measeurment *Solar Energy* **109** 95-104.

[20] Abderezak L, Aissa B, Hamza S 2015 Comparative study of three MPPT algorithms for a photovoltaic system *2015 World Congress on Information Technology and Computer Applications Congress (WCITCA)* 1-5.

IOP Conf. Series: Materials Science and Engineering **220** (2017) 012030 doi:10.1088/1757-899X/220/1/012030

Study on Temperature Characteristics of Frequency -domain Dielectric Spectrum of Capacitive Oil-paper Insulation Bushing Based on Extended Debye Model

X. W. Xu[1], G. Pan[2*], K. Wang[1] and W. D. Liu[2]

[1]Yunnan Electric Power Research Institute, Yunnan Power Grid Co., Ltd., Kunming 650217, China

[2]College of Electrical Engineering, Southwest Jiaotong University, Chengdu 610031, China

*E-mail:pangang0820@163.com

Abstract. In order to study the influence of temperature on the frequency domain spectroscopy (FDS) of the oil-paper insulation system, the 72.5kV bushing test model was fabricated according to the actual size, and the FDS of the bushing was tested at different temperatures. According to the test curves of the dielectric constant at different temperatures, the influence of temperature on FDS is analysed, and then the extended Debye equivalent circuit model is used to fit the FDS test curves at different temperatures. The characteristic parameters that could be used to characterize the temperature characteristics are extracted, and the relationship between the characteristic parameters and the temperature is established. The results show that dielectric constant tend to move in the high frequency with the increase of temperature. The parameters of the equivalent circuit model are sensitive to the temperature, the insulation resistance R_g and the maximum time constant branch parameter R_1 show the exponential function with the temperature, the minimum time constant branch parameter R_3 and the temperature show a power function relationship, so the variation of characteristic parameters can be applied to evaluate the influence of temperature on the FDS of the oil-paper insulation bushing.

1. Introduction

Capacitive oil-paper insulation bushing as the insulation and mechanical support equipment of the transformer. its insulation performance directly affects the reliability of the transformer operation [1-2]. The dielectric response based on the frequency-domain dielectric spectroscopy (FDS) used to assess and diagnose the transformer insulation state has been extensively studied [3]. The method of constructing function model to extract the characteristic parameters can be used to quantify the influence of temperature on FDS measurement results [4]. The resistance and capacitance parameters of the branch circuit of the extended Debye model can explain the loss characteristics of oil-paper [5-6].

Based on the extended Debye model, the influence of temperature on FDS of the bushing is studied in this paper. First, 72.5kV bushing test model was fabricated according to the actual size. The dielectric spectra of the bushing in different frequency were measured. And then use the extended Debye equivalent circuit model to fit the FDS test curve at different temperatures, then the relationship between the characteristic parameters and the temperature is extracted and established.

2. Experiment

IOP Conf. Series: Materials Science and Engineering **220** (2017) 012030 doi:10.1088/1757-899X/220/1/012030

2.1. *Preparation of sample*

The sample studied in this paper is based on a laboratory-made 72.5 kV bushing model which the thickness of the cable paper used to produce the capacitor core is 0.1 mm and the thickness of the aluminum foil plate is 0.03 mm, and degassed 45 # transformer oil. Insulated paper was placed in 101-1A electric blast drying oven under the environment of 110 ℃/50pa for 72h. The moisture content of the paper was reduced to below 0.5% through using the METTLER TOLEDO C20 Karl Fischer water titrator to titrate. Degassed 45 # transformer oil was placed in a vacuum drying oven under the environment of 80 ℃/50p for 48h, water titrator titrated the oil until the moisture content down to 10ppm or less.

The insulation layer of inner insulation core is set 17 layers, the capacitance core outside diameter is 30.4mm, aluminum foil and cable paper are cut according to the calculated size of the capacitor core, hollow aluminum tube is set as the center, the cable paper is tightly wrapped around the aluminum tube, as a certain thickness is wounded and then add a layer of aluminum foil as a capacitive screen. The wounded capacitor core model is shown in Figure 1.

2.2. *Test plan*

Before the test, the bushing was placed in the oven under the environment of 40 ℃/ 50pa for 48h to keep the insulating paper fully immersed in oil. Then adjust the temperature of the oven to test the FDS curve at different temperatures. The selected test temperature was 15 ℃, 25 ℃, 35 ℃, 45 ℃, 60 ℃. IDAX-300 as the test tool of FDS curve, the selected frequency range is 1mHz-1kHz. The moisture content of the oil-immersed bushing model is 0.53%. FDS test wiring is shown in Figure 2.

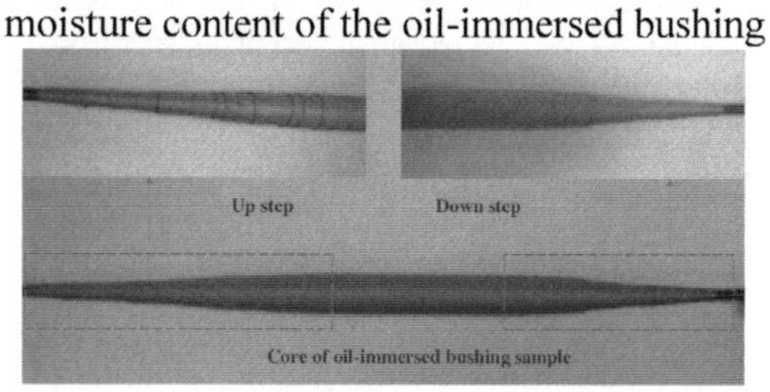

Figure 1. 72.5kV capacitor core model.

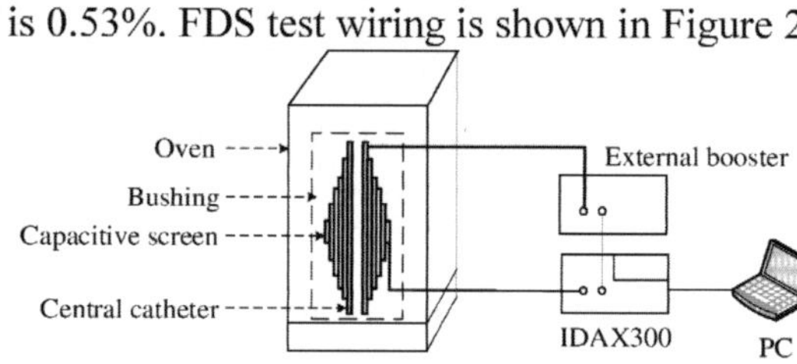

Figure 2. FDS test wiring diagram.

3. Result analysis

Figure 3 shows the 0.53% water content of oil -paper insulation of the bushing at different temperatures, the measurement results of the complex permittivity of the real part ε' and the imaginary part ε'' change with the change of frequency.

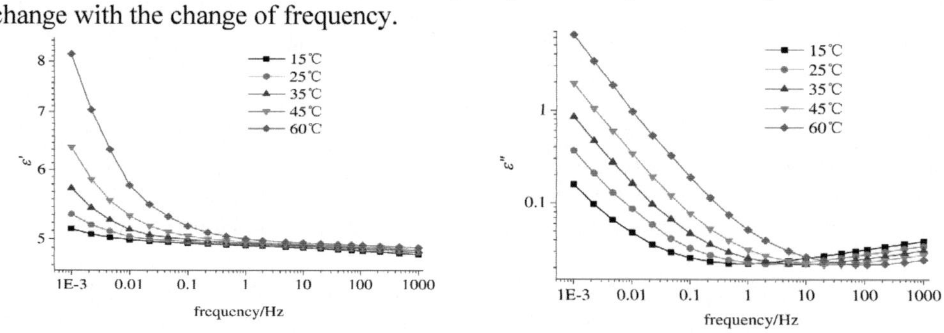

(a) ε' **(b)** ε''

Figure 3. the dielectric constant at different temperatures.

It can be seen from Figure 3 that the influence of temperature on the FDS test curve, in the range of 0.001-1000Hz, ε' increases with the increase of temperature when the frequency is less than 1Hz, and ε' does not change with the increase of temperature when the frequency is more than 1Hz. ε'' increases in the range of 0.001 to 10 Hz with the increase of temperature and decreases in the range of

10-1000 Hz with the increase of temperature. instantaneous polarization plays a leading role in the high frequency band results that ε' basically does not change with the change of temperature when the frequency is more than 1Hz. The main reason why ε'' changes is that when the temperature rises, the thermal motion of the conductive particles increases and the conductance loss becomes larger. The conductivity loss decreases with the increase of temperature, while the relaxation time constant associated with polarization decreases with the increase of temperature, the higher the temperature is, the shorter the settling time will be, and ε'' tends to shift to the right. In the high frequency band, ε'' appears to decrease with the increase of temperature.

4. Extended Debye equivalent circuit model of oil - paper insulation bushing

4.1. Extraction of the characteristic parameters

The main insulation system of bushing consists of multiple layers of insulating paper, the oil gap between paper and paper, aluminum foil that can make voltage and electric field uniform. oil-paper insulation system can be simplified to the extended Debye equivalent circuit model shown in Figure 4, where R_g represents the insulation resistance of the oil-paper insulation system, C_g is the geometric capacitance, and the R_i-C_i branch circuit represents the lossy relaxation polarization at different relaxation time [7].

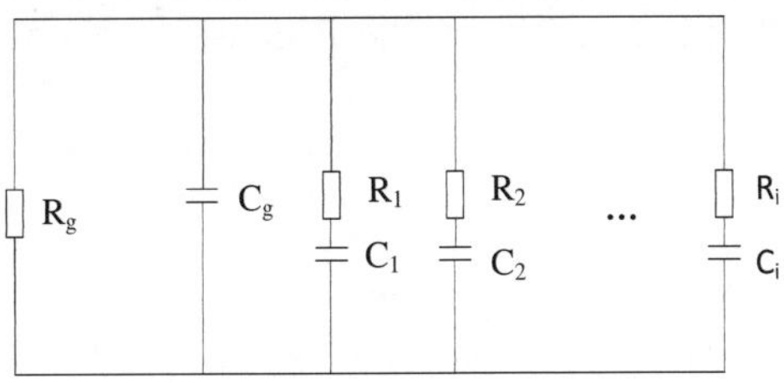

Figure 4. Debye equivalent circuit model of oil-paper insulation.

It is known that the Debye equivalent circuit model can be used to calculate the complex capacitance expression of the dielectric properties of oil-paper insulation of the bushing. Real part, imaginary part of complex capacitance can be expressed as:

$$C'(\omega) = \mathrm{Re}\{\frac{1}{j\omega Z}\} = C_g + \sum_{i=1}^{n} \frac{C_i}{1+(\omega R_i C_i)^2} \tag{1}$$

$$C''(\omega) = -\mathrm{Im}\{\frac{1}{j\omega Z}\} = \frac{1}{\omega R_g} + \sum_{i=1}^{n} \frac{\omega R_i C_i^2}{1+(\omega R_i C_i)^2} \tag{2}$$

It can be seen from Figure 4 that each branch circuit has the same characteristics of the parameters, in order to simplify the model, we take the above expression n = 3. According to the frequency domain dielectric curve of 0.53% water content of complex capacitance of bushing in different temperature conditions, least squares method is used to fit complex capacitor expression (1) , expression (2) and the frequency domain dielectric curve of complex capacitance. The obtained characteristic parameter values are shown in Table 1. The fitting curve is shown in Figure 5.

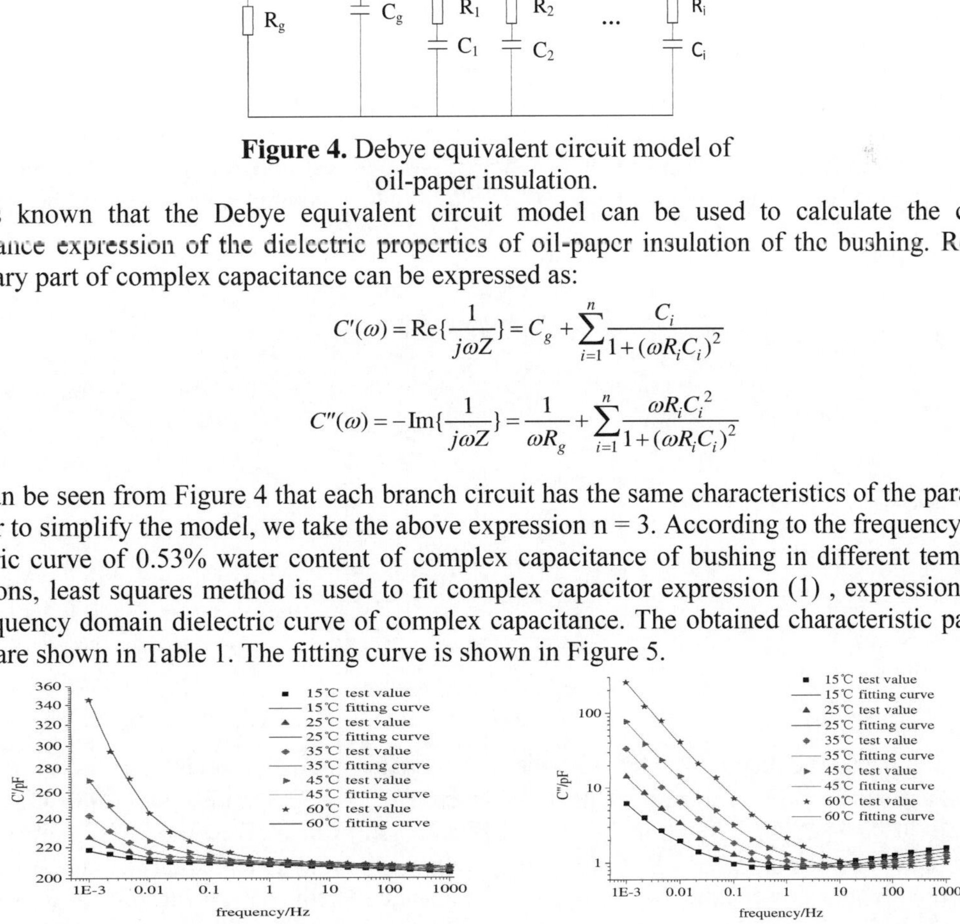

(**a**) C′ (**b**) C″

Figure 5. FDS test values and fitting curves of bushing at different temperature.

Table 1. Fitting parameter values of Debye equivalent circuit model at different temperatures

Temperature/℃	C_g/pF	R_g/GW	C_1/pF	R_1/GW	C_2/pF	R_2/GW	C_3/pF	R_3/GW
15	203.845	26.449	8.055	7.079	2.871	0.142	3.397	0.00068
25	205.039	11.479	6.613	3.989	2.427	0.292	3.2	0.00098
35	206.414	5.069	31.922	2.611	17.142	0.666	3.353	0.00364
45	207.62	2.207	57.488	1.656	15.299	0.511	4.457	0.01557
60	208.951	0.661	122.795	0.781	36.165	0.303	10	0.04245

Figure 5 shows that the frequency domain dielectric curve of complex capacitance and Debye dielectric response models have very good goodness of fit, It can be seen from Table 1, Ri, Ci each has a corresponding characteristic of the change with the increase of temperature, Where the insulation resistance R_g and the maximum and minimum time constant branch resistance R_i (i =1,3) are more sensitive to temperature, which is more obvious and regular with the change of temperature, so the relationship between these three parameters and temperature is selected as a research object.

4.2. Influence of temperature on characteristic parameters

Figure 5 shows the relationship between selected characteristic parameters and temperature. It can be seen from Figure 6 that the selected characteristic parameters and temperature separately show a quantitative relationship, the parameters and the corresponding test temperature are fitted separately, then the fitting formula shown in equation (3), equation (4) and equation (5) can be obtained.

$$R_g = A_0\, e^{(-T/B_0)} + C_0 \tag{3}$$

$$R_1 = A_1\, e^{(-T/B_1)} + C_1 \tag{4}$$

$$R_3 = a \times T^b \tag{5}$$

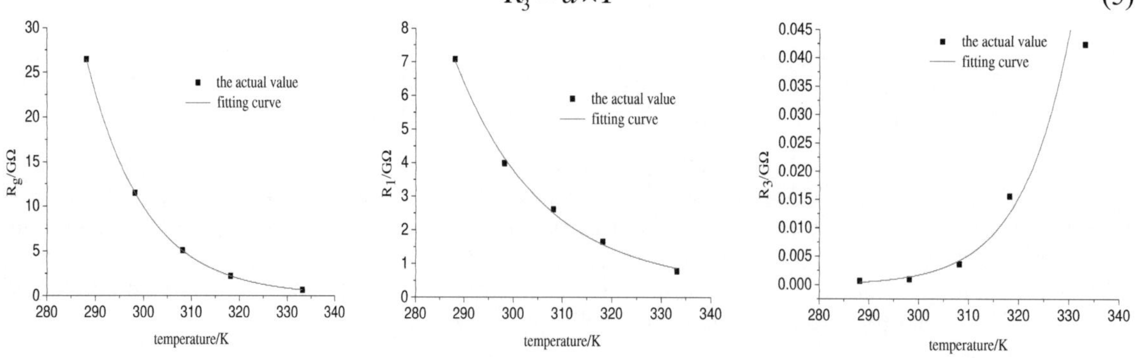

(**a**) R_g (**b**) R_1 (**c**) R_3

Figure 6. The relationship between the characteristic parameters and temperature.

The equation parameters obtained in this study are: A_0=7.3×1011, B_0=11.98, C_0=0.5542, A_1=7.89×107, B_1=17.7, C_1=0.3392, a=6.3×10-88, b=34.08. It can be seen from above formula, the fitting characteristic parameters R_g, R_1 and the temperature exhibit an exponential function, R_3 and the temperature exhibit a power function. Therefore, the change of temperature can reflect the variation law of the insulation resistance and the maximum and minimum time constant branch circuit resistance in the extended Debye model. It is shown that the extracted characteristic parameters can be used to evaluate the influence of temperature on the oil - paper insulation. The quantity relationships of equation (3), equation (4) and equation (5) show that the effect of temperature on FDS test results is significant. Therefore, the FDS test process must take into account the impact of temperature, it's necessary to ensure the same test temperature in assessing moisture content and aging state of the oil-paper insulation bushing.

5. Conclusion

In this paper, the FDS test results of oil-paper insulation of the bushing at different temperatures are analysed and studied, and the following conclusions are drawn:

1) At different test temperatures, ε' and ε'' tend to move in the high frequency with the increase of temperature. ε' is substantially free from temperature in the high frequency range.

2) Complex capacitance measured curve of Oil-paper insulation of the bushing and 3 branch Debye dielectric response function model show a good fitting effect. The fitting characteristic parameters R_g, R_1 and the temperature exhibit an exponential function, R_3 and the temperature exhibit a power function. These functional equations can be used to evaluate the influence of temperature on the FDS of the oil-paper insulation bushing.

References

[1] Zhou Y and Zhu X J 2006 Response analysis of frequency of high voltage oil-impregnated paper condenser bushing *High Voltage Engineering* **32** 163-6.

[2] Du Z B, Nie D X, Zhang L X, et al. 2013 Insulation evaluation of high voltage bushing based on frequency domain spectroscopy analysis *High Voltage Apparatus* **49** 6-11.

[3] Saha T K, Purkait P, Müller F 2005 Deriving an equivalent circuit of transformers insulation for understanding the dielectric response measurements *IEEE Transactions on Power Delivery* **20** 149-57.

[4] Wang S Q, Wei J L, Zhang G J, et al. 2012 Effect of temperature on dielectric response characteristics of oil impregnated pressboard *Transactions of China Electrotechnical Society* **27** 50-5.

[5] Baral A and Chakravorti S 2013 Assessment of non-uniform aging of solid dielectric using system poles of a modified debye model *IEEE Transactions on Dielectrics and Electrical Insulation* **20** 1922-33.

[6] Gao J, Liao R J, Wang Y Y, et al. 2016 Ageing characteristic quantities of oil-paper insulation for transformers based on extended debye model *Transactions of China Electrotechnical Society* **31** 212-7.

[7] Wang L, Zhou L J, Li X L, et al. 2013 Analysis of transformer oil-paper insulation dielectric response in time domain based on extended Debye *High Voltage Apparatus* **49** 39-43.

Study on optimized algorithm for mileage wheel of magnetic flux leakage detector

L Y Sun[1], Y B Li[2], Y T Wu[1], Q Y Xu[1], Y Cai[1]

[1]School of Control and Mechanical Engineering, Tianjin Chengjian University, Tianjin, 300384China

[2]State Key Lab. of Precision Measurement Technology and Instrument Tianjin University Tianjin, 300072, China

Abstract. Pipeline integrity is significant to safe operation of long-range pipeline. To avoid critical failure of the pipeline, which may lead to great loss of property and life, MFL_PIG is often used to detect the corrosion and leakage of the pipeline. To accurately locate the defects, mileage pulses emitted by the mileage wheel are used to and emit signal to single-chip microcomputer for position. This paper investigates the factors that may affect the precision of mileage wheel, an important part of pipeline corrosion and leakage detector (MFL_PIG), investigate its working principle and present an optimized algorithm for mileage wheel to increase the precision of detection.

1. Introduction

Long-range pipeline is important transportation tool in modern society and safety operation is crucial to pipeline. The pipeline corrosion and leakage detector (MFL_PIG) is one of the most widely used detectors to measure the integrity of the pipeline [1]. However, the situation in the pipeline is extremely intricate due to complex pipeline structure, transportation media, warp, and so on. Generally, the diameter is used to before MFL_PIG to clean the objective pipeline. There are many factors that may affect the positioning precision of detection of MFL_PIG, and the mileage wheel, as an important part of magnetic flux detector, plays significant role in the detection process. The mileage wheel is mileage measurement device, and its working condition is essential to measurement accuracy and reliability of MFL_PIG [2].

When MFL_PIG is marching in the pipeline, slippage or other failure may occur on the mileage wheel, which will lead to error of positioning. Therefore, 2 or more mileage wheels are often adopted. During the marching, each mileage wheel will emit a pulse signal when it rotates, which carries operating speed information of the diameter.

The signal which has the minimum error should be used as the distance measurement signal, that is, the tracking signal. Therefore, the signals which are received from different mileage wheel should be compared in real time to judge if it is suitable for tracking.

Selection of accurate tracking signal is important to ensure the mileage wheel's accuracy in the complex pipeline environment, and the design of the optimized algorithm for mileage wheel signals is discussed in detail in this paper.

After the introduction, the paper is organized in the following sections. Section 2 describes factors that affect the precision of the mileage wheel, and its working principle. Section 3 investigates the principle of optimized algorithm in detail, discusses the effect of different parameters on inverse time function K. Section 4 presents an experiment conducted on the pipelines to verify the optimized algorithm. Section 5 presents conclusions of the article.

2. Analysis of the factors affecting the structure and precision of the mileage wheel

2.1. Factors affecting the precision of the mileage wheel

There are many factors that may affect the measuring precision of mileage wheel [3-5].

A. Spring torque of the mileage wheel

Spring torque is one of the important factors which may affect the positioning accuracy of mileage wheel. The mileage wheels in the pipeline attached to the inner wall of the pipe firmly, and turn with the running of the detector. If the positive pressure between mileage wheel and pipe wall is large enough, the friction force should be increased to reduce the probability of slippage of the mileage wheel. Therefore, the torsion spring of the mileage wheel should be designed to maximum torque.

B. Tread pattern of the mileage wheel

The tread shape and depth of the mileage wheel is another factor affecting the positioning accuracy. The surface roughness of the mileage wheel directly affects coefficient of friction between the mileage wheel and the inner wall of the pipe. The rotation of mileage wheel is similar to a car driving on an icy road. Due to the effect of pipe wall wax, mileage wheel will slip. In order to increase the friction coefficient, with exception of increasing the positive pressure between mileage wheel and pipe wall, the mileage wheel surface is grooved and pressed with pattern, which can improve positioning accuracy of mileage wheel. In practical application, the space of mileage wheel surface groove is about 20mm, and the depth is about 3mm.

C. Wear of the mileage wheel

The diameter of steel mileage wheel is about 145mm. The wheel surface is treated with high-frequency quenching. Small magnetic blocks embedded in the spoke. When surface of the mileage wheel wear, the diameter of the mileage wheel decreases, result in changes in mileage pulse. Thus, errors occur in detection data. In practice, the error range of diameter mileage wheel is ±1.5 ram. If the error is greater than the range, the mileage wheel should be changed.

D. The rotation flexibility and swing error of the mileage wheel

Rotation flexibility and swing of mileage wheel are also one of the factors that affect the positioning accuracy of the mileage wheel. If the rotation of the mileage wheel is not flexible, the probability of mileage wheel slippage will be increased. When mileage wheel completes one turn, the number of mileage pulse is certain. If the gap between the mileage wheel and the arm is too large, the hall element installed on the arm of mileage wheel will not be able to perceive a signal caused by small magnetic blocks, resulting in less mileage pulse acquisition. Therefore, the gap between the swing range the mileage wheel and the arm should be less than 0.5mm.

Furthermore, the sundry goods in the pipe should be cleaned in case that the mileage wheel be wound by debris, which stuck the mileage wheel and lead to the loss of detection of data.

2.2. Working principle of mileage wheel

The When MFL_PIG operates in the pipeline, the mileage wheel firmly attaches to the inner wall of pipe rolling with the smooth running of the detector. A Hall element is installed on the arm of the mileage wheel, and small magnetic blocks mounted at even distance on mileage wheel. As the mileage wheel rotates, Holzer element generates a pulse signal, which will be acquired, multiplied, and sent to single-chip computer for optimization. The optimized signal will then be stored in the hard disk, and a square wave signal is formed (see Figure 1, above part).

The magnetic blocks are evenly distributed on the mileage wheel, and each two adjacent magnetic blocks corresponds to the distance of 10mm (mileage wheel rotation). In order to reduce the period of square wave cycle and improve the accuracy of the signal, another Hall element is placed in the middle of the magnetic blocks. Therefore, the distance between rising and falling part of the square wave is reduced from 10mm to 5mm (see Figure 1, low part).

In order to illustrate the working principle, the mileage wheel is extended into a straight line, as shown in figure 2. Assuming that there is only one Hall element C, the signal rolls when magnetic block 1 and 2 pass C respectively. The distance between block 1 and 2 is 10mm. When two Hall elements are provided, and the distance between the two components of A and B is 1.5 times of the distance of the magnetic blocks, the period of the signal decreased to half of the original. The distance between adjacent rising edge of the square wave reduced to 5mm, which increase the measuring precision.

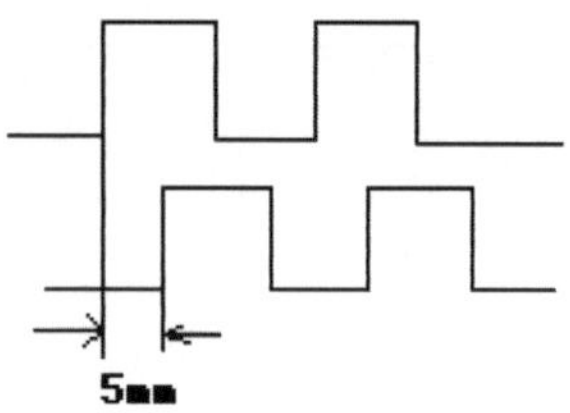

Figure 1. Schematic of Square wave signal of the mileage wheel.

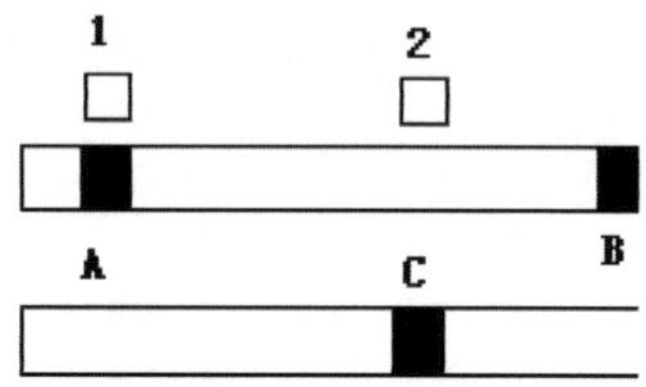

Figure 2. Schematic of the mileage wheel with magnetic blocks after straightening.

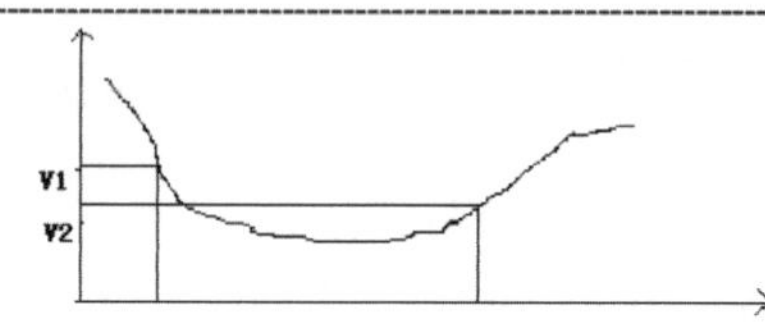

Figure 3. Relationship of speed V1 and V2.

3. Principle and implementation of the optimization algorithm for the mileage wheel [6-7]

3.1. Speed calculation algorithm of mileage wheel

When the mileage wheel is running normally or the speed is not particularly slow (that is, there is no stuck or serious slippage of the mileage wheel), the speed can be calculated as below,

$$V1=0.005/(T1-T2) \tag{1}$$

Where T1 and T2 are time of two adjacent rising edge of the mileage wheel signal.

However, when the mileage wheel is not running, or even stuck, the later rising edge is delayed for a long time, Equation (1) could not be used. At this time, another equation should be used instead.

$$V_2=0.005cn/T \tag{2}$$

In which n is decided according to the actual situation, 0.005*n means the distance which n rising edges passed, and T represents the time used for n rising edges.

V_1 in equation (1) is instantaneous speed of the mileage wheel and V_2 n describes the average speed of n rising edges. The relationship between V_1 and V_2 is shown in figure 3.

At the beginning of calculation, V_1 mode is used as reference speed, marked as V_1' (V_1' is determine in terms of actual situation of the inspection, varies between 0.5m/s to 5m/s.), and the following calculated speed will be compared with V_1'

If $V_1 > V_1'$, still use the V_1 as calculation speed;

If $V_1 < V_1'$, switch to V_2.

After switched to the V_2 calculation speed, the value of calculated speed V_2 will be compared with reference speed V_2',

If $V_2>V_2'$, switch to V_1 mode;

If $V_2<V_2'$, keep using the V_2 as calculation speed.

Here V_2' is determined according to the actual situation. The range of V_2' is 0.5m/s~ 5m/s.

3.2. Speed calculation algorithm of mileage wheel

Inverse time algorithm is used to construct the inverse function:

$$K=A/(BX^n+C) \tag{3}$$

Where A, B, and C are coefficients (the influence of these coefficients on K will be discussed in the following part) , and X is the difference between the speed of two mileage wheels (V_0 and V_S), that is

$$X=V_0-V_S \tag{4}$$

IOP Conf. Series: Materials Science and Engineering **220** (2017) 012031 doi:10.1088/1757-899X/220/1/012031

When the speed difference X is less than 0.01m/s, the system operate normally without switching. When the speed difference is over 0.1m/s, the algorithm switches immediately. When the speed is just in between 0.01m/s to 0.1m/s, inverse algorithm in equation (3) is used.

Therefore, When X=0.01, K approaches infinity in equation (3); when X=0.1, K approaches 0.

The influence of coefficient A, B, and C on function K will be investigated in this part.

A. 1st situation: the influence of parameter A on function K when parameter C and B are fixed, shown in figure 4 (a).It can be seen from figure 4 that K becomes larger with the increase of value of parameter A. And the curve becomes farther to the original of the coordinate ($A1<A2<A3$). The curve from left to right in turn is K3, K2, and K1。

B. 2nd situation: the influence of parameter B on function K when parameter A and C are fixed, shown in figure 4(b). It can be seen from figure 5 that K decreases with the increase of value of parameter B. And the curve becomes farther to the original of the coordinate ($B1<B2<B3$). The curve from left to right in turn is: K1, K2, and K3

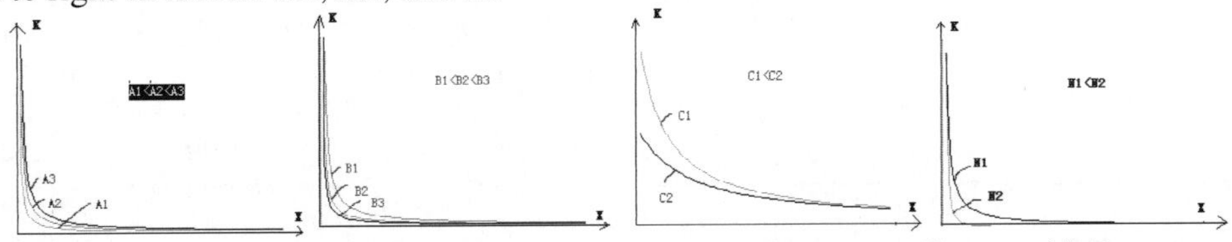

(a)Parameter A (b) Parameter A (c) Parameter C (d) Parameter N

Figure 4. Effect of parameter A, B, C and N on function K.

C. 3rd situation: the influence of parameter C on function K when parameter A and B are fixed, shown in figure 4(c). The influence of parameter C on function K is almost the same parameter B, the difference is that the curve of K becomes smoother with the rising of C. The curve from left to right in turn is: K2 and K1($C1<C2$)

D. 4th situation: the influence of parameter *n* on function K when parameter A, B and C are fixed, shown in figure 4(d). When parameters A, B, and C are fixed, the larger parameter n is, the steeper the curve of function K becomes. And the curve approaches to the original. The curve from left to right in turn is: K_2 and $K_1(n_1<n_2)$.

It can be known from the above that the parameters A, B, C and n have different effect on the shape of the curve of function K. In order to get the appropriate K, it is necessary to select A, B, C and N. The final expression of the K is as follows:

$$K=0.0001/ (X-0.01) \quad (5)$$

The final curve of K is shown in figure 5.

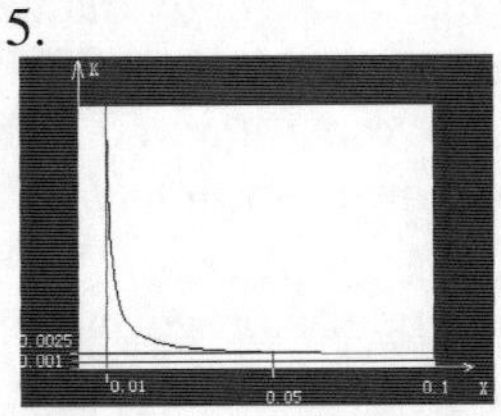

Figure 5. Final curve diagram of K.

The next step is to determine the switch time, a superposition parameter is established,

$$Z_{N+1}=Z_N+H/K \quad (6)$$

Equation (6) eliminated iterative and inverse operations. When $V_O<V_S$, H/K is negative, it is not possible to switch; only when $V_O>V_S$, Z increase, and gradually reach the required Z_0 to switch.

4. Experimental verification of the optimization algorithm of mileage wheel

In the experiment, three channels of signal generator or single-chip microcomputer system is used to emit three square wave pulse, which is to analog the mileage pulse signal sent by two mileage wheels. Signals are collected respectively and input to PCA module of the microcontroller, frequency and speed are calculated in real-time, and the above comparison rules are adopted. The experimental data based on three channels of analog pulse emitted by MCU system is shown in Table 1.

When frequency of the mileage square wave signal varies within 100Hz to 1MHz, the optimized algorithm can select the output pulse signal of the highest frequency correctly. By setting appropriate mileage accumulated error threshold s_e, the system can accurately calculate switching time needed and complete the channel switching. The sensitivity is less than 30Hz (which is independent with s_e).

In addition, when frequency of the three pulse signal from mileage wheel changes continuously, the system can determine the correct output signal by selecting the signal with highest frequency and the sensitivity is less than 10Hz. However, when the frequency of input signal is greater than 1MHz, false output channel may be selected and switch action may occur.

Table 1. Test data sheet for the optimization algorithm of mileage wheel

Time (second)	1	2	3	4	5	6	7	8	9	10	11
Pulse 1 (Hz)	100	200	300	500	700	1000	800	830	830	830	900
Pulse 2 (Hz)	100	200	200	500	700	1000	800	800	860	860	900
Pulse 3 (Hz)	100	200	200	500	700	1000	800	800	800	890	900
Output pulse	3	3	1	1	1	1	1	1	2	3	3
Result	√	√	√	√	√	√	√	√	√	√	√

5. Conclusion

There are many factors that affect the operation of mileage wheel, for example, the rotation flexibility, spring torque, swing angle, wear, or tread pattern of the mileage wheel of the mileage wheel. Of which, the most important is to determine the reference speed of the mileage wheel. The optimized algorithm is presented with detail working principle of mileage wheel and the parameters which have effect on inverse time function are investigated. Flow chart shows the judgment process of the system, and verification experiments were conducted to confirm the theoretical analysis of the algorithm.

Acknowledgment

The authors would like to thank foundation of Demonstration Project of Tianjin Marine Economic Innovation and Development Region (Grant No. 2015120024000473), and Applied Basic and Frontier Research Project of Tianjin (Tianjin Natural Science Foundation) (Grant No. 17JCYBJC19300).

References

[1] Sun L Y, Li Y B 2011 Hardware design of diameter analyzing system for liquid and gas pipeline *Applied Mechanics and Materials* 88-89 587-591.

[2] Wu X, Jin S J, Li Y B, et al. 2010 Above-ground marker system of pipeline internal inspection instrument based on geophone array *Nanotechnology and Precision Engineering* 8(6) 553-557.

[3] Sun L Y, Li Y B 2011 Large vertical storage tank bottom evaluation via Acoustic Emission signal analysis *Proceedings of the 2011 Chinese Control and Decision Conference* 3554-3558.

[4] Li Y, Zhou C, Liu S, Zhu H, Sun L 2015 An enhanced above ground marking method based on acoustic sensor array *Nanotechnology & Precision Engineering* 13 (2) 120-127.

[5] Li Y, Liu S, Dorantesgonzalez D J, Zhou C, Zhu H 2014 A novel above-ground marking approach based on the girth weld impact sound for pipeline defect inspection *Insight: Non-Destructive Testing & Condition Monitoring* 56 677-682.

[6] Li Y B, Cui Y Y, Sun L Y, Xiang H 2013 Azimuth Coefficient estimation of pipeline inspection gauge location system based on TDOA *Journal of Tianjin University (Science and Technology)* 46 (2) 186-189.

[7] Liu H, Liu M R, Zhang G J, et al. 2013 Study of a new scheme for MEMS vector sensor used for pipeline ground mark *Transducer and Microsystem Technologies* 32(11) 14-17.

IOP Conf. Series: Materials Science and Engineering **220** (2017) 012032 doi:10.1088/1757-899X/220/1/012032

A DFIG Islanding Detection Scheme Based on Reactive Power Infusion

M Wang[1], C Liu[1], G Q He[1], G H Li[1], K H Feng[2] and W W Sun[1]

[1]China Electric Power Research Institute, Haidian District Beijing, 100092 China
[2]China EPRI Zhangbei wind power research and detection company, Zhangbei, Hebei, 076450 China

Abstract. A lot of research has been done on photovoltaic (the "PV") power system islanding detection in recent years. As a comparison, much less attention has been paid to islanding in wind turbines. Meanwhile, wind turbines can work in islanding conditions for quite a long period, which can be harmful to equipments and cause safety hazards. This paper presents and examines a double fed introduction generation (the "DFIG") islanding detection scheme based on feedback of reactive power and frequency and uses a trigger signal of reactive power infusion which can be obtained by dividing the voltage total harmonic distortion (the "THD") by the voltage THD of last cycle to avoid the deterioration of power quality. This DFIG islanding detection scheme uses feedback of reactive power current loop to amplify the frequency differences in islanding and normal conditions. Simulation results show that the DFIG islanding detection scheme is effective.

1. General background information

Islanding is a condition in which a part of grid, which consists of local load and generator, continues to generate power even though it is isolated from main grid. Islanding can be harmful to equipments and cause safety hazards. For example, if a distributed generation does not have the capacity to adjust voltage or frequency, islanding will cause great voltage shift and frequency shift in the system, which may damage the grid and electric devices connected to the grid [1]. It is very necessary to detect and eliminate islanding rapidly and exactly.

In recent years, a lot of research has been done on islanding detection schemes for PV power system and a variety of detection schemes has been developed [2]. There are mainly two types of islanding-detection schemes for PV power system, which are power line carrier communications-based anti-islanding scheme and local islanding detection scheme. Power line carrier communications-based anti-islanding scheme is not widely used due to its large expenses and complexity in design and installment [3] [4]. While local islanding detection schemes, which rely on the PV converter to detect islanding, is widely used. Under this method, islanding is detected based on changes in output voltage and current signals monitored by PV converter. If output voltage and current signals drift from the presumed threshold, islanding can be detected and the converter will stop working [5]. Local islanding detection scheme can be classified into two types include passive anti-islanding scheme [6]-[8] and active anti-islanding scheme. While active anti-islanding scheme can also be classified into several types, among which, active frequency drift scheme [9] and reactive power disturbance scheme [10] are widely employed to detect islanding in PV power system.

2. Islanding in wind turbine generation system

Experiments and real cases show that islanding may also occur in wind farm. In 2012, the China Electric Power Research Institute carried out a 330kV short-circuit experiment in Jiuquan, China. This experiment shows that islanding occurs in a large number of wind farms when breaker trips at 330kV power lines and islanding condition lasts for 22s at most.

In the experiment, it is noticed that voltage still existed in the circuit after breaker was turned off at 26s, which shows that islanding can occur in wind farm. Figure 1 shows the experiment results of 1.5MW wind turbine. Waveforms of voltage, current, active power and reactive power of phase A are as shown.

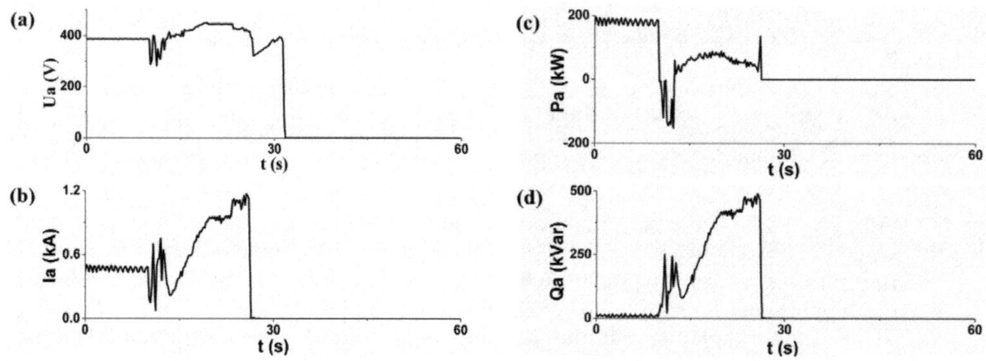

Figure 1. Diagram of phase A of 1.5MW wind turbine outlet.

In the experiment, voltage and current of wind turbines were observed. Islanding occurred at 10s, outlet voltage dropped to 0 at 32s, current drops to 0 at 26.4s and islanding lasted for 22s.The experiment shows that islanding can occur in wind farms and can last for a period of time, which may lead to equipment damage and pose great danger to utility workers if they are not aware of it. Hence, it is very meaningful to carry out the study on islanding detection for wind turbine.

3. Islanding detection scheme of DFIG

3.1. Wind turbine system

Figure 2 shows a wind turbine generation system. Where $P_0 + Q_0$ is the output power of DFIG, $P_{load} + Q_{load}$ is the local load, $P_G + Q_G$ is the power transported to the grid, and U_0 is the voltage at the point of common coupling (the "PCC").

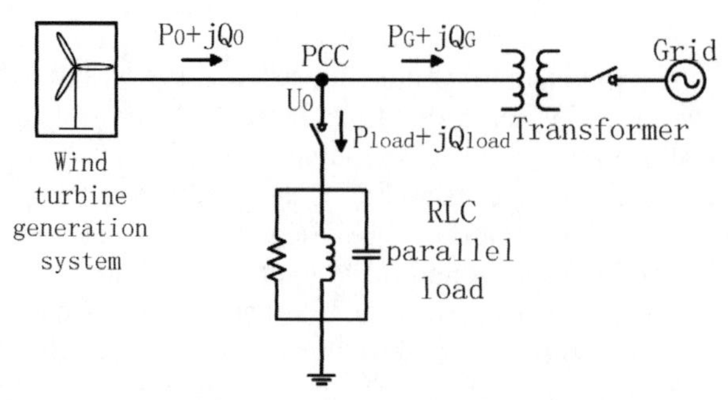

Figure 2. Islanding happens in wind turbine generation system.

If there is a big mismatch between the output voltage and local load in islanding condition, the voltage at PCC and angular frequency can be obtained as follows.

$$U = U_0\left(1 + \frac{P_G}{P_{load}}\right) \tag{1}$$

$$\omega = \frac{2Q_f P_{load}\omega_0}{Q_G + \sqrt{Q_G^2 + 4Q_f^2 P_{load}^2}} \tag{2}$$

Where U_0 is the voltage of PCC andω_0 is the angular frequency before the islanding is formed.

If there is a big mismatch between the output voltage and local load, P_G and Q_G will be large, and the change in U andω before and after the islanding is formed will also be large. Hence, it is easy to trigger the voltage protection and frequency protection of wind turbines. While if the mismatch between output voltage and local load is small. The change in voltage and frequency may be too small to trigger the abovementioned protection system. Passive islanding detection scheme may fail to detect islanding. When the mismatch between the active/reactive power and load is small, the PLC load shown in Figure 2 will still satisfy (3) and (4).

$$Z \approx R + j\frac{\omega L}{1-\omega^2 LC} \tag{3}$$

$$U^2 = P_{load}R = Q_{load}\frac{1}{\frac{1}{j\omega L} - \omega C} \tag{4}$$

According to (3), the imaginary part of the impedance Z of the RLC parallel load

$$Im(Z) = \frac{\omega L}{1-\omega^2 LC} \tag{5}$$

And according to (4), the imaginary part of the conjugation of Z, which is expressed as Z^*

$$Im(Z^*) = \frac{P_{load}R}{Q_{load}} \tag{6}$$

Hence,

$$Im(Z) = -Im(Z^*) = \frac{P_{load}R}{Q_{load}} = -\frac{\omega L}{1-\omega^2 LC} \tag{7}$$

From (7),

$$\omega = \frac{Q_{load}}{2Q_f P_{load}\sqrt{LC}} \pm \frac{1}{2}\sqrt{\frac{Q_{load}^2}{CP_{load}} + \frac{4}{LC}} \tag{8}$$

Where$Q_f = R\sqrt{\frac{C}{L}}$ is the quality factor of the RLC parallel circuit.

Considering that power generated by wind turbines in normal conditions is mainly active power while reactive power is approximate 0. And the value of ω is positive, $\frac{Q_{load}^2}{CP_{load}}$ can also be ignored, then

$$\omega \approx \frac{Q_{load}}{2Q_f P_{load}\sqrt{LC}} + \frac{1}{\sqrt{LC}} \tag{9}$$

During islanding, $P_{load} = P_0$, $Q_{load} = Q_0$. Hence, the angular frequency in islanding is affected by the active power and reactive power generated by the wind turbine. In normal conditions, since $P_0 \gg Q_0$, the change in reactive power will have a much greater effect on the frequency than the change in active power. Hence, this paper uses reactive power as the cause of change in frequency.

3.2. Control theory of DFIG islanding detection scheme based on reactive power infusion

Under the traditional reactive power infusion scheme, continuous reactive power disturbance will be applied to the wind turbine system, which may lead to the system to become unstable.

To improve stability of wind turbine system, this paper presents a new wind turbine islanding detection scheme, in which the change rate of voltage harmonic distortion and the reactive power disturbance are used to detect islanding. In this scheme, voltage harmonic distortion rate of the voltage

IOP Conf. Series: Materials Science and Engineering **220** (2017) 012032 doi:10.1088/1757-899X/220/1/012032

waveform output by wind turbines is to be monitored. At the moment when islanding is formed, a large proportion of harmonics can occur and the THD can increase dramatically. After islanding is formed, THD increases slowly due to the output characteristics of the wind turbine. Hence, THD can be monitored continuously and a trigger signal can be obtained by dividing the voltage THD by the voltage THD of last cycle. Use K as the trigger signal for reactive power infusion. If K is greater than the threshold value, then compare the frequency of the turbine and the threshold value. If the frequency is greater than the threshold value, then shut down the wind turbine. If not, start to infuse reactive power into the system. If voltage frequency is greater (less) than 50Hz, then infuse reactive power by$\Delta Q = 0.01\text{pu.}$ ($\Delta Q = -0.01\text{pu.}$). During the infusion, monitor the voltage frequency for 10 subsequent cycles. If the change of frequency stays within the threshold value, increase reactive power infusion by $\Delta Q = 0.1\text{pu.}$(when the frequency increases) and $\Delta Q = -0.1\text{pu.}$ (when the frequency decreases). During the infusion, continue to monitor the voltage frequency. If the frequency drifts from the threshold value, shut down the wind turbine; if not, stop the infusion and start to monitor the voltage THD.

To testify the rationality of using K as trigger signal, this paper uses the experiment data of the abovementioned 1.5MW wind turbine islanding experiment. And to make it more convincing, long term data of phase A voltage waveform are used. The THD and the trigger signal K are as shown in Figure 3.

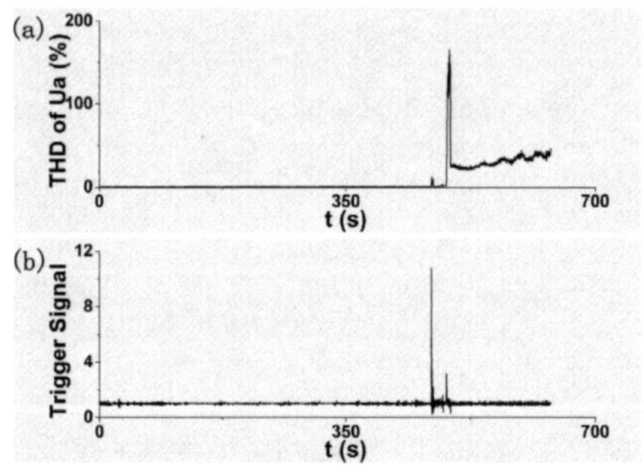

Figure 3. THD and Trigger signal of U_a in experiment.

As shown in Figure 3, when the system is in normal operation, phase A voltage THD is approximate 0.8% and the trigger signal is 1. At the moment islanding is formed, THD increases to 10% and the trigger signal rapidly increases to 11. During islanding, THD increases slowly and the trigger signal decreases. Hence, it is rational to use K as the trigger signal for reactive power turbulence.

Compared to PV converter and direct-driven wind turbine generator, the structure of DFIG is much more complex. In DFIG, stators are directly connected to grid, rotor is excited by three-phase PWM converter, electromagnetic power is transferred via both stator and rotor routes [11]. The reactive power of DFIG is generated by stator and controlled by the rotor excitation current. Hence, rotor current is used to control the reactive power of DFIG. Figure 4 shows the control block diagram of reactive power infusion of DFIG which is controlled by grid voltage vector orientation.

Where P_s is stator active power, Q_s is stator reactive power, ω_{slip} is slip angle frequency, $\psi_{sdq}$is d and q component of stator flux linkage vector, i_{rdq} is d and q component of rotor current vector, U_s is stator voltage vector, U_{sdq} is d and q component of stator voltage vector, L_s is stator inductance and L_m is mutual-inductor of stator and rotor.

Monitor the THD and frequency of three-phase voltage stators. K equals to THD divided by THD of the last cycle. If K is greater than the threshold value, reactive power infusion will be triggered.

Reactive power disturbance$\Delta Q_1 = \pm 0.01$pu. will be infused toouter reactive power loop to affect the three-phase PWM converter of rotor side and then to control the reactive disturbance power of DFIG. If voltage frequency increases (decreases), increase the infusion of reactive disturbance power by $\Delta Q_2 = \pm 0.1$pu.. Continue to monitor the output voltage frequency. If voltage frequency becomes greater than the threshold value, shut down DFIG.

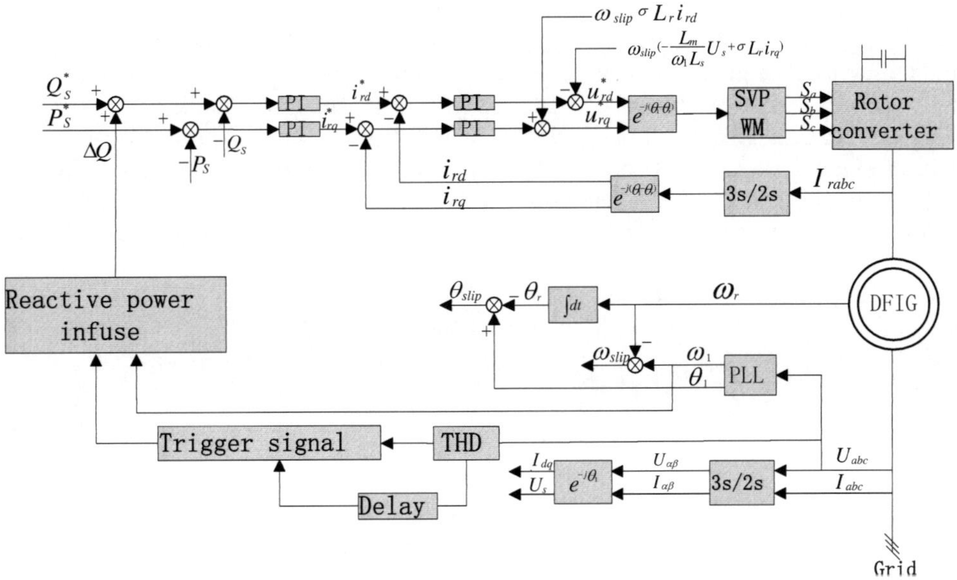

Figure 4. Schematic diagram of island detection and control for DFIG.

3.3. Simulation

To examine the feasibility of the proposed scheme, a simulation model of electromagnetic transient process in 1.5MW DFIG is established by MATlab/Simulink. The model consists of wind speed model, DFIG model, RLC parallel load model, as well as power system model. It is as shown in Figure5.

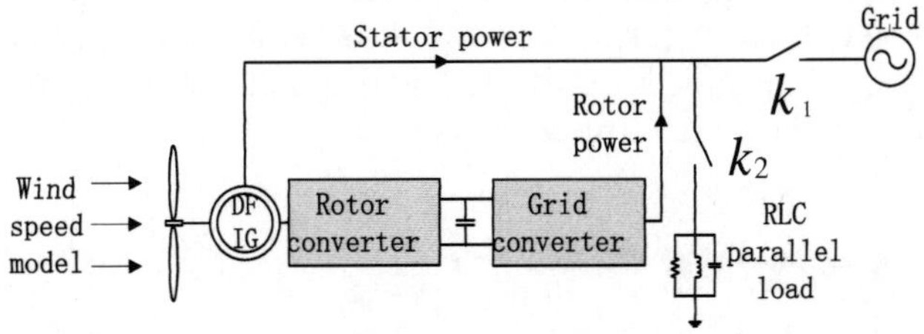

Figure 5. Structure diagram of islanded DFIG simulation.

The simulation test shows the situation in which islanding detection fails due to lack of active anti-islanding scheme. In this test only frequency protection and voltage protection are used. In accordance with the technical rule for connecting wind farm to power system under GB/T-19963, the allowable range of voltage is $0.9\text{pu.} < U < 1.1pu.$ and the allowable range of frequency is $48\text{Hz} < f < 50.5\text{Hz}$. DFIG will be shut down if the voltage amplitude or frequency is beyond the allowable range.

In accordance with the test method of the anti-islanding testing code for inverter of utility-interconnected photovoltaic power station of NB/T 32010-2013, switch on breaker k1, switch off

IOP Conf. Series: Materials Science and Engineering **220** (2017) 012032 doi:10.1088/1757-899X/220/1/012032

breaker k2 and monitor the DFIG output value of active power and reactive power. The value of active power is $P_0$and the value of reactive power is Q_0. Then, adjust the RLC load capacity to P_0 and Q_0. Hence, the DFIG output of the simulation model is almost the same with the local load.

When wind speed is 15m/s, DFIG power reaches the power rating. In this model, constant wind speed is set at 14m/s. Then$P_0 = 1.325\text{MW}$, $Q_0 = 0$。

The simulation period is 6s and islanding is formed at 2s. Breaker k1 is switched on and breaker k2 is switched off at the beginning of the test. At 2s, breaker k1 is switched off and breaker k2 is switched on. Islanding lasts for 4s. As shown in Figure 6, under the condition that the active power and reactive power outputted by DFIG equates to the local load, the voltage and frequency stay within the range of $0.97\text{pu.} < U < 1.03pu.$ and $49.9\text{Hz} < f < 50.1Hz$. In this situation, neither passive frequency protection nor passive voltage protection can detect islanding.

As shown in Figure7, at 2s, THD zooms from 3% to 12% and trigger signal zooms from 1 to 3.6. Although THD has a large shift after islanding is formed, the value of trigger signal remains less than 3 after 2s. Hence, it is proved that change rate of voltage harmonic distortion can be used as the trigger signal.

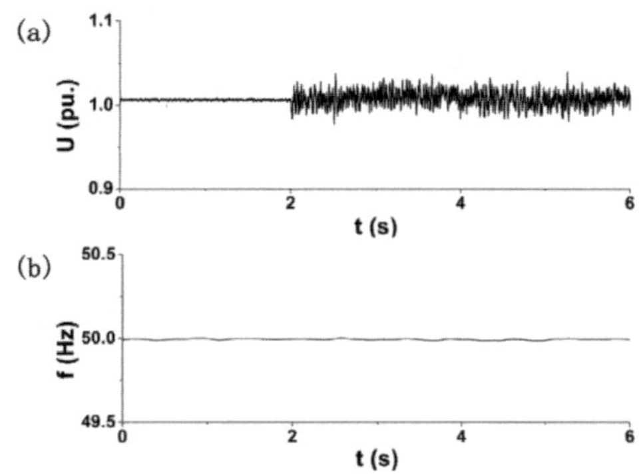

Figure 6. Voltage amplitude and frequency of DFIG after islanded.

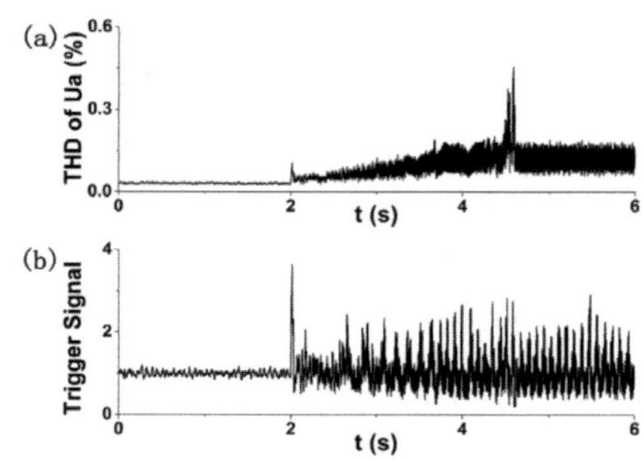

Figure 7. THD of U_a and Trigger signal in simulation.

The second stimulation test is to examine the feasibility of reactive power infusion scheme in this simulation model. Reactive power infusion model is added to this stimulation model. The simulation period is still 6s. Breaker k1 is switched on and breaker k2 is switched off at the beginning of the test. At 2s, breaker k1 is switched off and breaker k2 is switched on. Islanding lasts for 4s. The threshold of reactive power infusion trigger signal is set as K=3. The diagrams of DFIG output voltage, frequency, active power of this test are as shown in Figure8.

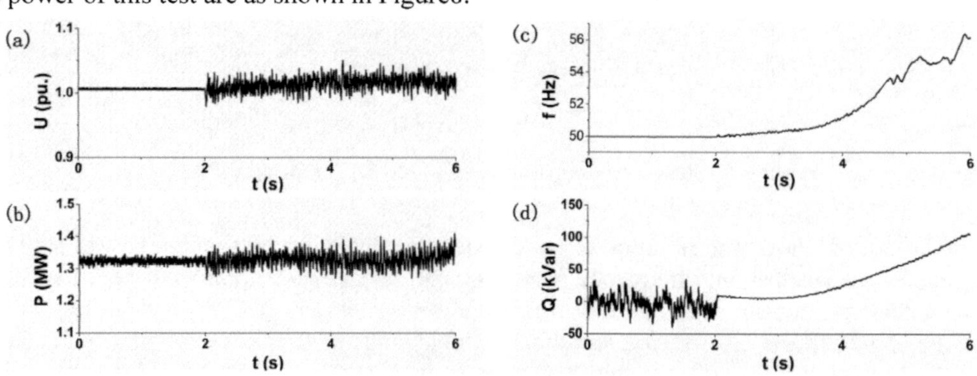

Figure 8. Diagram of voltage amplitude and frequency, active and reactive power with reactive power disturbances.

As shown in Figure8, in islanding, the shift of active power and voltage amplitude of DFIG is very small. When islanding is formed, reactive power increase from 0 to 10kVar (0.007pu.) in 1s and increases to 110kVar (0.073pu.) in 4sfrequency increases gradually and reaches the frequency threshold of 50.5Hz in1.51s and reaches 56Hz in 4s. Islanding lasts for 1.51s in this test.

4. Conclusion

Islanding not only exists in PV system, but also wind turbine system. More work should to be done to solve this problem. When islanding happens, the mismatch between the output reactive power of the turbine and local reactive load will cause frequency shift of the isolated grid. So it is easy to detect the islanding by infusing a small reactive disturbance into the system and monitoring the frequency of the system, if the frequency drifts from the threshold value, then shut down the wind turbine. However, the continuous infusion of reactive power may cause grid voltage instability, especially when the number of the turbine is large. So a trigger signal was proposed. The rate of harmonic distortion which can be calculate as $K = THD(N)/THD(N-1)$ is a reliable and simple value to be this trigger signal and it will not do harm to the grid during the process of calculation.

Acknowledgments

This work was supported by National Science and technology support program (2015BAA06B01) and State Grid Corporation of China Science & Technology Project: Study and Application of Centralized Anti-Islanding on Large-Scale renewable energy.

References

[1] Xie D 2014 Research on islanding detection technology for multi-grid connected inverter based distributed generation Hefei Univ. of Technology Anhui China.

[2] Chao Y, Wu G and Zeng X J 2009 Protection technology for distributed generation systems *Power System Protection and Control* **2** 99.

[3] Ropp M, Larson D and Meendering S 2007 Discussion of a power line carrier communications-based anti-islanding scheme using a commercial automatic meter reading system *Proceedings of the 4th IEEE Conference on Photovoltaic Energy Conversion* 2351.

[4] Dysko A, Booth C, Anaya-Lara O and Burt G M 2007 Reducing unnecessary disconnection of renewable generation from the power system *IEEE Renew. Power Gener.* **1** 41.

[5] Redfern M A, Usta O and Fielding G 1993 Protection against loss of utility grid supply for a dispersed storage and generation unit *IEEE Trans. on Power Delivery* **8(3)** 948.

[6] Francesco De M, Marco L and Antonio D A 2006 Overview of anti-islanding algorithms for PV systems Part 1: Passive Methods *12th International Power Electronics and Motion Control Conference* 1878.

[7] Vieira J C M, Freitas W and Wilsun X 1993 Performance of frequency relays for distributed generation protection *IEEE Trans. on Power Delivery* **3** 1120.

[8] Jang S I, Kim K H 2004 An islanding detection method for distribution generation using voltage unbalance and total harmonic distortion of current *IEEE Trans. on Power Delivery* **19** 745.

[9] Ropp M E, Begovic M and Rohatgi A 1999 Analysis and performance assessment of the active frequency drift method of islanding prevention *IEEE Trans. on Energy Conversion* **3** 810.

[10] Jeong J B, Kim H J and Ahn K S 2005 A novel method for anti islanding using reactive power *27th International Telecommunications Conference* 101.

[11] He Y K, Hu J B and Xu L 2012 *Grid-connected Operation of Wind-turbine Driven Double Fed Induction Generators* (Beijing: China Electric Power Press) 278.

IOP Conf. Series: Materials Science and Engineering **220** (2017) 012033 doi:10.1088/1757-899X/220/1/012033

Mining Research on Vibration Signal Association Rules of Quayside Container Crane Hoisting Motor Based on Apriori Algorithm

Chencheng Yang, Gang Tang and Xiong Hu

Logistics Engineering College, Shanghai Maritime University, Shanghai 201306, China

ychencheng@126.com

Abstract. Shore-hoisting motor in the daily work will produce a large number of vibration signal data,in order to analyze the correlation among the data and discover the fault and potential safety hazard of the motor, the data are discretized first, and then Apriori algorithm are used to mine the strong association rules among the data. The results show that the relationship between day 1 and day 16 is the most closely related, which can guide the staff to analyze the work of these two days of motor to find and solve the problem of fault and safety.

1. Introduction

In order to adapt to more and more ships, shorten loading and unloading time and improve efficiency, the quay crane, which plays an important role in the port operation, is developing towards high-speed operation and heavy lifting capacity, so the hoisting motor is easy to produce the problem of large load, big impact and big sway [1]. Hoisting motor vibration signal data is very valuable and analytical value, but these data are often disorganized and there is no obvious feature, so the need for a hidden relationship between data mining methods Apriori algorithm is a frequent itemset mining algorithm for mining association rules, which uses join and pruning methods to generate a small number of candidate sets and then generate frequent itemsets. Finally, strong association rules are generated, which are widely used in various fields of social life the Apriori algorithm can mine the association rule in the motor vibration signal data, analyze the mechanical state of the motor, and find out the motor fault and safety hazard [2].

2. The principle of Apriori algorithm

There are two events A and B. The support of association rule $A \Rightarrow B$ refers to the proportion of all transactions in the database that contain the event AB, that is the probability $P(A \cup B)$.The confidence of association rule AB is the number of transactions that contain both AB and include A ratio of the number of transactions, that is the probability $P(B/A)$.

Apriori algorithm is a priority algorithm. Every time the database is scanned, it calculates all the support in the database. The items that support less than the minimum value are set to form a frequent itemset L1. Then it scans the database and generates a frequent itemset L2. The kth scan generates a frequent itemset

IOP Conf. Series: Materials Science and Engineering **220** (2017) 012033 doi:10.1088/1757-899X/220/1/012033

Lk of length k and the (k+1)th scan generates a frequent itemset Lk+1 of length k+1. The items in the candidate frequent itemset Ck+1 are reduced according to the pruning properties of the algorithm and scan the database to generate frequent itemsets according to the items in Ck+1 [3-4].

Apriori algorithm is an iterative way. First, it draws C1 from each item in the database. Then it draws the frequent itemset L1 from the minimum support [5-6]. Then it finds the combination Ck+1 of Lk and L1. Next it finds Lk+1 according to the minimum support and repeats this cycle until k-order. Finally, the confidence of itemset Lk is calculated.If the confidence is greater than the minimum confidence, it is a qualified rule. Method flow chart is shown in Figure 1 [7].

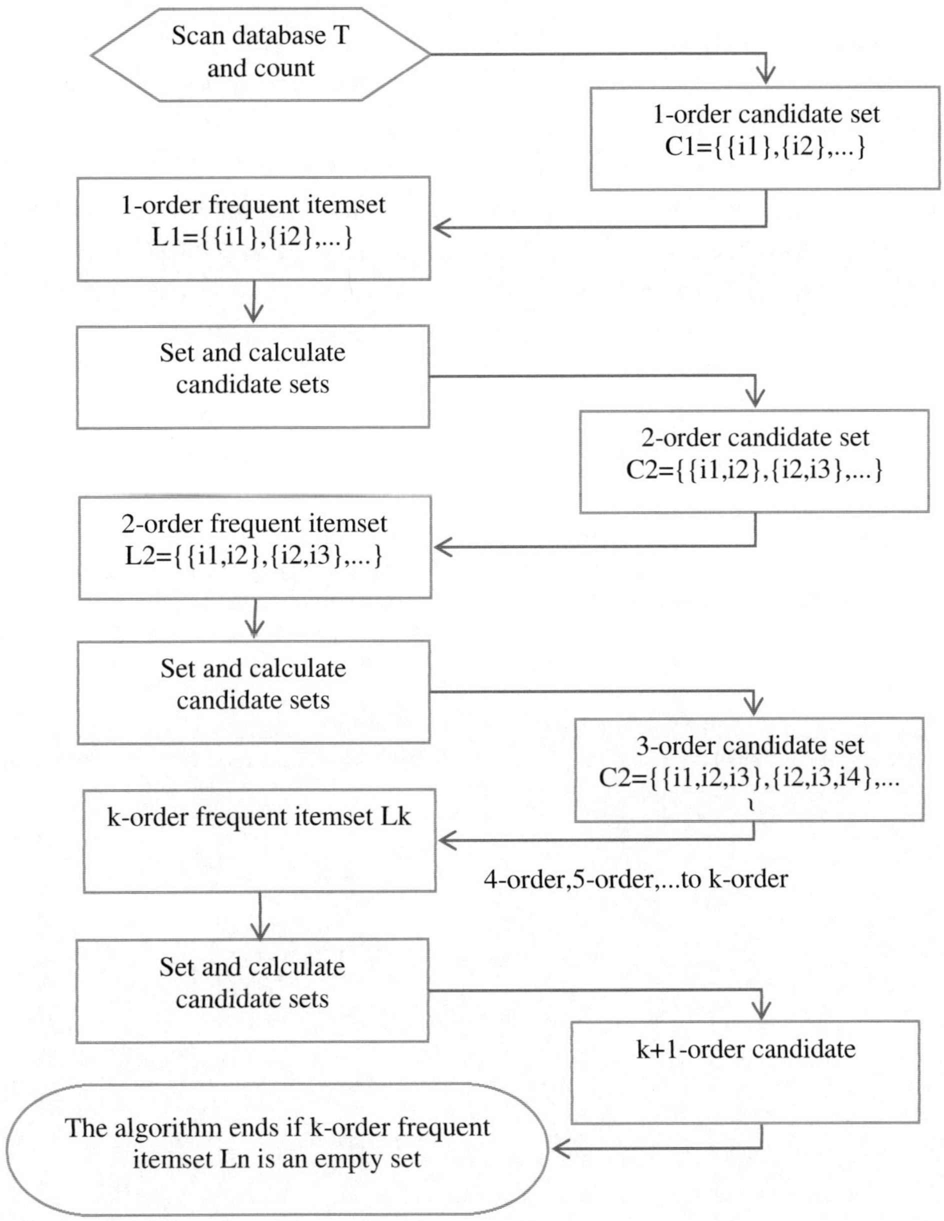

Figure 1. Apriori algorithm data mining schematic.

3. Data association rule mining process

Data association rules mining process includes preparing data, discretizing data, reading Matlab and running programs, generating frequent itemsets, generating strong association rules, analyzing strong association rules and so on.

The vibration signal data of 28 days were selected from 2009.12.28-2010.01.03 and 2010.01.18-2010.02.07, and these data were used as the original data of association rule mining. Because of the large amount of data, in order to facilitate the mining association rules and procedures for the operation, all data is divided into 28 columns (6 bits, respectively, hours, minutes and seconds), part of the data shown in Table 1:

Table 1. Part of the raw data

time/day	09.12.28	09.12.29	09.12.30	09.12.31	10.01.01	10.01.02
000001	1.348226717	1.461543966	0.147767787	1.732512912	1.988745204	0.215233879
000021	0.557416152	1.111776506	0.07700468	1.456330582	2.260567322	0.183927996
000031	1.051686803	1.776334604	0.079273979	0.809488764	1.830401507	0.174495186
000041	1.641297123	1.450069932	0.07770084	1.545590266	1.296293009	0.213564406
000051	0.978158806	1.24278345	0.110310373	1.262294948	1.843240024	0.209127551
000101	2.488720126	1.292546795	0.134116793	0.967015838	2.608663835	0.141376765
000121	1.4078731	1.850598325	0.106696628	0.610512535	0.549463057	0.273129161
000131	2.304847972	1.144550309	0.114081028	1.45902719	0.561034852	0.236489105
000141	1.671902493	1.253938726	0.073675301	1.268312074	6.87629457	0.251650358
000151	6.599175619	0.807598386	0.115374575	1.465584009	1.120279761	0.106547819

Because the association rules analysis algorithm discretizes the data, all transactions need to be pretreated, and the data is discretized by the definition of association rules. The commonly used methods are Boolean discretization and multi-valued discretization [8-9]. Boolean discretization, the data in the normal range of data is mapped to 0, in a non-normal range of the mapping is 1 (that is, the emergence of early warning state is 1, otherwise 0), where the data is greater than or equal to 1, The data of less than 1 becomes 0. The discretization result is shown in Table 2:

Table 2. Boolean discretization results

time/day	09.12.28	09.12.29	09.12.30	09.12.31	10.01.01	10.01.02
000001	1	1	0	1	1	0
000021	0	1	0	1	1	0
000031	1	1	0	0	1	0
000041	1	1	0	1	1	0
000051	0	1	0	1	1	0
000101	1	1	0	0	1	0
000121	1	1	0	0	0	0
000131	1	1	0	1	0	0
000141	1	1	0	1	1	0
000151	1	0	0	1	1	0

Put the data of Boolean discretization into Matlab and save it as a file that can be transferred directly. Finally, the association rule mining based on Apriori algorithm can calculate the degree of support and

IOP Conf. Series: Materials Science and Engineering **220** (2017) 012033 doi:10.1088/1757-899X/220/1/012033

confidence between different time points of each day, and get the frequent itemsets and strong association rules[10-11].

Set a support and confidence threshold min_sup = 0.35, min_conf = 0.6. The itemsets whose support degree is greater than the threshold are frequent itemsets, and the strong itemsets whose confidence is greater than the threshold are strong association rules.Through Matlab calculating support The frequent itemsets are shown in Table 3.

Table 3. Frequent itemsets results

Number	Frequent itemsets	Support
4	1 13	0.3500
6	1 16	0.4164
12	13 16	0.3504

And then calculate the confidence of each frequent itemset and compare it with the minimum confidence threshold. The result shows that the frequent itemsets of number 7 do not meet the requirements, so 13 strong association rules are finally obtained.

From the strong association rules, we can see that the probability of early warning signals of two-day lifting motors with strong association rules is higher at the same time, and there is a strong correlation between them, among them, the first and the 16th day (2009.12.28 and 2010.01.26) are most closely related.

The above results show that the two days the motor is in a similar working environment and working conditions, the two days may have a bad weather conditions or the motor is in high-load working condition, which can re-analysis of these two days of motor work, Draw conclusions and guide the relevant personnel in a timely manner to resolve failures and security risks.

4. Conclusion

In order to verify the results, using Matlab to draw the 1st day and 16th day of the data fitting diagram, as shown in Figure 2:

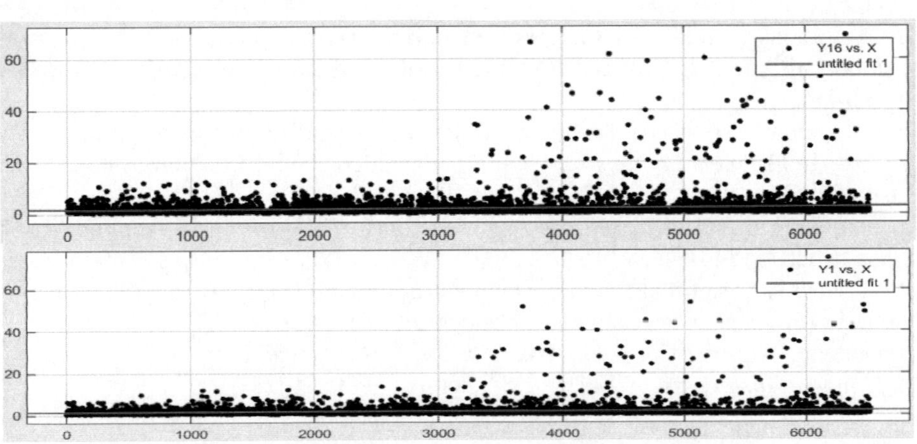

Figure 2. Data fitting diagram

It can be seen from the figure that the distribution of the data in these two days is similar, which proves the correctness of the mining association rules.

The vibration signal data of the hoisting motor of the quayside crane is an important reference for analyzing the mechanical state of the motor, and the association of the vibration signal data with the association rules is an important method to evaluate the working status of the motor, and to prevent and reduce the fault. The association rules between the mechanical data of the motor provide a direction for processing the huge data of the motor, and can monitor and evaluate the working state of the motor. It can also find out the potential faults and safety hazards ahead of time, and has important reference to the relevant staff significance, in later studies can be more than a number of properties of the measurement points to analyze the data to be more accurate and comprehensive association rules.

Acknowledgements
This work was supported by the National Natural Science Foundation of China (No. 31300783), China Postdoctoral Science Foundation (No. 2014M561458), Doctoral Fund of the Ministry of Education Jointly Funded Project (No. 20123121120004), the Shanghai Maritime University Research Project (No. 20130474), the Shanghai Top Academic Discipline Project- management science & engineering, and the high-tech research and development program of China (No. 2013A2041106).

References

[1] Wang Zhi-xin. *Data Mining and Condition Recognition on Characteristic Data of Quayside Container Crane*[D]. Shanghai: Shanghai Jiao Tong University, 2008:15-23.
[2] Xiaoyan Cui, Shimeng Yang, Darning Wang. *An Algorithm of Apriori Based on Medical Big Data and Cloud Computing*[C]. 2016 4th International Conference on Cloud Computing and Intelligence Systems. IEEE Conference Publications, 2016: 361-365.
[3] Liu Hua-ting, Guo Ren-xiang, Jiang Hao. *Research and Improvement of Apriori Algorithm for Mining Association Rules*[J]. Computer Applications and Software, 2009, 26(1) : 146-149.
[4] Cui Guan-xun, Li Liang, Wang Ke-ke, et al, *Research and Improvement of Apriori Algorithm for Mining Association Rules*[J]. Computer Applications, 2010, 30(11) : 2952-2955.
[5] Huang Chang-hai, Gao De-yi, Hu shen-ping, et al, *Association Rule Analysis of Vessel Traffic Accidents Based on Apriori Algorithm*[J]. Journal of Shanghai Maritime University, 2014, 35(3) : 19-22.
[6] Wang Yu, Zhang Wei-hong, Liu Yu. *Application of Association Rules in Information Accessibility WebsiteBased on Apriori Algorithm*[J]. Journal of Jilin University, 2013, 31(1) : 101-106.
[7] Jon Hills, Anthony Bagnall, Beatriz Iglesia, et al, *BruteSuppression: A Size Reduction Method for Apriori Rule Sets[J].* Journal of Intelligent Information Systems, 2013, Vol.40 (3), pp.431-454.
[8] Lamine M. Aouad, Nhien-An Le-Khac, Tahar M. Kechadi. *Performance Study of Distributed Apriori-like Frequent Itemsets Mining*[J]. Knowledge and Information Systems, 2010, Vol.23 (1), pp.55-72.
[9] Dong Juan Gu, Lei Xia. *A Novel and Improved Apriori Algorithm*[J]. Applied Mechanics and Materials, 2015, 543-546.
[10] Guo Min Zhuang. *Simulation Environment of Virtual Visualization Technology Based on Apriori Algorithm*[J]. Applied Mechanics and Materials, 2014, 1131-1135.
[11] Niu Zhendong, Nie Yaoxin, Zhou Qian, et al, *A Brain-region-based Meta-analysis Method Utilizing the Apriori Algorithm*[J]. BMC neuroscience, 2016, 23.

IOP Conf. Series: Materials Science and Engineering **220** (2017) 012034 doi:10.1088/1757-899X/220/1/012034

Load Weight Classification of The Quayside Container Crane Based On K-Means Clustering Algorithm

Bingqian Zhang[1, a], Xiong Hu[2, b], Gang Tang[2, c] and Yide, Wang[3, d]

[1] Logistics Research Center, Shanghai Maritime University, China
[2] Logistics Engineering College, Shanghai Maritime University, China
[3] IETR CNRS UMR6164, University of Nantes, France

[a]bingqianzhang1116@126.com, [b] huxiong@shmtu.edu.cn, [c] gangtang@shmtu.edu.cn, [d] yide.wang@univ-nantes.fr

Abstract. The precise knowledge of the load weight of each operation of the quayside container crane is important for accurately assessing the service life of the crane. The load weight is directly related to the vibration intensity. Through the study on the vibration of the hoist motor of the crane in radial and axial directions, we can classify the load using K-means clustering algorithm and quantitative statistical analysis. Vibration in radial direction is significantly and positively correlated with that in axial direction by correlation analysis, which means that we can use the data only in one of the directions to carry out the study improving then the efficiency without degrading the accuracy of load classification. The proposed method can well represent the real-time working condition of the crane.

1. Introduction

Nowadays, the demand for crane is growing. Due to the lack of scientific basis, there are still problems in the evaluation of service life of the crane. [1] There is a strong association between the service life of the crane and the actual load of each operation. The traditional life assessment methods usually pay attention to the methodological research and lack actual data and experiment verification [2-4].

Even we can figure out the total load during a period through the record book, the load weight for each individual operation is hard to know. The vibration signal of the hoist motor has a nonlinear relationship with the actual load of each operation of the crane. [5] The analysis of the real-time vibration of the motor allows estimating the load weight of each operation which provides reliable data for the further evaluation of crane service life and improves the accuracy of the assessment.

Data analysis methods are essential for analyzing the ever-growing massive quantity of data. [6] One of the most popular and efficient clustering methods is the K-means method [7,8]. This paper proposes a container crane load classification method based on K-means clustering algorithm. The proposed technique allows discriminating and classifying a set of vibration signals of the hoist motor corresponding to different classes of the loads.

The paper is organized as follows. Section 1 describes the data acquisition and raw data pre-processing. Section 2 analyzes the acquired data and presents the crane's load classification. Section 3 validates the classification method with real-time data. Section 4 concludes the paper.

2. Data processing

IOP Conf. Series: Materials Science and Engineering **220** (2017) 012034 doi:10.1088/1757-899X/220/1/012034

2.1. Data acquisition

NetCMAS (network condition monitoring and assessment system) has been effectively used for equipment state analysis and management in big container companies. [9] NetCMAS acquires and processes real-time signals of different equipment. It can also save the data. As the data acquisition is in real-time, the signal contains two kinds of states including working and non-working in addition of interference signals.

As mentioned previously, the load is directly related to the vibration signals which can be required directly. The data analyzed in this paper are acquired by NetCMAS from real-time vibration signals in radial and axial directions of a measuring point of hoist motor of the crane (refer with: Figure 1, equipment installed on the crane in a Shanghai port). We select the data from 2pm January 18, 2010 to11pm February 7, 2010 for a period of three weeks which are comparatively representative.

Figure 1. left : crane; right : hoist motor.

2.2. Data preprocessing

The real-time data collected by the system, which usually contain abnormal data, repeated data or missing data, cannot be used directly to their classification and they should be preprocessed.

2.3. Data normalization preprocessing

The data are limited to the range of 0~1 through their normalization which is a convenient way for the future data processing and reducing the operation time.

First of all, we get the data, representing the radial and axial vibration signals of the hoist motor, of each hour and take the average hourly. There are 490 sets of data totally. The maximum value of the radial direction is 3.14, and the maximum value along the axial direction is 4.93. They are used for the data normalization (the data are normalized by dividing each value with the maximum value). The range (the max value minus the min value of the data) for radial data is 3.08, and the range for the axial data is 4.87 which means that the magnitude of change of vibration signals is not big. So it meets the conditions for the K-means clustering analysis. The standard deviations along the axial and radial directions are 0.61 and 0.87 respectively which indicate that the dispersion of the data is relatively small.

3. Load classification

3.1. Crane load and motor vibration

The motor of crane is generally a three-phase asynchronous model. The load of quayside crane is proportional to the input power of the motor according to the mechanical characteristics of three-phase asynchronous motor. What's more, the input power is proportional to the electromagnetic torque. The input power is inversely proportional to the slip.

After the motor stator being connected to the power supply network, the fundamental wave of magnetic motive force produces the rotating magnetic field with same speed and same directions. The amplitude is proportional to the stator current which is proportional to the input power.

Generally, with the increase of the load of the hoist motor, the vibration of the hoist motor will also increase [10]. Therefore, in order to investigate the load category, the radial and axial vibration intensity data can be used. However, they are not two independent signals. To investigate their relationship, the correlation analysis is carried out.

3.2. Correlation analysis

We use R language and SPSS (Statistical Product and Service Solutions) to analyze the correlation between the axial and radial vibration intensities. SPSS is a series of software products and related services for statistical analysis, data mining, predictive analysis and decision support tasks promoted by IBM. In this study, the axial vibration intensity, denoted by xi, is taken as independent variable and the radial one, denoted by yi as dependent variable. There is a significant positive correlation (refer with: Figure 2) between the axial and radial vibration signals for a measuring point as shown in figure 2. Let X represent the radial direction, and Y the axial direction. In figure 2, a point represents the data of one hour and different colors correspond to different days. This correlation can be quantified by the following correlation coefficient (refer with: (1)):

$$r=\frac{N\sum x_i y_i-\sum x_i\sum y_i}{\sqrt{N\sum x^2_i-(\sum x_i)^2}\sqrt{N\sum y^2_i-(\sum y_i)^2}} \tag{1}$$

where r is the Pearson correlation value whose absolute value is limited by one; N is the number of the data samples; xi is the ith radial vibration intensity; yi is the ith axial vibration intensity.

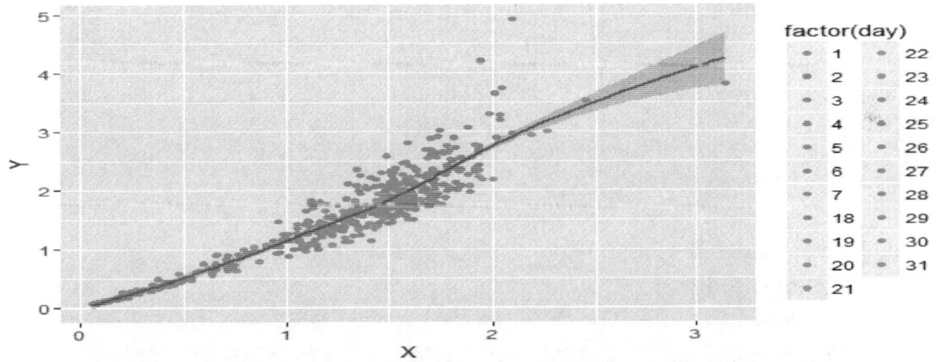

Figure 2. Axial vibration intensity in terms of radial vibration intensity

After calculating, the Pearson correlation value between the radial and axial vibration intensities is 0.964. So, there is a significant positive correlation between the radial and axial intensities. In order to get a quantitative model of the axial vibration intensity (y) and radial vibration intensity (x), we establish the following linear regression model (refer with: (2)) between them:

$$y=ax+b \tag{2}$$

By applying the least squares method to the data, we get a = 0.895 and b = 0.151. Finally, the mathematical model (refer with: (3)) of the relation between the radial and axial vibration intensities can be written as:

$$y=0.895x+0.151 \tag{3}$$

This model shows that we can predict the radial vibration intensity values from the axial value through Equation 3. Consequently, the data of one of the two directions can be used for the load classification, which will reduce the memory requirement and calculation complexity.

3.3. Classification for the crane's load

In practice, the load of the hoist motor can be divided into 5 categories representing the 5 typical working conditions, and the working cycle of the hoist motor in this paper is defined as:

1. non-load: idle;
2. small-load: lifting few rated load and generally lifting small load;
3. medium-load: occasionally lifting rated load and generally lifting medium load;
4. high-load: usually lifting rated load;
5. over-load: continually lifting rated load.

4. Load classification based on K- means cluster

By comparing the hierarchical clustering and K-means clustering, the most common classification methods, it is shown that the K-means clustering is more suitable for this study due to the large amount of data. The output results by K-means are easier to understand than the results by hierarchical clustering.

This paper classifies 21 sets of data by K-means clustering and sets time as categorical variables and vibration signals in radial and axial directions as the independent variables. Six categories are shown in Table 1.

Table 1. Hoist motor vibration intensity K-means clustering

Categories	**1**	**2**	**3**	**4**	**5**	**6**
Time	0127	0125	0119、0120	0121	0122、0128	0118、0123
			0124、0130	0131	0129、0203	0126、0201
			0205、0206		0204	0202、0207

From Table1, we observe that the load state of the hoist motor can be defined to five different categories which is summarized in Table 2.

Table 2. Classification standard of hoist motor load

Load Condition	**Y**	**X**
Non-Load	$0<y\leq 0.4356$	$0<x\leq 0.3763$
Fractional-Load	$0.4356<y\leq 0.5199$	$0.3763<x\leq 0.4197$
Medium-Load	$0.5199<y\leq 0.6101$	$0.4197<x\leq 0.5293$
High-Load	$0.6101<y\leq 0.7069$	$0.5293<x\leq 0.5963$
Over-Load	$0.7069<y\leq 1$	$0.5963<x\leq 1$

In order to verify the reliability of this classification result, we classify the vibration intensity of the radial and axial directions separately. The results, in Table3, show that the categorized results in axial direction are almost matched to those in radial direction. Hence, we can only use the data in one of the directions to carry out the study.

Table 3. Categorized results in radial and axial directions separately

Categories	1	2	3	4	5	6
Percentage Of Matching Degree [%]	100	85	85	100	100	90

The axial vibration and radial vibration intensities are positively correlated, and their classification results are almost the same, therefore we only need to measure the vibration intensity in one direction in practical application, which will greatly reduce the amount of calculation and improve the calculation speed.

In the following, we calculate the number of times of each working state of the quayside container crane in one day by coding through Java. We can learn from the real-data in NetCMAS that the data interval is about 15s. The statistical results are shown in Table 4. The data displayed in this table can be used to accurately evaluate the service life of the crane which is such a great improve. As we all know, the over-load working states usually cause the damage to the crane.

Table 4. Proportion of load time in January 26th

Load Condition	Time Percentage [%]	Operation Times
Non-Load	16	115
Small-Load	44	316
Medium-Load	4	29
High-Load	17	123
Over-Load	19	137

According to the statistical results, the hoist motor is occasionally in the condition of lifting rated load and usually in the state of small and medium load.

5. Conclusion

From the above analysis, this paper made the following conclusions:

1. This paper analyzes the real-time vibration data of the hoist motor of the crane in order to classifying the load of crane. This method has many advantages such as: selecting few characteristic parameters, simple operation, high accuracy and strong practicality.
2. The classification results are well adapted to the engineering application. According to the classification results, this paper calculates the number of times and proportion of operation of each type of the load of the crane in one day. That method provides helpful data for the evaluation of service life of the crane in application.
3. We have come to the conclusion that there is significant positive correlation between radial and axial vibration signals. This conclusion has practical significance for engineering application, which can simplify the monitoring work and release memory.

Acknowledgements

This work was supported by the National Natural Science Foundation of China (No. 31300783), China Postdoctoral Science Foundation (No. 2014M561458), Doctoral Fund of the Ministry of Education Jointly Funded Project (No. 20123121120004), the Shanghai Maritime University Research Project (No. 20130474), the Shanghai Top Academic Discipline Project- management science & engineering, and the high-tech research and development program of China (No. 2013A2041106).

References

[1] Fan X. Evaluation Method of Remaining Fatigue Life for Crane Based on the Acquisition of the Equivalent Load Spectrum by the Artificial Neural Network[J]. Journal of Mechanical Engineering, 2011, 47(20):69.

[2] Yin Y, Hu X, Lin T R. A practical approach to analyze the non-stationary signals of a quayside container crane motor using a combined empirical mode decomposition and wavelet packet quantization technique[J]. Noise Control Engineering Journal, 2016, 64:págs. 126-133.

[3] Xu X, Hu X, Jiang S. Statistic analysis and predication of crane condition parameters based on SVM [C]// IEEE International Conference on Automation and Logistics. IEEE, 2010:109-113.

[4] LeiDong. Some fatigue life prediction method research [D]. University of science and

technology of China, 2006.

[5] Nakazono K, Ohnishi K, Kinjo H, et al. Vibration control of load for rotary crane system using neural network with GA-based training[J]. Artificial Life and Robotics, 2008, 13(1):98-101.

[6] Zuber N, Bajrić R. Application of artificial neural networks and principal component analysis on vibration signals for automated fault classification of roller element bearings[J]. Eksploatacja i Niezawodnosc - Maintenance and Reliability, 2016, 18(2):299-306.

[7] Ding C, He X. {\it K-means clustering via principal component analysis [C]// International Conference on Machine Learning. ACM, 2004:29.

[8] Kanungo T, Mount D M, Netanyahu N S, et al. An efficient k-means clustering algorithm: analysis and implementation[J]. IEEE Transactions on Pattern Analysis & Machine Intelligence, 2002, 24(7):881-892.

[9] Yin Yinan, Hu Xiong, li yong. Heavy lifting motor vibration research and health evaluation standard to determine [J]. Journal of vibration and shock, 2013, 32 (12) : 67-71.

[10] Kuehne S. Model assisted computing of three-phase asynchronous motor torque[J]. 1999.

2nd International Conference on Design, Materials, and Manufacturing IOP Publishing
IOP Conf. Series: Materials Science and Engineering **220** (2017) 012035 doi:10.1088/1757-899X/220/1/012035

State Identification of Hoisting Motors Based on Association Rules for Quayside Container Crane

Q. Z. Li, T. Gang*, H. Y. Pan and H. Xiong
Logistics Engineering College, Shanghai Maritime University, Shanghai, China

gangtang@shmtu.edu.cn

Abstract: Quay container crane hoisting motor is a complex system, and the characteristics of long-term evolution and change of running status of there is a rule, and use it. Through association rules analysis, this paper introduced the similarity in association rules, and quay container crane hoisting motor status identification. Finally validated by an example, some rules change amplitude is small, regular monitoring, not easy to find, but it is precisely because of these small changes led to mechanical failure. Therefore, using the association rules change in monitoring the motor status has the very strong practical significance.

1. Introduction

In the past, most of the researches on association rules mining in engineering field only analyzed the association rules and did not analyze and analyze the long-term evolution and minor changes of association rules [1-3]. Association rules are used to monitor the change of mechanical state of shore-bridge hoisting motors, so it is necessary to carry out association rules mining for long-term monitoring information of quayside bridge. Therefore, the proposed partition association rule-similarity [4-5], that is, The Similarity Degree of Mechanical Information Association Rules of Hoisting Motors.

2. Similarity criteria

The similarity between association rules and the formula is as follows:

$$S_{ij}=\begin{cases} \dfrac{l_{ij}\sum_{p} X_{ijp}}{|A_{ij}|}\times\dfrac{h_{ij}\sum_{q} Y_{ijp}}{|B_{ij}|} & \text{if } |A_{ij}|\neq 0 \text{ and } |B_{ij}|\neq 0 \quad (1\text{-}1)\\ 0 & \text{if } |A_{ij}|\neq 0 \text{ or } |B_{ij}|=0 \quad (1\text{-}2)\end{cases}$$

Among $l_{ij}=\dfrac{|A_{ij}|}{\max(|M_i^{t_1}|,|M_j^{t_2}|)}$,

$$H_{ij}=\frac{|B_{ij}|}{\max(|N_i^{t_1}|,|N_j^{t_2}|)} \quad (1\text{-}3)$$

L_{ij} and $H_{ij}$are used to represent the axial similarity of the regular lift motor and the hoist motor respectively. The similarity S_{ij} is the general rule similarity of the motor, and its range is [0,1]. After the rule similarityity is calculated, we can use the maximum similarity of rit1 and rjt2 to measure the rule change at time t1 and t2 [6-10].

The maximum similarity formula is:

$$S_i^{t1}=\max(S_{i1}^{t1},S_{i2}^{t2},\ldots,S_{iRt2}^{t1})$$

$$S_j^{t2}=\max(S_{j1}^{t2},S_{j2}^{t2},\ldots,S_{jRt1}^{t2})$$

If the maximum similarity $S_{ij}=1$ of the association rule, the association rule exists at any time, so it is called stationary rule. If the support of association rules is increasing, the rules are becoming stronger and more stable. On the contrary, if the support is decreasing, the rules are becoming weaker and there is a tendency to disappear.

If the maximum similarity of association rule $0 < S < 1$, it means that the left and right association rules are partially similar. (RMT) is set. If the maximum similarity S> RMT and disappears at time t2, the rule is the disappearance rule. If the maximum similarity of the association rule S <RMT, and the time t2 is different from the time t1, the rule is a new rule.

3. Categories of association rules

In this paper, a large number of association rules mining for lifting motors of a bridge crane from January to June of 2015 are carried out, and some interesting results are obtained (GA - kmeans result is taken as an example). But some of these rules have always existed, some have disappeared over time, and others have changed over time. Therefore, according to the regularity of these rules, into the similarity of the formula, the association rules will be divided into four categories, including the stability of the rules, changes in the rules, the disappearance of the rules and new rules.

3.1. The stability of the rules

During the months of January through June 2015, there have been some fixed rules that do not change much in the numerical range or number of projects, as shown in Table 1.

Table 1. Steady Rules (January and February)

Lifting Motor Radial Frequent Itemsets	Lifting Motor Axial Frequent Itemsets	Support (%)	Confidence (%)

L[0.0532 0.8655]	R[0.0513 34.4633]	65.16 56.35	70.41 68.86

The left side of the association rule is the vibration intensity in the radial direction of the hoist motor, and the right side is the vibration intensity in the axial direction. When the time t12 in January and February, respectively, the support of the rule is 65.16 and 56.35, and the confidence is 70.41 and 68.86, the changes are small. According to the formula to calculate the maximum similarity and are 1, and the support will be in the range of 32 - 38%. The rule states that the mechanical condition of the hoist motor remains stable during the two months without any major change. Therefore, this rule is called a stable rule.

3.2. The rules of change

Container crane hoisting motor with the continuous wear and tear of mechanical components, various parts of the state continue to occur in subtle changes, so the overall state of the bridge crane is always changing, it will produce a lot of fine rule changes.

Some of the rules found in January are shown in Table 2.

Table 2. Rules for Changes (January)

Lifting Motor Radial Frequent Itemsets	Lifting Motor Axial Frequent Itemsets	Support（%）	Confidence（%）
L[0.0532 0.8655]	R[0.0513 24.4633]	62.22	66.87

Some of the rules found in March are shown in Table 3.

Table 3. Rules for Changes (March)

Lifting Motor Radial Frequent Itemsets	Lifting Motor Axial Frequent Itemsets	Support（%）	Confidence（%）
L[0.0570 0.8578]	R[0.0513 35.4633]	50.93	59.20

Some of the rules found in May are listed in Table 4.

Table 4. Rules for Changes (May)

Lifting Motor Radial Frequent Itemsets	Lifting Motor Axial Frequent Itemsets	Support（%）	Confidence（%）
L[0.0586 1.3763]	R[0.0622 22.0970]	35.50	54.28

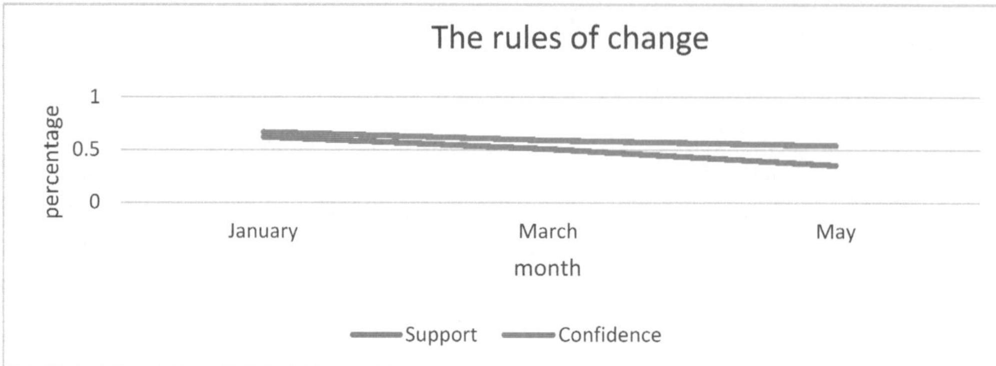

Figure 1. The rules of change

From the above 3 months, the same radial and axial vibration intensity,as shown in Figure 1. Support and confidence were high in the January rule, but in the May change, support and confidence fell by about 10%. During this period, a new rule interval was found in which the range of values from R [0.0513 24.4633] to R [0.0513 35.4633] in March was larger, but the number remained unchanged, indicating a change rule.

3.3. The rules of disappearance

From January to June 2015, some rules will suddenly disappear, the phenomenon often occurs in a part of the parts change, and then rule disappears. As shown in Table 5:

Table 5. Rules for Disappearance (February)

Lifting Motor Radial Frequent Itemsets	Lifting Motor Axial Frequent Itemsets	Support（%）	Confidence（%）
L[0.0.640 1.0621]	R[0.0663 15.7336]	30.19	47.26

The rule of disappearance shown in the above table is obtained in February. From the table, the association rules of the motor radial and axial vibration can be obtained, but the rule disappears in March, and the maximum similarity S = 0, Set RMT = 0.5. That is, the rule is disappears.

3.4. New rules

Usually some rules will disappear after the new rules will appear with the emergence of new rules in June as shown in Table 6:

Table 6. Rules for New Students (June)

Lifting Motor Radial Frequent Itemsets	Lifting Motor Axial Frequent Itemsets	Support (%)	Confidence (%)
U1[0.0534 0.9177]	U2 [0.0399 0.5484]	63.78	79.82

Prior to June, the association rules did not exist, the rule of support and confidence are high, indicating that the new rules are significant. The maximum similarity of the rule S = 0, so this is a new rule.

4. Summary

In this paper, we first propose to identify the change of hoist motor state of the quayside bridge by using the change of association rule. The similarity degree is used as the measure of association rule partitioning. According to this standard, the association rules are divided into: stable rules; disappearing rules, changing rules and new rules. In order to make better use of association rules to identify the mechanical state of hoisting motor of quayside crane, the real-time data of field hoisting crane is analyzed, which further proves the above viewpoints. This view can find the change of the state of the quay crane from the subtle changes of the monitoring data, which is very important for the condition monitoring, diagnosis and evaluation of the quayside container crane.

Acknowledgements

This work was supported by the National Natural Science Foundation of China (No. 31300783), China Postdoctoral Science Foundation (No. 2014M561458), Doctoral Fund of the Ministry of Education Jointly Funded Project (No. 20123121120004), the Shanghai Maritime University Research Project (No. 20130474), the Shanghai Top Academic Discipline Project- management science & engineering, and the high-tech research and development program of China (No. 2013A2041106).

References

[1] Jinag X C, Wang D ZH ,Ning Y. 2016.Analysis method of power grid fault diagnosis based on association rule, TP311.13; TM727

[2] Wang L, Wu L L, Xiao X Y.2015.A method and system for prediction of rolling product quality based on association rule tree: CN104298778A [P].

[3] Zhao D M Zhang X, Wei J, Liang W CH, Zhang D Y.2014Research on fault diagnosis of power grid for fault recovery process [J]. *Proceeding of the CSEE.*

[4] Chen M C, Chiu Ai-lun, Chang H-H .2005.Mining changes in customer behavior in retail marketing, *Expert system with applications,* 28, 773-781

[5] V Radhakrishna, SA Aljawarneh, PV Kumar, V Janaki. 2017. A novel fuzzy similarity measure and prevalence estimation approach for similarity profiled temporal association pattern mining[J]. *Future Generation Computer Systems.*

[6] Cao Z-W, Qian J, Zhang W M . 2007.A Synthetic Concept Similarity Computing Method [J] .*Computer Science*

[7] Tan R, Lu Zh Q Yan X P.2008.Study on Association Rules Mining Model Based on Similarity in Distributed Environment [J] .*Application Research of Computers,* , 25 (3): 695-697.

[8] Rodrguez MA, 2003, Egenhofer MJ.Determining semantic similarity among entity classes from different ontologies. *IEEE Transactions on Knowledge and Data Engineering.*

[9] Formica A.Ontology2based concept similarity in Formal Concept Analysis. *Journal of Information Science.*

[10] Weeds J. 2002.The Reliability of A Similarity Measure， *Proceedings of the Fifth UK Special Interest Group for Computational Linguistics.*

IOP Conf. Series: Materials Science and Engineering **220** (2017) 012036 doi:10.1088/1757-899X/220/1/012036

Co-simulation on vibration characteristics of uniaxial shaker based on AMESim and ADAMS

W Liu[1], D B Zhang[2] and M Su[1]

[1]Simulation Laboratory of Mechanical & Control Engineering, Guizhou Normal University, Guiyang Guizhou 550014, China;
[2]School of Mechanical Engineering, Guizhou University, Guiyang Guizhou 550025, China

Abstract. In this paper, we propose a selection method of the increased amplitude stability parameter for the unstable motion problem of the uniaxial shaker before its stable operation. On the basis of single-axis shaker dynamics equations, an ADMAS and AMESim combined simulation model is established. According to this model, the vibration characteristics under different parameters are solved by using the control variate method. The simulation results show that the motor speed, the eccentric mass and the inclination of screen surface are the three main factors which affect work starting state of uniaxial shaker. The working efficiency of uniaxial shaker is controlled by the motor speed while the amplitude is affected by the eccentric mass. Moreover, the inclination of screen surface plays a decisive role in the distribution of the vibration track before uniaxial shaker reaching the stable operation state. The relatively stable movement is obtained by optimizing the parameters, which provides a new way to improve the stability of uniaxial shaker.

1. Introduction

Uniaxial shaker[1] widely used in modern industry, its controllability projectile angle, adjustable amplitude, diversified hierarchical levels and other screen surface vibration characteristics of the vibrating screen performance play a decisive factor. Shaker work, if the shaker vibration amplitude is too large, the body will lead to instability, increasing the impact on the ground; and if the vibrating screen to reach a stable state for a long time, will affect the life of the body. Li Bin-hui[2] adopts PID control strategy to optimize the parameters, to achieve the stability of the control, but this method is complicated, and the workload of the experiment is large when the parameters are optimized; Zhang La-la[3] discusses the influence of amplitude, throwing index, screen angle on the average speed of screening material particles, the speed change reflects the fluctuation amplitude from the side, but in practice, these three parameters can not be precise control, parameter selection is limited; Chen Xiao-qiang[4] analysis of the response of a shaker system, through numerical calculation and experimental procedures modified exciting force, stable amplitude ideal, computational complexity theory, applicable to verify the optimized parameters results. In fact, the time from start to hold steady state is the key point to study the vibration characteristics of the vibrating screen, and the output vibration direction of the ellipse can be more intuitive to analyze the stability of the amplitude.

This research is based on the dynamic equation of uniaxial shaker, and the AMESim and ADAMS software are combined co-simulation [5,6]. A control system is established in AMESim, and the mechanical vibration system is established in ADAMS. The speed of the drive motor, the quality of the eccentric block and the inclination angle of the screen face are simulated. According to the displacement time curve, the eccentric block angular velocity curve and the displacement synthesis curve, the steady state of the vibrating screen is analyzed, and it introduced the concept of throwing index and verify, then compare the simulation results under different conditions, get the factors that affect the vibration state, optimize the structure of the vibrating screen, and output the optimized curve.

2. Structure and working principle of uniaxial shaker

The basic structure of uniaxial shaker is shown in figure 1. Mainly consists of vibration exciter (including 2 sets of eccentric block, rotating shaft, bearing), box, screen surface, support springs, bearings, conveyor belt (or coupling), vibration motor and other parts of the vibration. Wherein, the eccentric block is fixed with rotating shaft and the motor transmits energy through the transmission belt, the eccentric device of uniaxial shaker obtains the rotational speed, thereby generating the exciting force and driving the box body under the action of the supporting spring, and making the material entering the sieve surface to generate vibration and reaching the screening effect.

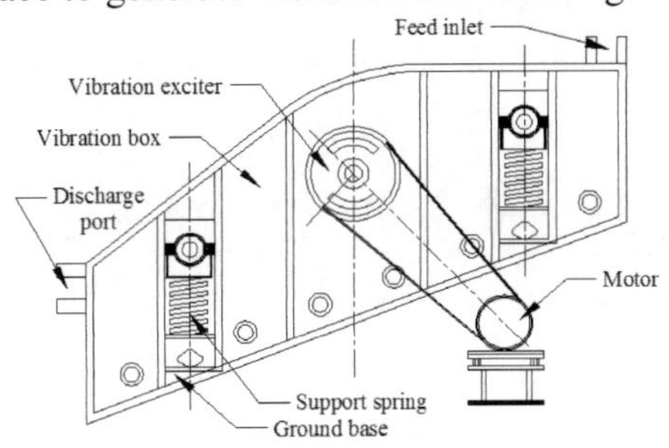

Figure 1. Schematic diagram of single shaft vibrating screen.

3. Modeling of uniaxial shaker

3.1. Mathematical modeling of uniaxial shaker

According to the Darren Bell theorem of the motor rotor, the dynamic equations of the vibrating screen mainly include: the characteristics of the vibration motor, load torque, friction torque, etc.

According to the electric drive theory, the practical expression of the mechanical characteristics of the exciting motor is [7]:

$$T = \frac{2T_m}{\frac{s}{s_m} + \frac{s_m}{s}} \tag{1}$$

Among: $s = \frac{n_0 - n}{n_0}$

Where: T_m—Maximum torque of vibration motor, N.m; T—Motor output torque, N.m; s—The vibration motor slip; s_m—The maximum vibration motor slip; n_0—Synchronous speed of vibration motor, r.s^{-1}, n—Vibration exciter speed, r.s^{-1}, $n = \frac{\omega}{2\pi}$; ω—Vibration exciter angular velocity, rad.s^{-1}.

According to the single shaft driving device, the vibration of the axis line direction and the friction force between the vibration exciter and the box body are not considered, the gravity of the eccentric block provides energy, prompting the vibration exciter rotated around the axis line to obtain the track motion. Vibrating screen to overcome the gravity of the eccentric block to produce the load torque of[8]:

$$T' = \sum_{i=1}^{4} m_i r_i g \sin\varphi \tag{2}$$

Where: φ—Eccentric block angular displacement of rotating shaft, rad; $\sum_{i=1}^{4} m_i r_i$—product of mass radius of four eccentric blocks, $\sum_{i=1}^{4} m_i$ is the total mass of four eccentric blocks, kg, i=1, 2, 3, 4; g—Acceleration of gravity, 9.8m.s^{-2}.

In the study of the characteristics of the excitation motor drive system, the friction torque of the motor rotor is [9]:

$$T_f = T_v + f\omega + \sum_{i=1}^{4} m_i r_i f \omega^2 \frac{d}{2} \tag{3}$$

Where: T_v—Motor rotor constant resistance moment, N.m; f—Friction coefficient; d—bearing bore diameter, m.

Combined with the above parameters, the control equation of the vibrating screen drive system can be obtained:

$$\frac{2T_m n_z}{\dfrac{n_0\pi - 30\omega n_z}{n_0\pi s_m} + \dfrac{n_0\pi s_m}{n_0\pi - 30\omega n_z}} - T_v - f\omega - \frac{1}{2}\sum_{i=1}^{4} m_i r_i f \omega^2 d - \sum_{i=1}^{4} m_i r_i g \sin\omega t = J\frac{d\omega}{dt} \tag{4}$$

Where: n_z—Conveyor belt between the motor and the vibrator; J—Moment of inertia, kg.m^2.

By the dynamic equation, the research on the vibration characteristics of vibrating screen contains the amplitude, frequency, the angle of the screen surface and so on[10]. In Liu Huan-sheng's paper[11], it obtained a comprehensive evaluation index, that is throwing index, it is used in the definition of single axis vibrating screen is:

$$D = \frac{\omega^2 H \sin(\alpha + \beta)}{g\cos\beta} \tag{5}$$

Where: H—Vertical component of amplitude, m; α—The initial projection angle, rad; β—Screen surface inclination, rad.

3.2. Vibration system model of uniaxial shaker based on ADAMS

According to the working principle of uniaxial shaker[12], the simple vibration system model is established by UG8.0 and imported into ADAMS[13], as shown in Figure 2. The inside of the screen box is simplified, and the quality is controlled by the cylinder on both sides of the symmetrical distribution. At the same time, the connection between the box and the vibration exciter is connected with the rotary pair, the support is connected with the four supporting springs, and the bottom of each bearing and the ground are fixed with a fixed pair.

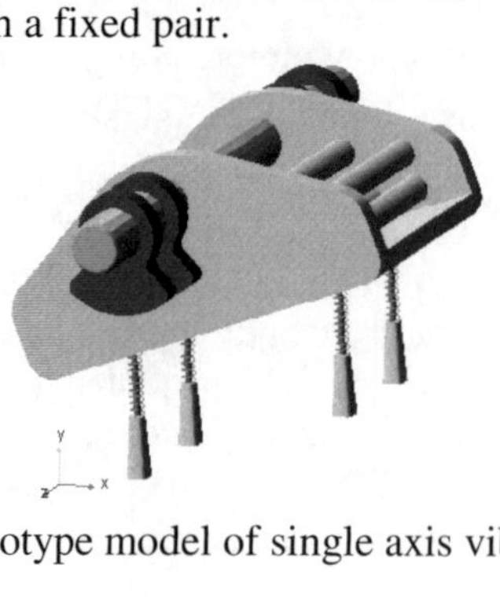

Figure 2. Prototype model of single axis vibrating screen.

In ADAMS, the model direction is established as shown in figure 2, the direction of gravity is the negative direction of the Y axis, and the Z direction is parallel to the horizontal direction of the cylinder in the box, and the X direction is perpendicular to the horizontal direction of the cylinder.

Some material properties of vibrating screen are defined in table 1.

Table 1. Material properties of vibrating screen

Name	Density[kg.m^3]	Name	Spring coefficient[N.m]	Damping coefficient[N.s.m^{-1}]
Box	8545.0	Support spring	700	0.58
Eccentric block	1.9222E+04	X direction bushings force	5500	0.06
Rotary shaft	1.9222E+04	Z direction bushings force	450000	0.68

Due to the joint simulation, ADAMS needs to get the torque energy from AMESim, while the angular velocity of the eccentric device system vibration transmitted to the AMESim, so as to exchange the simulation data, the study needs to establish the input and output variables, the specific content as shown in table 2.

Table 2. ADAMS input and output variable table

Variable name	Variable type	Physical meaning
Torque	Output variable	Motor torque, acts on the exciter
AngularV	Input variable	the angular velocity of Shaker eccentric
XDis	Input variable	X direction displacement of vibrating screen box
YDis	Input variable	Y direction displacement of vibrating screen box

3.3. Model of uniaxial shaker control system based on AMESim

According to the dynamic equation, the control system model based on AMESim[14] is established in Figure 3. AMESim is mainly to provide torque energy for the ADAMS vibration system, at the same time to receive the angular velocity of its output, and feedback to the torque, forming a closed loop. According to the working principle of the shaker, the system is divided into the vibration system and the control system. The joint simulation diagram is shown in Figure 3.

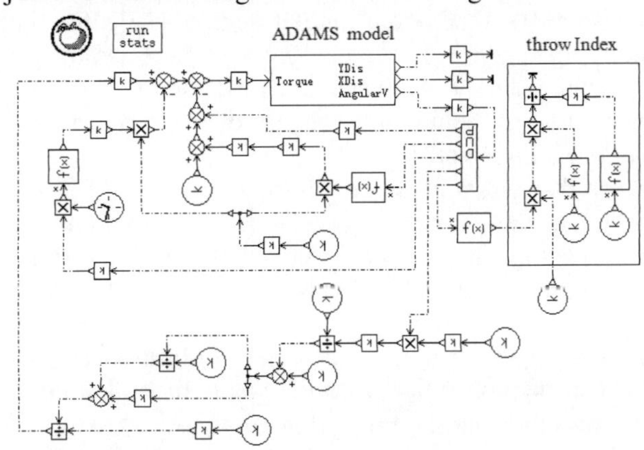

Figure 3. Joint simulation of AMESim and ADAMS.

The main parameters in AMESim are set as shown in table 3:

Table 3. Main parameter settings for AMESim

Parameter name	Parameter character	Value	Units
Motor maximum torque	T_m	3	N.m
The maximum motor slip	s_m	0.152	/
Conveyor belt transmission ratio	n_z	0.8	/
Motor rotor constant resistance moment	T_v	0.01	N.m
Bearing bore diameter	d	0.3	m
Friction coefficient	f	0.002	/

4. Sections, subsections and subsubsections

To ensure the consistency of the two software, the interface of the two model is adjusted by several gain modules. The simulation is set to Discrete co-simulation, The two software respectively solve the problem, develop the advantages of each solver and exchange the obtained data [15].

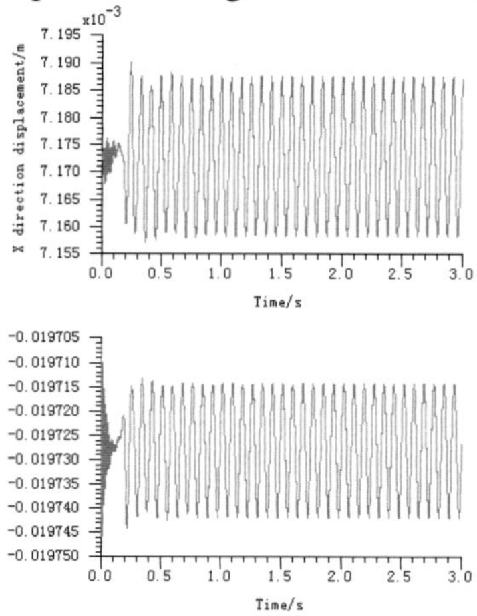

Figure 4. X and Y direction displacement curves of single axis vibrating screen.

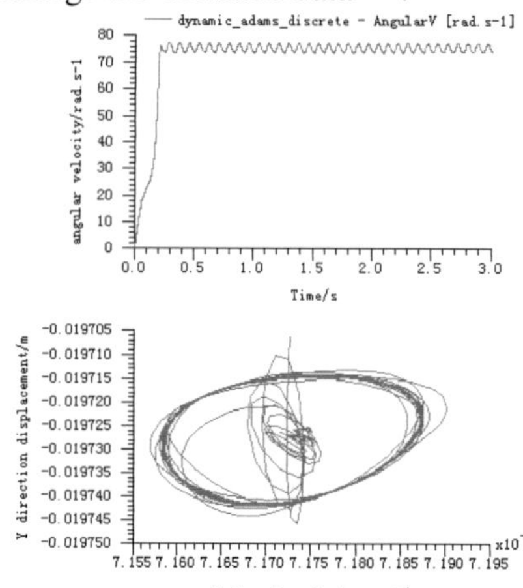

Figure 5. Velocity and displacement curve of eccentric angle.

Setting the time interval is 0.001s, the system simulation time is 3s, the step size is 0.0005s. Setting the motor speed is 900 r.min^{-1}, the angle between the screen surface and the ground is 20°, the material density and size remain unchanged. The displacement time curve of the X and Y direction of the vibrating screen, the change curve of the angular velocity of the eccentric block and the trajectory of the displacement are shown in Figure 4 and 5.

By the graph, you can see that vibrating screen in the direction of Y and X made positive cosine curve movement after reaching a stable state, in a short period of time can be achieved basically stable, the formation of elliptical trajectory, the stability is better. At the same time, the size of the angular velocity is stable and the regularity of the circulation is maintained.

5. Simulation of factors affecting vibration characteristics of uniaxial shaker

Through the analysis and experiment of initial projectile angle, the initial projection of uniaxial shaker angle between 70°~74°[11]. This study set the projection angle $\alpha = 72^{\circ}$. In guarantee the normal work of vibrating screen case, throwing index size is generally between 3~5. From the definition of throwing index can know its size and amplitude, eccentric angular velocity, the inclination is related, the specific model is shown in figure 3. In the experiment, study on the effect of motor speed, eccentric quality, screen angle and other factors change on the throwing index, so as to optimize the internal structure of vibrating screen, achieve the ideal screening effect.

5.1. Effect of motor speed

For different motor speed simulation, the results are as shown in figure 6.

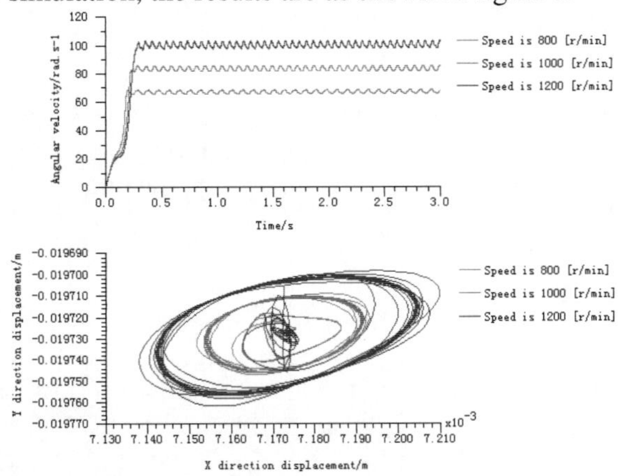

Figure 6. Effect of motor speed on the throwing index.

The speed of the three simulations is 800r.min^{-1}, 1000r.min^{-1}, 1200r.min^{-1}. As can be seen from Figure 6, the motor speed determines the angular speed of the eccentric device, when the speed increases, the formation of the elliptical trajectory more standardized, the time corresponding to the steady state of the vibrating screen is increased, and the numerical value of angular velocity and the amplitude of vibration are also larger. Comparison of the three groups of simulation images, you can get: when the speed is greater, the same time, the formation of the trajectory is more intensive, the better the efficiency of the work. Three sets of vertical component data are obtained by simulation: 2.15×10^{-5}, 3.61×10^{-5}, 5.48×10^{-5}. Using the formula (5) can be calculated from their throwing index are: 1.05, 2.78, 5.95. According to the requirements of the throwing index, we can see: with the increasing of speed, throwing index increased. When the speed is less than 800r.min-1 and greater than 1200 r.min-1, the throwing index does not meet the conditions, so the appropriate rotational speed in the range of 800~1200 r.min-1.

5.2. Effect of eccentric mass

For different eccentric mass, because the eccentric device size in ADAMS is fixed, for the sake of simplicity, the quality can be changed by changing the properties (density) of the material, the simulation results are obtained as shown in figure 7.

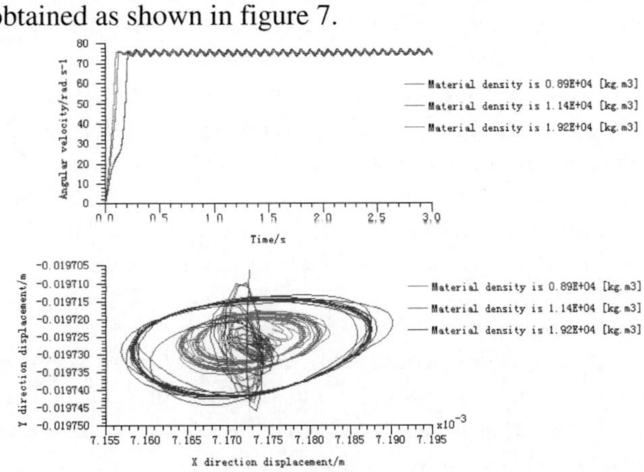

Figure 7. Effect of eccentricity on the quality of the throwing index.

The speed of the three simulation is 900r.min^{-1}, and the material density is 0.89×10^4 kg.m^3, 1.14×10^4kg.m^3, 1.92×10^4kg.m^3. Figure 7 can be seen, with the density (eccentric mass) of the larger, the angular velocity reaches the equilibrium state by the time to grow more, stable time will be longer, the fluctuation range of the vibration is small, and the amplitude of the stability is increased. Three sets of vertical component data are obtained by simulation: 1.64×10^{-5}, 2.24×10^{-5}, 3.61×10^{-5}. Using the formula (5) can be calculated from their throwing index are: 1.26, 1.72, 2.78. Due to the limitation of speed, the three simulation results need to be further improved. According to the requirements of the throwing index, we can see that with the density of the material increases, throwing index also increased, third groups of experimental data is close to the normal work requirements. Therefore, the appropriate choice of materials with large density as the experimental tool can be more close to the ideal experimental results.

5.3. Effect of screen surface inclination

For the different screen surface inclination, due to the change of inclination angle, the center of vibration box is changed and the experimental track diagram's centre is biased. Therefore, the synthetic trajectory of each displacement is listed separately, and the results are shown in Figure 8.

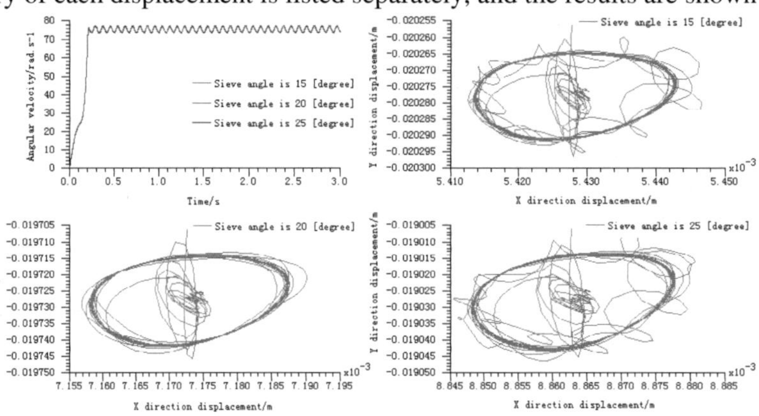

Figure 8. Screen angle effect on the throwing index.

The speed of the three simulation is 900r.min^{-1}, and the screen surface inclination is 15°, 20°, 25°. Can be seen from figure 8, In the three groups, the angular velocity is almost identical, and the same time to achieve stability. It can be obtained that with the change of the angle of the screen surface, there is a great difference in the vibration state before reaching the steady state. In the case of 20°, the trajectory is relatively concentrated, the vibration amplitude is small, and it is in line with the ideal working state, the vibration of 15° and 25° is more dispersed, and it is not practical. Three sets of vertical component data are obtained by simulation: 3.32×10^{-5}, 3.61×10^{-5}, 3.73×10^{-5}. Using the formula (5) can be calculated from their throwing index are: 2.47, 2.78, 2.94. According to the requirements of the throwing index, the latter two groups of test data are more close to the normal range of data. However, when the inclination angle is 25°, the amplitude of the vibration state is too large to reach a stable state, which can not be included in the reference range, therefore, under normal operating conditions, the inclination angle is more favorable for the 20° state.

6. Parameter optimization simulation

Through simulation and analysis, motor speed, eccentric quality, screen angle has an effect on the throwing index of vibrating screen, especially in the design, selection of suitable motor speed is necessary. On the basis of above experiments, considering the influence of various parameters on the amplitude stability and throwing index, using Exploration Design module in AMESim to optimize the parameters, set the constraints for the motor speed in the range of 800~1200r.min^{-1}, the material density is maintained at 1.92×10^4kg.m^3, the screen angle between 15°~25°. Through the optimization and simulation of the motor speed, the main effect of the throwing index curve and iteration of the results shown in figure 9, the simulation shows that when the motor speed is 1100 r.min^{-1}, the throwing index is the best. Finally, the material density is 1.92×10^4kg.m3, the screen surface is 20°. The simulated image is shown in figure 10.

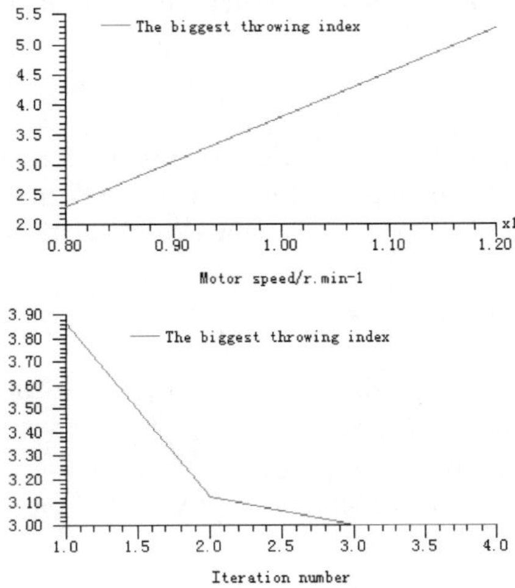

Figure 9. Simulation of motor speed optimization.

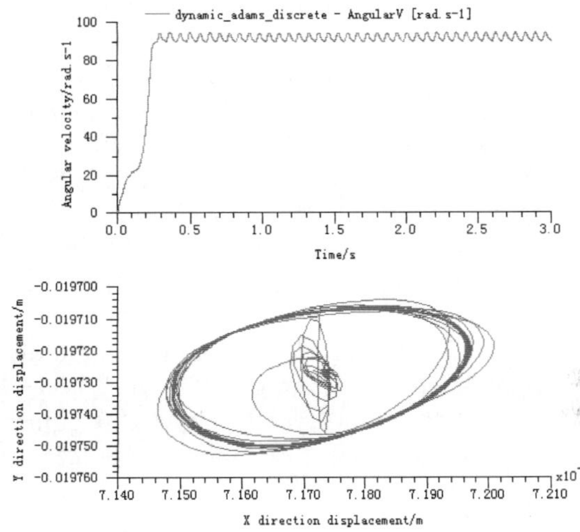

Figure 10. The velocity and displacement of the optimized eccentric angle.

Figure 10 compared with the previous curve, we can see that when the speed is the same, the angular velocity of the eccentric angle is still the same as the stability of the numerical value, and the vibration amplitude is smaller, which is in line with the actual situation before the steady state is reached. The vertical component of the amplitude of the simulation is: 4.39×10^{-5}. Using the formula (5), throwing index calculated is: 3.89. The throwing index is in the range of normal, and in accord with the design requirements of uniaxial shaker. Therefore, it is significant to study the vibration characteristics of vibrating screen by selecting the rotational speed of the motor, the eccentric mass and the dip angle of the screen, also provides reference for the improvement of vibrating screen.

7. Conclusion

In this study, the mathematical model of uniaxial shaker and the combined simulation model of AMESim and ADAMS are established, the vibration characteristics of small vibration sieve are compared and analyzed. Through the simulation results, the following conclusions are drawn:

(1) In uniaxial shaker, the speed of the motor, the quality distribution of the eccentric device, the angle between the screen surface and the ground influence the vibration characteristics of the vibrating screen, meanwhile, they are also an important factor in the ideal working condition of vibrating screen.

(2) The speed of the motor mainly influences the angular velocity of the eccentric device, which directly affects the throwing index size and vibration sieve to achieve stable working efficiency.

(3) The greater the quality of eccentric device, the greater the vibration amplitude of the uniaxial shaker, the time will be extended to reach a stable state. Mainly because of the increase of the eccentric mass, the exciting force will be increased, so that the box is more intense.

(4) Screen angle in the range of vibrating screen is 15°~25°, select the appropriate angle, the inclination of 20°, can make the vibration sieve to vibration state stability before remained stable, with regularity, and step by step to achieve stable operating condition.

In the process of designing and analyzing the vibrating screen, we need to consider all the factors that affect the vibration characteristics of the vibrating screen, and constantly try to get the best result.

Acknowledgments
This work was finically supported by the Foundation of Science and Technology Innovation Talents Team of Guizhou Province ([2014]4013).

Reference

[1] Banaszewski T 1985 Vibration forms in screening machines with single mass unbalanced weight drive *Aufbereit tech* **26(7)** 415-421.
[2] Li B, Gao H, Zhao B, et al. 2006 Vibration amplitude stability control for an antiresonant vibrating griddle *Journal of Liaoning Technical University* **25(3)** 429-431.
[3] Zhao L, Zhao Y, Liu C, Li J, Dong H 2011 Simulation of the screening process on a circularly vibrating screen using 3D-DEM *Mining Science and Technology (China)* **21(5)** 677-680.
[4] Zhang N, Hou X, Wen B 2008 Research for Design of Vibration Screen Based on Dynamic Optimization Design *Coal Mine Machinery* **29(3)** 12-15.
[5] Chen X, Peng T, Xu E, et al. 2012 Shearer Drum Height Adjustment System Based on Co-simulation of ADAMS and AMESim *Coal Mine Machinery* **33(11)** 40-42.
[6] Wang N, Li J, Zhang H, et al. 2015 Co-simulation Study of Broken Belt Protector Based on ADAMS and AMESim *Coal Technology* **34(12)** 238-240.
[7] Wu H 2002 *Motor and electric drive* Chongqing: Chongqing university press 46-48.
[8] Banaszewski T, Liu P 1991 Vibrating Screen starting torque calculation *Mining & Processing Equipment* **1** 38-39.
[9] Li G 1996 Research on the characteristics of single excited motor drive system *Mining & Processing Equipment* **8** 38-40.
[10] Zhou J, Zhou Y, Zhang H 2013 Simulation Analysis of Circular Vibrating Screen Motion Based on ADAMS *Coal Mine Machinery* **34(4)** 101-103.
[11] Liu H 1999 Circular vibrating screen material mass ejection Angle and the design of screen surface inclination *Mining & Processing Equipment* **7** 43-44.
[12] Wen B, Liu F, Liu J 1989 *Vibrating screen vibrating feeder vibration conveyor design and debugging* Beijing: Chemical Industry Press 65-67.
[13] Chen F 2013 *ADAMS 2012 virtual prototyping technology from entry to the master* Beijing: Tsinghua University press.
[14] Fu Y, Qi H 2011 *LMS Imagine. Lab AMESim system modeling and simulation-tutorial examples* Beijing: Beihang University press.
[15] Shen J, Liu L, Tang H 2009 Simulation and analysis on vibratory hydraulic system on tandem rollers using AMEsim and ADAMS *Chinese Journal of Construction Machinery* **7(1)** 31-35.

A new configuration method for accelerometers in rotating accelerometer-based gravity gradiometer

Hongwei Wei[1] and Meiping Wu[1*]

[1]College of Mechatronics and Automation, National University of Defense Technology, Changsha 410073, China;

*E-mail: meipingwu@263.net

Abstract. As a sensor for direct measurement of gravity gradient signals, accelerometer plays an important role in the rotating accelerometer-based gravity gradiometer. Therefore, the way to position the accelerometer has an important influence on the gravity gradient measurement. In this article, we have analysed the principle of gravity gradient measurement, and points out some defects in the configuration of traditional gravity gradients. Finally, we proposed a new way to position the accelerometers in gravity gradiometer, in which, the sensitive axis of the opposite accelerometers are in the same direction. And the results of simulation indicate that the measurement error of the new way to position the accelerometers is about 9.29%, while the traditional configuration is about 39.68%, then the the efficacy of the new configuration method could be proven.

1. Introduction

Gravity gradient reflects the variations of the gravity in space. The gravity gradient is a very weak signal. To detect this weak signal, many kinds of high precision instruments were manufactured, based on different principles, such as cold atomic interferometer-based gravity gradiometer[1-3], the rotating accelerometer-based gravity gradiometer [4, 5], the superconducting gravity gradiometer[6, 7], MEMS gravity gradiometer[8] and some others. Among them, the rotating accelerometer-based gravity gradiometer is the unique one that has passed through the flight test and achieved the desirable sensitivity in the field surveys.

The most important component of the gravity gradiometer is Gravity Gradient Instrument (GGI) [4]. To our knowledge the structure of each GGI is as shown in figure 1.

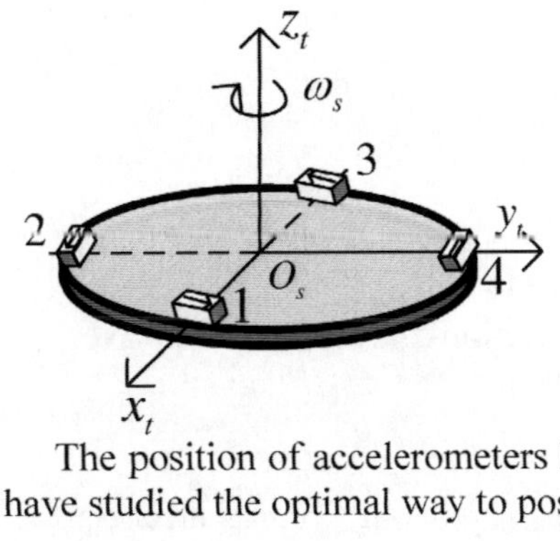

Figure 1. The schematic diagram of GGI. It consists of four accelerometers (1, 2, 3, 4); the arrows represent the direction of the sensitive axis of the accelerometers; and ω_s is the rotation speed of the disc.

The position of accelerometers have an important influence on the measurement. In this paper, we have studied the optimal way to position accelerometers in GGI.

The following is the structure of this article. In the second section, the principle of gravity gradient measurement based on four accelerometers and rotating disk is introduced. In the third section, we have proposed a new way to position accelerometers. In the fourth section, we used simulation data to prove that the new configuration works better than traditional configuration.

2. Measuring principle

As shown in figure 1, the combination of the accelerometers' output signals contains the gravity gradient information. The gravity gradient signal is modulated to the second harmonic of the rotational speed. Thus, the gravity gradient signal can be obtained by demodulating the combined signal.

The relationship between the outputs of accelerometers and the gravitational gradient components is as follows:

$$\frac{f_2+f_4-f_1-f_3}{2r_s}=2\left(\Gamma_{xy}+N_1\right)\cos\left(2\omega_s t\right)+\left(\Gamma_{xx}-\Gamma_{yy}+N_2\right)\sin\left(2\omega_s t\right) \tag{1}$$

where Γ'_{xx}, Γ'_{yy}, Γ'_{xy} are components of gravity gradient tensor, f_1, f_2, f_3, f_4 are outputs of accelerometers, N_1, N_2 are other known coupling signals.

With the help of equation (1), we can demodulate $f_2+f_4-f_1-f_3$ at $\sin\left(2\omega_s t\right)$ yields $\Gamma_{xx}-\Gamma_{yy}$, and demodulate it at $\cos\left(2\omega_s t\right)$ yields Γ_{xy}.

3. A new configuration of accelerometers

From equation (1) we know that the gravity gradient signals contained in the differences of two paired accelerometers. In other words, when outputs of accelerometers on opposite sides of the table were differenced, we may achieve values of the gravity gradient signals at the midpoint of the accelerometers. To our knowledge, accelerometers on GGI are arranged as shown in figure 1. Actually, accelerometers are not perfectly working, so the position of accelerometers will affect the measurements accuracy.

In this article, we have proposed a new configuration, as shown in figure 2.

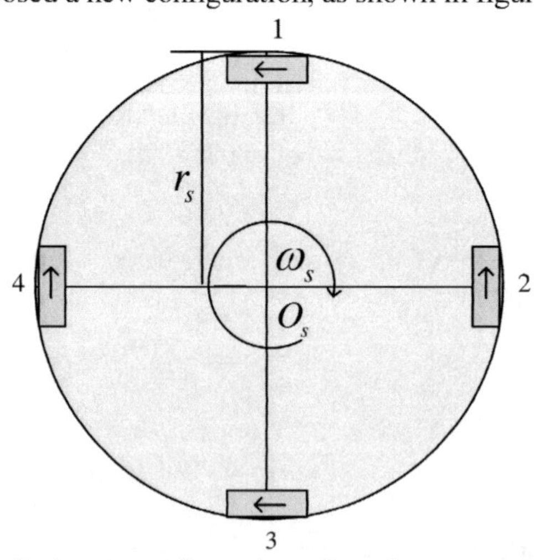

Figure 2. A new configuration of accelerometers in GGI.

The difference between figure 2 and figure 1 is that the direction of the sensitive axes of accelerometer 3 and 4 are different.

3.1. Measuring model of accelerometer

Ideally, the specific force measuring model of the accelerometer is generally represented by the following formula:

$$N_a = kf_{in} \tag{2}$$

where N_a is the output of the accelerometer, k is the scale factor of the accelerometer, and f_{in} is the specific force as input of the accelerometer.

But in fact, due to the interference of the external environment, the limitation of the processing technology and the differences of device's performance, the actual measurement model is different from equation (2). In inertial navigation system, equation (3) is often used as the mathematical model of the accelerometer:

$$N_n = k_0 + k_1 f_{in} + \varepsilon \tag{3}$$

where N_n is the output of the accelerometer, k_0 is bias of the accelerometer, k_1 is the scale factor of the accelerometer, f_{in} is the input of the accelerometer, and ε is the measurement noise.

The accuracy of the model of equation (3) is sufficient to meet the accuracy requirements of inertial navigation. However, for gravity gradient measurement, this model is still too rough. Next, we use equation (4) as the mathematical model of the accelerometer:

$$N_g = k_0 + k_1 f_{in} + k_2 f_{in}^2 + \varepsilon \tag{4}$$

in which k_2 represents the parameter factor associated with the quadratic term.

Because the main purpose of this paper is to analyze the effect of the accelerometers configuration on the gravity gradient measurement, the influence of the installation error is not considered in the following analysis.

3.2. Comparison of two configurations

As previously mentioned, the gravity gradient signals are implied in the difference of accelerometers' outputs.

Next, we have analyzed the accuracy of measurement for difference specific force under the two configurations.

It is assumed that the accelerometers in the accelerometer group are symmetrically mounted on both ends of the turntable diameter (as shown in figure 3), the mounting points are denoted as A and B. The specific force at A and B are denoted as f and $f + \Delta f$, respectively.

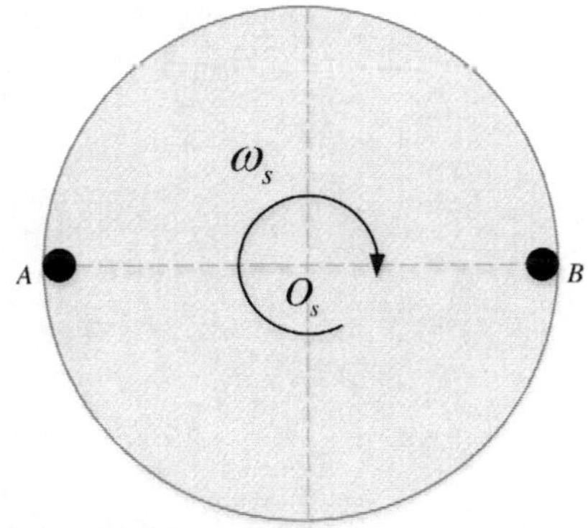

Figure 3. The position of the accelerometers on the disc.

The output of accelerometer 1 at A is recorded as:

$$N_{Ay} = k_0 + k_1 f + k_2 f^2 + \varepsilon_{A1} \tag{5}$$

The output of the accelerometer 2 at B, which has the same sensitive axis direction with the accelerometer 1, is:

$$N_{By'} = k_0' + k_1'(f + \Delta f) + k_2'(f + \Delta f)^2 + \varepsilon_{B2} \tag{6}$$

The output of the accelerometer 3 at B, which has the opposite sensitive axis direction with the accelerometer 1, is:

$$N_{By''} = k_0'' - k_1''(f + \Delta f) + k_2''(f + \Delta f)^2 + \varepsilon_{B3} \tag{7}$$

In order to obtain the difference of specific force Δf between *A* and *B*, N_{Ay} and $N_{By'}$ are differentiated:

$$\begin{aligned} N_{By'} - N_{Ay} &= (k_0' - k_0) + (k_1' - k_1) f + (k_2' - k_2) f^2 + k_1'\Delta f + 2k_2' f \Delta f + k_2'\Delta f^2 + \varepsilon_m \\ &= \Delta k_0 + \Delta k_1 f + \Delta k_2 f^2 + k_1'\Delta f + 2k_2' f \Delta f + k_2'\Delta f^2 + \varepsilon_m \end{aligned} \quad (8)$$

where $\varepsilon_m = \varepsilon_{B2} - \varepsilon_{A2}$.

Similarly, to obtain the difference of specific force Δf, N_{Ay} and $N_{By'}$ are summed:

$$\begin{aligned} N_{By''} + N_{Ay} &= (k_0'' + k_0) - (k_1'' - k_1) f + (k_2'' + k_2) f^2 - k_1''\Delta f + 2k_2'' f \Delta f + k_2''\Delta f^2 + \varepsilon_a \\ &= 2k_0 + 2k_2 f^2 + \delta k_0 - \delta k_1 f + \delta k_2 f^2 - k_1''\Delta f + 2k_2'' f \Delta f + k_2''\Delta f^2 + \varepsilon_a \end{aligned} \quad (9)$$

where $\varepsilon_a = \varepsilon_{B3} + \varepsilon_{A1}$.

Since $k_0, k_0', k_0'', k_1, k_1', k_1'', k_2, k_2'$ and k_2'' are the results of calibration, they are known quantity. Therefore terms $\Delta k_0 + \Delta k_1 f + \Delta k_2 f^2$ and $2k_0 + 2k_2 f^2 + \delta k_0 - \delta k_1 f + \delta k_2 f^2$ can also be calculated. Finally, by equation (8) and equation (9), we can achieve Δf.

Compare equation (8) and equation (9), we can find that equation (9) requires more compensation. This means that when the parameter factor of the accelerometer drifts, using the configuration shown in figure 1 will result in a greater error.

Next, we will use simulation results to prove this conclusion.

4. Simulation

On the basis of the hypothesis that scale factors of accelerometers drift in the same trend, the accuracy and the robustness of the proposed configuration method are verified by the simulation test.

The simulation is carried out according to the following steps: first, setting up the simulation environment; second, using outputs of the accelerometers to calculate Δf; third, compare the results of above two configurations, and determine which configuration is better.

The steps of the simulation are shown in figure 4.

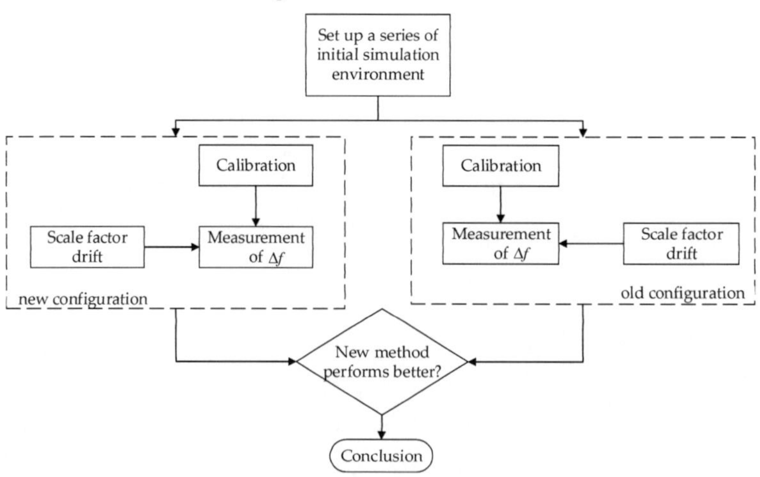

Figure 4. Simulation steps.

4.1. Initial Simulation Parameter

The initial parameters of the simulation are as follows:

Noise intensity of each accelerometer: $1\mu g / \sqrt{Hz}\,(1\sigma)$

Specific force at *A*: $f = 35 m/s^2$

Specific force at *B*: $f+\Delta f=(35+0.00001)m/s^2$

The parameter factors of each accelerometer are shown in table 1.

Table 1. Parameter factors of each accelerometer

Parameter Factors	k_0	k_1	k_2	k_0'	k_1'	k_2'	k_0''	k_1''	k_2''
Values	96	4800	0.0960	49	4900	0.0490	49	4900	0.0490

The parameter factors' drift of the accelerometers are shown in figure 5.

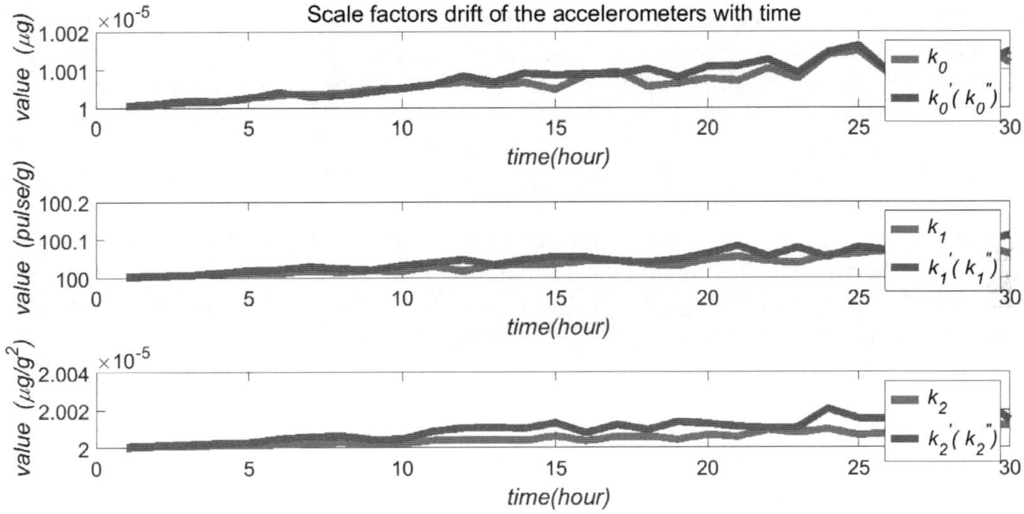

Figure 5. Parameter factors' drift of the accelerometers with time.

4.2. Simulation Results

In the two configurations, the accuracy of the measured Δf are shown in table 2 and table 3.

Table 2. Δf , measured using the old configuration

True value	Measured value	Relative measurement error
0.0001	0. 000138685	39.68%

Table 3. Δf , measured using the new configuration

True value	Measured value	Relative measurement error
0.0001	0. 000090713	9.29%

It is clear from table 2 and table 3 that the measurement error of the new configuration method is about 9.29%, while the old configuration is about 39.68%.

5. Conclusion

In this paper, we proposed a new accelerometer configuration method in gravity gradiometer, which helps to improve the accuracy of gravity gradient measurement. The improved configuration has the following advantages:

1) When there is a large drift in the parameter factors of the two accelerometers, as the change of accelerometer parameters have the same trend, the difference between the two accelerometers' parameter factors is relatively stable. Therefore, the performance of the new configuration is relatively stable.

2) The coefficient of Δf in equation (9) is $2k_2''f-k_1''$, less than $k_1'+2k_2'f$ in equation (8), this makes the resolution and anti-interference ability decrease when using the old configuration to measure Δf .

IOP Conf. Series: Materials Science and Engineering **220** (2017) 012037 doi:10.1088/1757-899X/220/1/012037

References

[1] Snadden M J, McGuirk J M, Bouyer P, Haritos K G, and Kasevich M A 1998 Measurement of the Earth's Gravity Gradient with an Atom Interferometer-Based Gravity Gradiometer *Physical Review Letters* **81(5)** 971-974.
[2] De Angelis M, Bertoldi A, Cacciapuoti L, Giorgini A, Lamporesi G, Prevedelli M, et al. 2008 Precision gravimetry with atomic sensors *Measurement Science and Technology* **20(2)** 022001.
[3] Sorrentino F, Bodart Q, Cacciapuoti L, Lien Y H, Prevedelli M, Rosi G, et al. 2014 Sensitivity limits of a Raman atom interferometer as a gravity gradiometer *Physical Review A* **89(2)** 023607.
[4] Lee J B 2001 FALCON gravity gradiometer technology *Exploration Geophysics* **32(4)** 247-250.
[5] Chan H A, Moody M V, and Paik H J 1987 Superconducting gravity gradiometer for sensitive gravity measurements. II *Experiment. Physical Review D* **35(12)** 3572-3597.
[6] Moody M V, Paik H J, and Canavan E R 2003 Principle and performance of a superconducting angular accelerometer *Review of Scientific Instruments* **74(3)** 1310-1318.
[7] Moody M V 2011 A superconducting gravity gradiometer for measurements from a moving vehicle *Review of Scientific Instruments* **82(9)** 094501.
[8] Flokstra J, Cuperus R, Wiegerink R J, and van Essen M C 2009 A MEMS-based gravity gradiometer for future planetary missions *Cryogenics* **49(11)** 665-668.

IOP Conf. Series: Materials Science and Engineering **220** (2017) 012038 doi:10.1088/1757-899X/220/1/012038

On-line Monitoring for Cutting Tool Wear Condition Based on the Parameters

Fenghua Han[1*], Feng Xie[1]

[1]School of Electrical Engineering & Automation, Anhui University, Hefei 230601, China

*E-mail: 1617972207@qq.com

Abstract. In the process of cutting tools, it is very important to monitor the working state of the tools. On the basis of acceleration signal acquisition under the constant speed, time domain and frequency domain analysis of relevant indicators monitor the online of tool wear condition. The analysis results show that the method can effectively judge the tool wear condition in the process of machining. It has certain application value.

1. Introduction

With the wide use of numerical control machine tools, the modern manufacturing system is developing towards the direction of the highly automated. The real-time monitoring for the state of cutting tool is needed. According to the statistics [1], NC machine tools equipped with cutting tool monitoring system can make its productivity increased by ten percent to sixty percent. The utilization rate of machine tools are increased by fifty percent, so the real-time monitoring of machine tool is great significance.

Wang Ming used wavelet theory to extract the vibration signal and analysised the change of the signal amplitude and frequency vibration to determine the state of tool wear. Gao Weijia put forward a kind of method based on artificial neural network after blade abrasion value prediction and used the universal tool microscope to measure the abrasion value.

2. Cutting tool wear monitoring method analysis

The method of cutting tool in machining condition monitoring contains the processing of signal, analyze signal, extract feature and diagnose wear fault. Generally the cutting tool vibration signal contains the valid information of the tool wear. The tool wear state is judged by picking up processing vibration signal from the acceleration sensor and analyzing the relevant indicators[2].

2.1. The time domain analysis

Time domain waveform is usually the most intuitive analysis method. With obvious characteristics of the wear, the time domain waveform directly correspond to different degree of wear. The vibration waveforms show the different shapes[3].

Amplitude parameter analysis of indicators are divided into dimensional parameter and the dimensionless parameter. The dimensional parameters contain the mean, root amplitude, kurtosis, etc. Dimensionless parameters have waveform, pulse, tolerance index and so on. The following is a calculation formula of each index:

Variance $\sigma^2 = \frac{1}{N}\sum_{i=0}^{N-1}(x_i - \overline{x})^2$ (1), Kurtosis index $x_q = \frac{1}{N}\sum_{i=0}^{N-1} x_i^4 / x_a^2 - 3$ (2),

Power spectral centroid $FC = \frac{\sum_{i=1}^{N} f_i p_i}{\sum_{i=1}^{N} p_i}$ (3) ,The mean square frequency $MSF = \frac{\sum_{i=1}^{N} f^2{}_i p_i}{\sum_{i=1}^{N} p_i}$ (4)

The x_i for collection of data, x_p as the peak, N as the collection data points, f_i the power spectrum of the frequency, p_i the power spectrum amplitude.

2.2. *The frequency domain analysis*

Frequency domain analysis of tool wear monitoring is based on the analysis of time domain waveform. It is to further find out the vibration internal relations of acquisition signal of normal and wear parts.

3. The design for tool wear experiments

The tool wear test is to use new knife from early state to the late continuous processing. In order to study the different levels of wear, the wear mechanism of cutter and the influence law of wear, the different parameters of the time frequency domain are analyzed. The experiment is mainly to collect the acceleration signals under the constant speed. Acceleration sensors are adsorbed to the tool spindle lateral X and Y direction. It will be collected on the stability of the dynamic signal by the acceleration sensors. The signals are transmitted to the acquisition card and imported into the MATLAB software to analyze in the computer [4].

In order to apply vibration test method in tool wear monitoring, as shown in figure 1 is collected vibration signals of tool wear monitoring system.

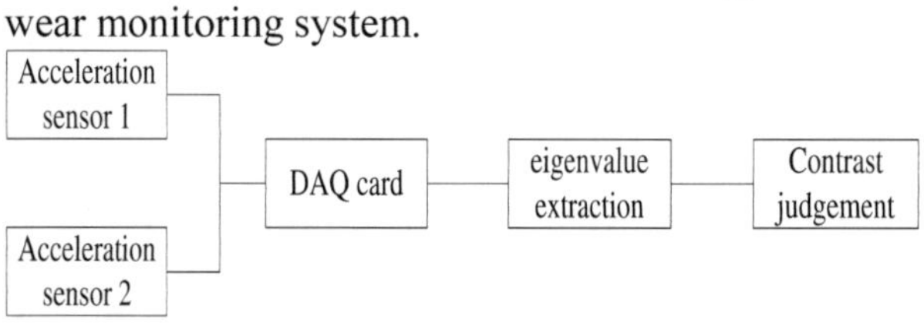

Figure 1. Collection and analysis of vibration signal monitoring figure.

The system consists of milling drilling machine, the acceleration sensor, a four-channel data acquisition card (NIUSB4431) and PC software MATLAB analyzer. The experiment of specific machine of ZXJ7016 is the speed regulating pole change motor machine. The cutter material is a straight shank end mill with two teeth. The sampling frequency is 20000 Hz. The sampling time is 2 seconds and the signal acquisition is taken from the new state to severe wear.

4. Results of data analysis

The spindle speed of 750 r/min, feeding rate of 70 mm/min, turning quantity is 0.5 mm and the processing thickness is 5 mm.

To contrast the different level tool wear state are based on the analysis of the time domain and frequency domain parameters. The different parameters have different sensitive of the wear degree. It is need to select a number of effective parameters as the characteristics of fault diagnosis [5].

IOP Conf. Series: Materials Science and Engineering **220** (2017) 012038 doi:10.1088/1757-899X/220/1/012038

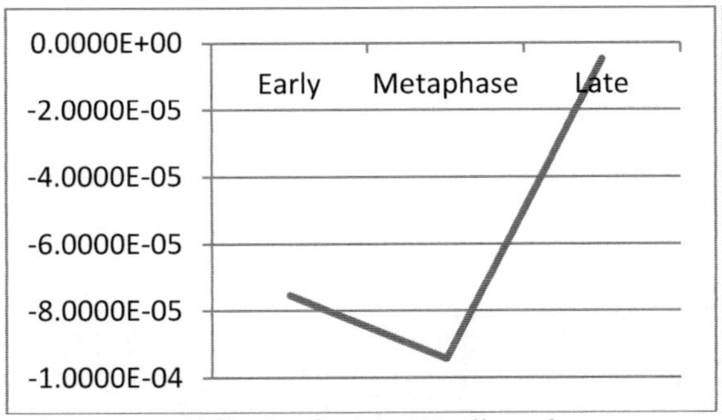

Figure 2. Average line chart

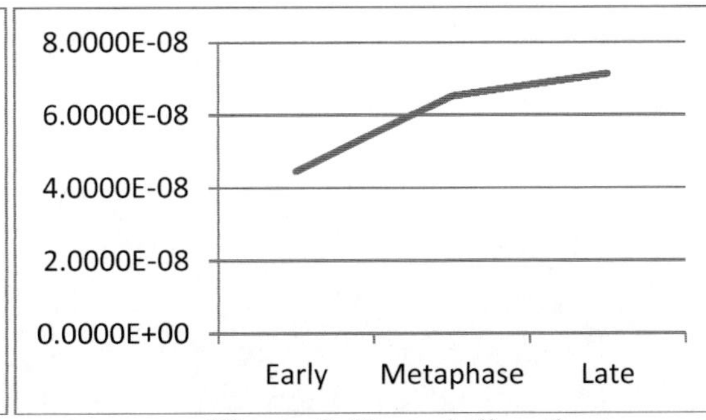

Figure 3. Kurtosis line chart

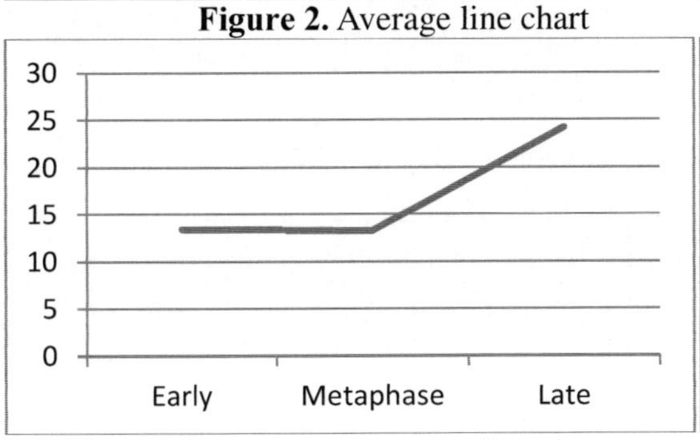

Figure 4. Impulsion index line chart

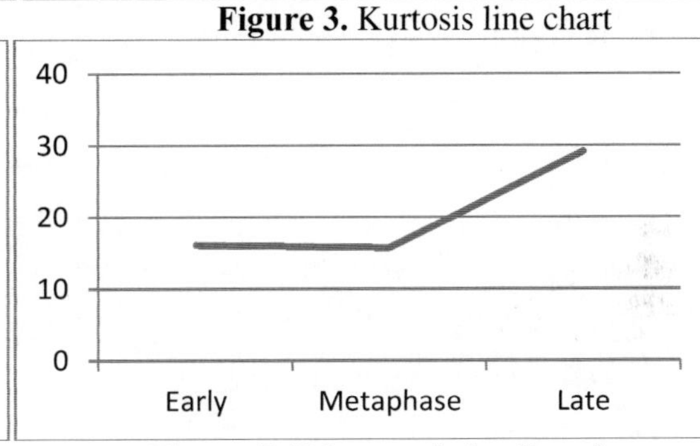

Figure 5. Tolerance index line chart

From figure 2 to figure 5 in the line chart, the average, kurtosis, impulsion index and tolerance index is more sensitive indicator of tool wear signals. These can be used as mixed characteristics of tool wear monitoring.

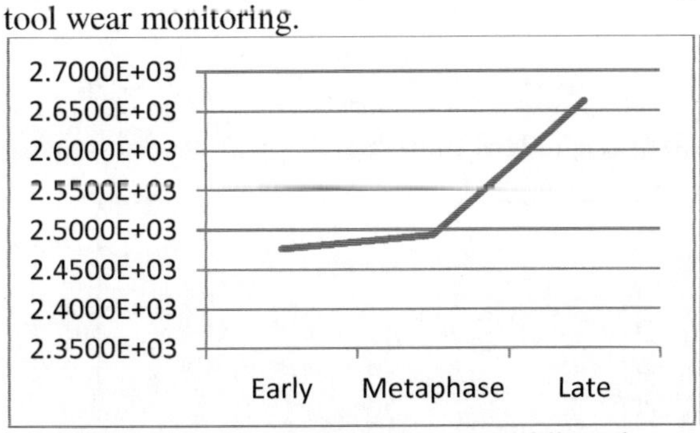

Figure 6. Power spectral centroid line chart

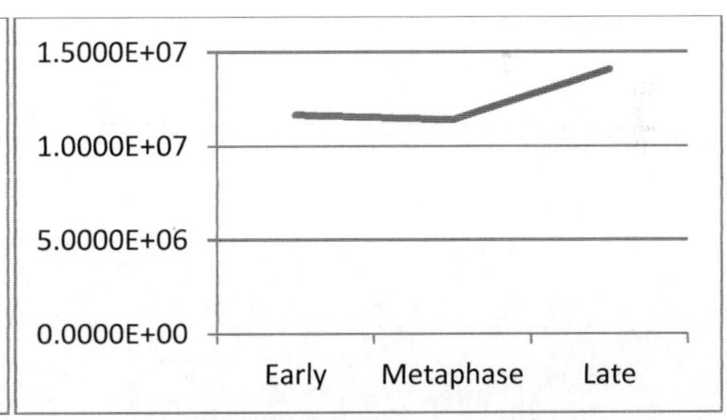

Figure 7. Mean square frequency

The above two line chart about the frequency domain index, both the mean square frequency and the power spectral centroid are sensitive to the vibration signal of the tool wear and have better signal indication. Here are all wear period of time domain and frequency domain spectrum analysis[6].

IOP Conf. Series: Materials Science and Engineering **220** (2017) 012038 doi:10.1088/1757-899X/220/1/012038

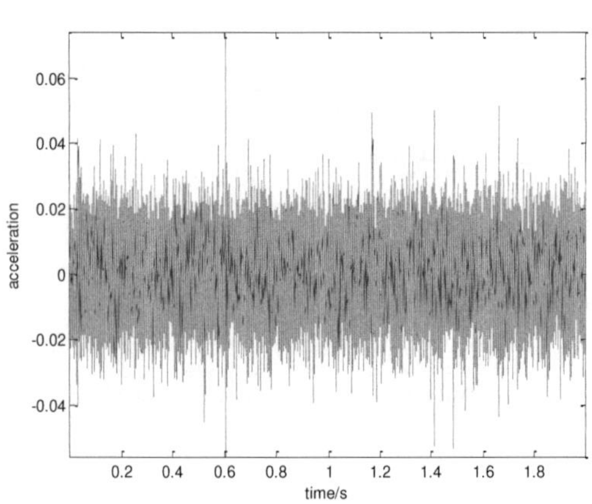

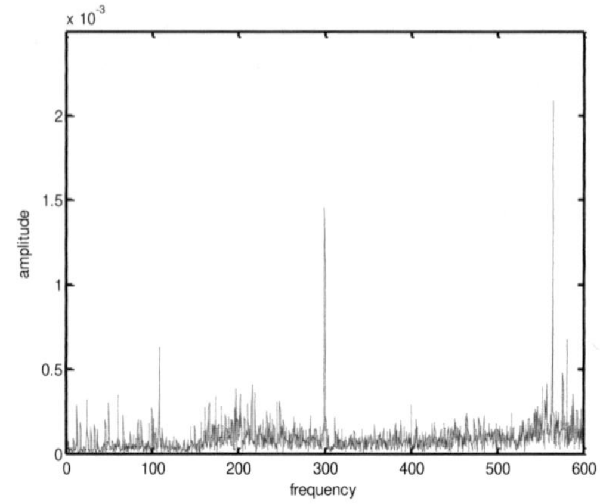

Figure 8. Early tool wear time domain diagram. **Figure 9.** Early tool wear frequency domain diagram.

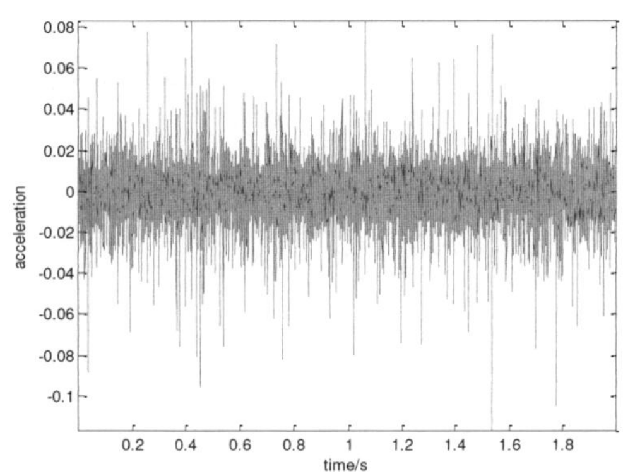

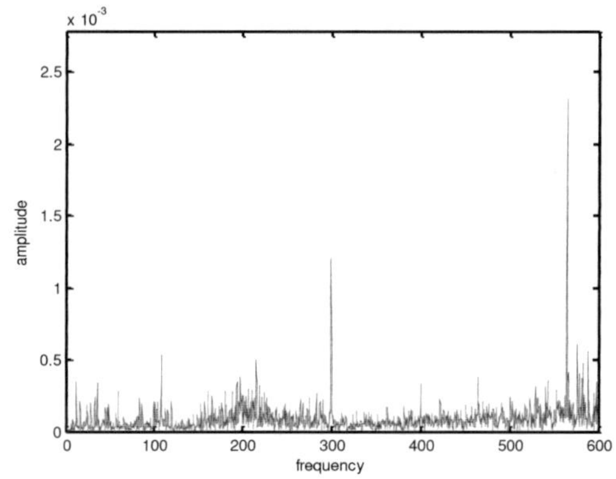

Figure10. Mid tool wear time domain diagram. **Figure11.** Mid tool wear frequency domain diagram.

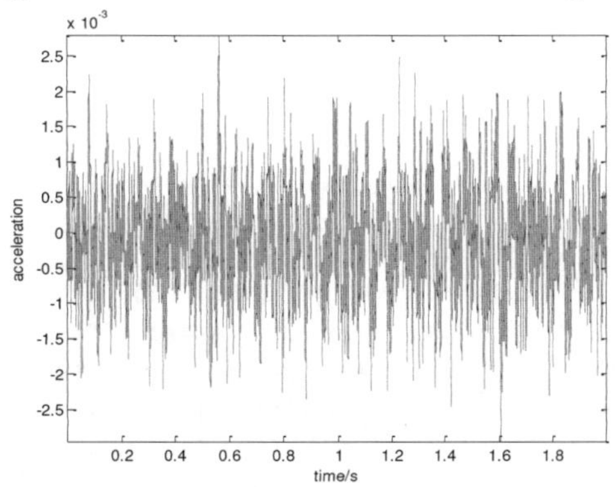

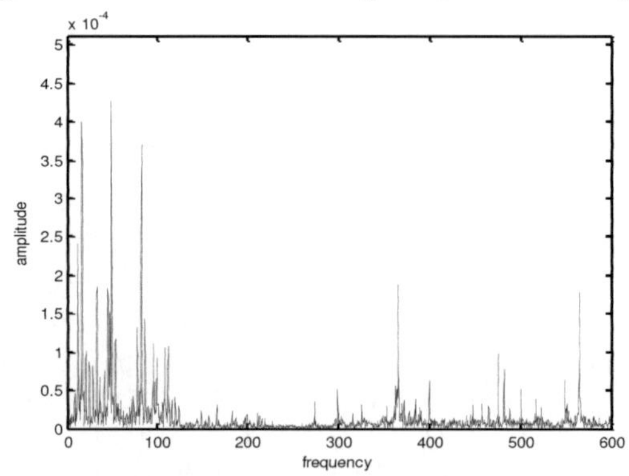

Figure12. Late tool wear time domain diagram. **Figure13.** Late tool wear frequency domain diagram.

The vibration signal can get varying degrees of cutting tool wear from the time domain and frequency domain. Above a few of frequency domain diagrams are typical. Due to the spindle speed switch frequency is 12.5 Hz, the experiment of the cutter material has two teeth and the switch frequency is 25 Hz. It can be seen the frequency of 125Hz or 200 Hz and 550 Hz, in the early or mid time of the tool wear degree difference is not big, but have great peak. Contrasting with the wear of late, the frequency of the tool's wear is not as high as the frequency of 25 Hz or more than the set value, that's said the tool wear badly. These as shown in figure 8 ~ 13 stages time frequency domain[7].

IOP Conf. Series: Materials Science and Engineering **220** (2017) 012038 doi:10.1088/1757-899X/220/1/012038

5. Conclusion

Through the analysis of the relevant, the vibration signal can be observed from the time domain and frequency domain in the figure. In the early or middle time of the tool wear degree, the difference is not big, but the middle is relatively stable. The late vibration is more intense. According to the different vibration signal, the cutting tool indicates different wear state[8]. In the time domain parameters, the mean square amplitude, waveform of index and kurtosis distinguish from the cutting tool wear state. In the frequency domain parameters, the mean square frequency and power spectrum centroid of the cutting tool wear have obvious corresponding characteristic. The example proves that the time domain and frequency domain analysis method can effectively extract the characteristics of the cutting tool and identify the quality problems of the cutting tool wear. To a certain extent, the method improves the accuracy of tool wear monitoring in real time.

Reference

[1] Xu M 2009 CNC machine tool wear monitoring data processing method *Journal of combination machine tools and automatic processing technology*.

[2] Zhuang Z 2009 Based on the method of acoustic emission and vibration state of tool wear detection research Shanghai: Shanghai Jiaotong University.

[3] Zhang C 2010 CNC machine tools typical fault analysis and diagnosis system design Qingdao: Qingdao University of Science and Technology 16-22.

[4] Dominik I, Iwaniec M and Lech L 2013 Low frequency damage analysis of electric pylon model by fuzzy logic application *J Low Frequency Noise Vib Active Control* **32** 239–252.

[5] Dimla D E 2000 Sensor signals for tool -wear monitoring in mental cutting operrations-a review of mental *International Journal of Machine Tools & Manufacture* **40(8)** 1073-1078.

[6] Ghani J A, Rizal M, Nuawi M Z, Haron C H C, Ramli R 2010 Statistical analysis for detection cutting tool wearbased on regression model in *Proceedings of the International MultiConference of Engineers and Computer Scientists*.

[7] Scheffer C, Kratz H, Heyns P S, Klocke F 2003 Development of a tool wear-monitoring system for hard turning *International Journal of Machine Tools and Manufacture* **43(10)** 973-985.

[8] Chiu M C 2014 Numerical assessment of rectangular side inlet/outlet plenums internally equipped with two crossed baffles usingan FEM, neural network, and GA method *Journal of Low Frequency Noise, Vibration and Active Control* **33** 271–288.

IOP Conf. Series: Materials Science and Engineering **220** (2017) 012039 doi:10.1088/1757-899X/220/1/012039

Study unique artistic lopburi province for design brass tea set of bantahkrayang community

V Pliansiri[1] and S Seviset[1]

[1]Program in Technology of Industrial Product Design, Department of Architectural Education and Design, Faculty of Industrial Education and Technology, King Mongkut's Institute of Technology Ladkrabang, Bangkok 10520, Thailand

Abstract. The objectives of this study were as follows: 1) to study the production process of handcrafted Brass Tea Set; and 2) to design and develop the handcrafted of Brass Tea Set. The process of design was started by mutual analytical processes and conceptual framework for product design, Quality Function Deployment, Theory of Inventive Problem Solving, Principles of Craft Design, and Principle of Reverse Engineering. The experts in field of both Industrial Product Design and Brass Handicraft Product, have evaluated the Brass Tea Set design and created prototype of Brass tea set by the sample of consumers who have ever bought the Brass Tea Set of Bantahkrayang Community on this research. The statistics methods used were percentage, mean ($\bar{X}$ =) and standard deviation (S.D.) 3. To assess consumer satisfaction toward of handcrafted Brass tea set was at the high level.

1. Introduction

Bantahkrayang Community, Tahlaechupsorn District, Lopburi Province, one of villages inherits the intellect of creating the local handcraft. Brass is a metal handcraft and always a popular local handcraft of Bantahkrayang community that is inheritance from the past. The manufacture of Brass is evolved from skills of what the body of knowledge is collected via inheritance, improvement and development. Then it transforms into the Brass at present. It inherits from intellectual to intellectual and generation to generation.

The production of Brass has got a beautiful shape, unique and very interesting. The Brass casting of Lopburi Province has been originated for long time. The finding of historical evidence was shown that the northern of Bantahkrayang, at the back of Tong Pu Temple has ever been the village of smith and some unburned remains of antique kiln were found. According to the evidence, the metal casting of Bantahkrayang is a heritage of ancestor by taking the metal to cast into equipment what is suitable for any era. Moreover, it cast into the artistic and animal shapes [1].

The Occupational Group of Brass Caster, Bantahkrayang is facing with the issue of product shape that the design development isn't various, the number of consumers are decreased, the purchases order of exportation are also decreased due to inbound competitors who manufacture the Brass instead. Moreover, the national competition of producing the Brass is increased for provinces at present. The cost of Brass and fuel oil is more expensive [2].

The Brass Tea Set is one of a popular product among the consumers who love making of tea. The shape of Brass Tea Set of Bantahkrayang Community can buy anywhere and it's not unique. This has an effect on the decreased choice or taste of buying because it's special for consumers who love Brass Handicraft only.

The detail above, researcher is interested in studying the unique artistic Lopburi Province in order to design the Brass Tea Set of Bantahkrayang. It's to add the value of product and the distribution channel. It's also the everlasting conservation of Brass casting which is the intellectual of Thai people.

2. Objective

- To study on identity and the production process of handcrafted Brass Tea Set
- To design the handcrafted Brass Tea Set
- To assess consumer satisfaction toward of handcrafted Brass tea set

3. Research methods

The methodologies of this study were as follows.

3.1. Study the production process and the style of Handcrafted Brass Tea Set

The researcher studied the production methods of handcrafted Brass Tea Set by document, book, interview and the related study. Then all the data were analyzed and approached to the development of Handcrafted Brass Tea Set.

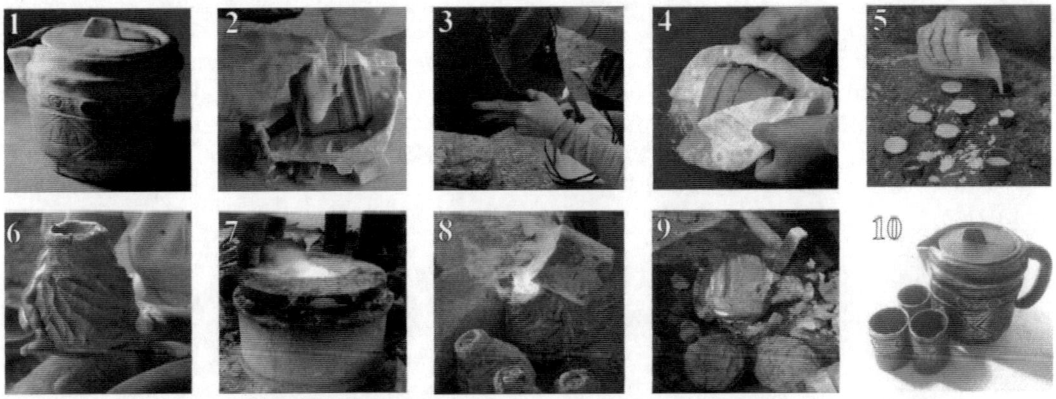

Figure 1. Diagram of Production Process of the Handcrafted Brass Tea Set

There are several steps for production process as follows.

1. Clay is modeled for plaster model, 2. Remove the silicone from primary model to pour the wax inside, 3. Pour wax into the model, 4. Remove the model in water with normal temperature to get wax model, 5. Plaster is poured by the wax model, 6. Daub the clay on the plastered model, 7. Melt the Brass at a temperature of 1500°C for 4 hours, 8. When the Brass is boiled, pour it in the model that you prepared, 9. Get the model for the first step, smash the Brass model when it is cool, 10.Polish the Brass model as you want. Then it finishes.

3.2. Design the Handcrafted Brass Tea Set

Design the Handcrafted Brass Tea Set that based on a theory of Wiroon Tungcharoen [3]. This theory was applied to the design relating to the quality of product, and it is a conceptual product development, The instruments used were primary data collection, interviews and questionnaires. the designing process then started by using collected data, through analytical processes with conceptual framework for product design; The Quality Fucnction Deployment, the theory of inventive Problem Solving, Principles of Craft Design, and the Resverse Engineering Principle. The experts from both the fields of Industrial Product Design and Brass handicraft product, whom concern Brass Tea Set format, were included to evaluate the design and creats prototype of brass tea set, the researcher have drafted 30 models of sketch Design and by using "Reverse engineering" to analyze, the best 3 forms were selected and assessed by experts and creats prototype of Brass Tea Set.

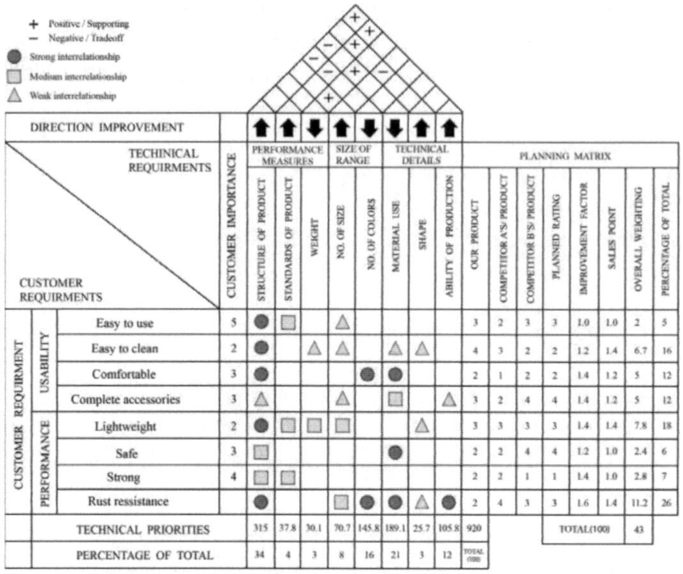

Figure 2. shows the thinking process of Quality Function Deployment framework From the Theory of Inventive Problem Solving framework, the "structure of product" is a technical conflict. According to the 40 of TRIZ. The 10th principle is "Prior action"
Source: Vuttipong Pliansiri Somchai Seveset

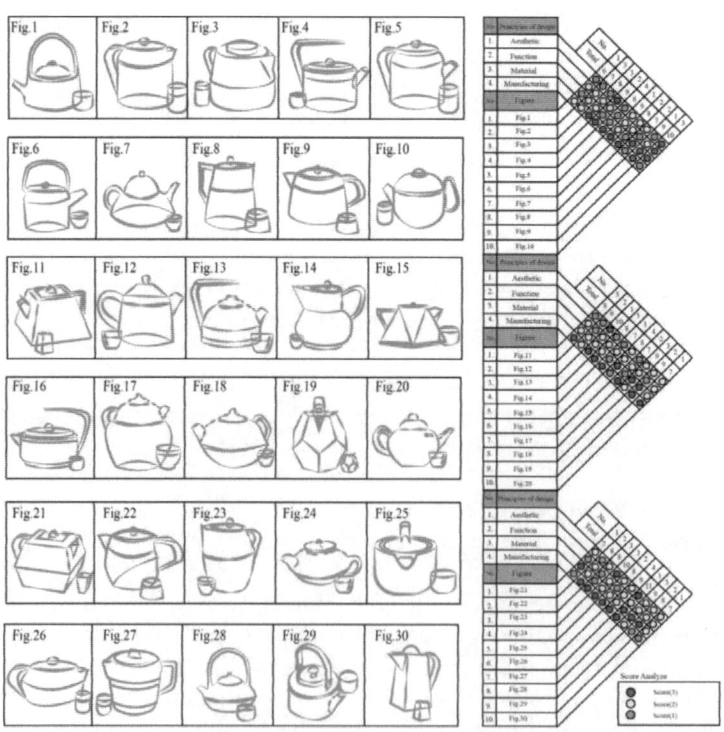

Figure 3. Shows drafted 30 models of sketch Design and by using "Reverse engineering" to analyze, the best 3 forms (Fig.27 Fig.13 Fig.10) were selected and assessed by experts and creats prototype of Brass Tea Set.

IOP Conf. Series: Materials Science and Engineering **220** (2017) 012039 doi:10.1088/1757-899X/220/1/012039

Table 1. The optimum scores on considering model selection are for Brass Tea Set Development

No.	Considerable Criteria	Design 1 		Design 2 		Design 3 	
		$\bar{x}$	S.D.	$\bar{x}$	S.D.	$\bar{x}$	S.D.
1	Aesthetic	4.33	0.38	3.33	0.38	3.55	0.19
2	The matching with usability	4.11	0.71	3.55	0.57	2.88	0.76
3	The optimum of production material	3.88	0.57	3.44	0.57	3.33	0.38
4	The matching with production	4.00	0.38	3.77	0.38	3.11	0.71
	Total Mean	4.10	0.51	3.52	0.48	3.22	0.51
	Optimum	Highest		Medium		Medium	

Table 1. Researchers have drafted 30 models of sketch Design and by using "Reverse engineering" to analyze, the best 3 forms were selected and assessed by experts. the handcrafted expert gave an opinion to the design of Brass Tea Set as follows; Design 1 was the most favorable of 3 designs by means ($\bar{x}$ = 4.10, S.D.=0.51). When each aspect was considered, it found that (1) Total mean of Aesthetic ($\bar{x}$ = 4.33, S.D.=0.38); (2) Total mean of the matching with usability ($\bar{x}$ = 4.11, S.D.=0.71); (3) Total mean of the optimum of production material ($\bar{x}$ = 3.88, S.D.=0.57) ; and(4) Total mean of the matching with production ($\bar{x}$=4.00, S.D.=0.38)

3.3. Assess consumer satisfaction toward Brass tea set

Assess consumer satisfaction toward Brass tea set (1) Total mean of Aesthetic ($\bar{x}$=4.33, SD=0.38); Total mean of matching with usability ($\bar{x}$=4.11, SD=0.71); (3) Total mean of the optimum of production material ($\bar{x}$=3.88, SD=0.57); and (4) Total mean of the matching with production ($\bar{x}$=4.00, SD=0.38)

4. Research's result

4.1. The result from the study of the identity and process of brass tea set

The result from the study of the identity and process of brass tea set craft is found that, the identity of Lop Buri province is its ancient land since prehistoric period, which has many archaeological evidences, such as ancient community, ancient remains, antiques, and the Lop Buri's animal which is monkey living with the resident for a long time. Nowadays, brass tea set has varied process, which are 1. Forging shape forming, 2. Turning shape forming. These processes are complex and need manufacturing only; the community cannot produce by itself. Thus this research is applied "Lost Wax" to produce Brass Tea Set so that the community is supported to produce Brass Tea Set by itself.

4.2. The design and development of handicraft products, brass tea sets

The design and development of handicraft products, brass tea sets Ban Tha Kra Yang community, Lop Buri Province, found that the identity of Lopburi province was related to the Slogan of the province. It was inspired to design pattern include monkey, wang Narai, phra prang sam yod and pikul flower of

Lop Buri Province, By analysis the handcrafted expert gave an opinion to the design of Brass Tea Set as follows; Design 2 was the most favorable of 3 designs by means ($\bar{x}$ = 4.61, S.D.=0.42). When each aspect was considered, it found that (1) Total mean of Aesthetic, Results the design study Brass tea set of Bantahkrayang it was inspired by shape relating to Historical remains of Lopburi Province, and designed in the framework of the core of the industrial design Udomsak Saributr [4] consider all *4* sides Aesthetic, Function, Manufacturing, The matching with production Researchers have drafted 30 models of sketch Design and Analytical principles "Product Design Theory" To select the appropriate model of the product.

4.3. Consumer satisfaction assessment results for Brass Tea Set Handicraft

Consumer Satisfaction Assessment Results for Brass Tea Set Handicraft The results of the study on identity and production process, as well as the design and development of brass tea set in Bantahkrayang Community to consumer satisfaction was to assess consumer satisfaction toward Brass tea set, (1) Total mean of Aesthetic($\bar{x}$ = 4.33, S.D.=0.38)

5. Discuss and summarize

According to the study on Handcrafted Brass Tea Set of Bantahkrayang Community divided into 3 parts.

The study of Handcrafted Brass Tea Set production process and design, it was found that the casting of wax models could be applied with other metal casting, for example, silver, copper and accessory molding work, which is in accordance with the research from Phensri Tongnoppakhun for developing the process of molding in accessory industry, which is careful process in order to reduce the damages and mistakes of works to the least [5].

To design the handcrafted of Brass Tea Set. The instruments used were primary data collection, interviews and questionnaires. The designing process then started by using collected data, through analytical processes with conceptual framework for product design; The Quality Function Deployment, the theory of inventive Problem Solving, Principles of Craft Design, and the Reverse Engineering Principle. The experts from both the fields of Industrial Product Design and Brass handicraft product, whom concern Brass Tea Set format, were included to evaluate the design and creats prototype of brass tea set.

To assess consumer satisfaction toward of handcrafted Brass tea set was at the high level.

6. Suggestions

6.1. Suggestion on this research

1. The Brass Tea Set is too heavy to reduce the thickness, 2. The handle should be wider for easy to use.

6.2. Applying the reseach's result

6.2.1. Study the production process and design of handcrafted Brass Tea Set. The researcher aimed to study the production process and design of handcrafted Brass Tea Set of Bantahkrayang Community, Lop Buri. Thus the reseach's result should be applied as a guideline of developing the handcrafted Brass Tea Set that can be also applied to other castings; silver, copper and others. Each casting is varied itself.

6.2.2. The development of handcrafted Brass Tea Set. The development of handcrafted Brass Tea Set Bantahkrayang Community must be mainly considered on the style because there are no technological manufactures for every process to help the production. The style is mainly on handicraft. Thus the development and design should be focused on the possibility of shape forming or casting those handicrafts.

6.2.3. Consumer satisfaction assessment results for Brass Tea Set. Consumer Satisfaction Assessment Results for Brass Tea Set Handicraft was shown that there were 2 consumer groups, one have bought to use daily and another have bought for home decoration. Thus, the product design should be varied for usability so that the consumer groups can be more varied and increasingly. These suggestions could be applied to improve the data on the design of Brass home decoration product development.

References

[1] Lopburi Fine Arts 1997 Identity and Wisdom in Lopburi vol 4 chapter 4 pp 293-294.
[2] Phanitsupa T 2013 Development a group of technicians And marketing strategies for Brass Products 41-50.
[3] Wiroon T 1983 Design., Bangkok, Visual Art Publishing 158-159.
[4] Udomsak S 2006 Technology of Industrial Product, O.S. Printing House p 19.
[5] Phensri T 2011 The research for developing the molding process in accessory Industry pp 1-2.

Research on squeal noise of tread brake system in rail freight vehicle

Zhang Jun[1*], **Li Yong-hua**[1], **Fang Ji**[1], **Zhao Wen-zhong**[1]

[1]College of Traffic and Transportation Engineering, Dalian JiaoTong University, Dalian 116028, China

*E-mail: zhj@djtu.edu.cn

Abstract. Brake squeal is a result of a unstable flutter from brake system, it results to the noise pollution in railway side and excessive wear of wheel tread. A finite element model of brake system for rail freight vehicle is set up, the contact and friction between the brake shoe and wheel tread is considered, the complex modals of brake system are calculated, the possibility of happening chatter and squeal noise are analyzed. The results show that the pressure angle or the brake force direction have a important influence on the unstable chatter and squeal noise, the more greater the pressure angle deviates from the wheel center, the more greater the possibility of happening chatter and squeal noise is, and the possibility of happening chatter and squeal noise is also increased along with the addition of friction factor.

1. Introduction

The thread brake is widely used because of simple structure, lower cost of manufacture and convenient control in rail freight vehicle, currently, the overwhelming majority of rail freight vehicle use the thread brake, it is that the brake force is directly acted on the wheel thread [1-2]. The thread brake is same as the disc brake, if the structure is designed unreasonably or running condition is poor, the unstable chatter and squeal noise may be induced[3-5]. The unstable chatter not only induce the high frequency squeal noise, produce the noise pollution in road side, but also cause excessive wear of wheels, wheel tread scratch and polygon wheel, this will seriously influence on the vehicle stability and steady in running[6]. The research show that the frequency domain of squeal noise produced by brake are wide, the noise is named the chatter noise if its frequency is below the 500Hz, in contrast the noise is named the high frequency squeal noise if its frequency is above 500 Hz. The high frequency squeal noise is a research focus and is difficulty in control[7]. Currently, the basic method of forecasting the instability chatter and squeal noise of brake system is to calculate complex modal of brake system, if real parts of complex modals exist any positive values, then the brake system has a lager possibility to happen the chatter and squeal noise in braking[8-9]. The chatter and squeal noise of brake system is related with the friction state[10], the possibility of inducing chatter and squeal noise can be depressed by improving the friction condition between the brake shoe and wheel tread , or improving the structure shape of brake system[11-12]; Moreover, the state of brake force is also major factor to influence on the chatter and squeal noise of brake system [13].

In this paper, the finite element model of brake system for rail freight vehicle is built, its complex modal is calculated, the effect of pressure angle and friction factor on squeal noise of brake system are analysed.

2. The analysis theory of complex modal

The brake noise belongs to frictional noise, the mechanism includes glue-sliding, friction-velocity negative slope and modal coupling, etc. The researches show that the squeal noise is accompanied by a vibration with a same frequency, this means that the squeal noise is induced by a unstable vibration because of friction. The method of researching squeal noise is that the whole brake system is chosen as a object, its finite element model of brake system considering the friction and contact is set up, then its complex modales are calculated, the complex eigenvalues are used to forecast that if the unstable vibration may be induced or the squeal noise may be aroused, the corresponding frequency is just right the squeal frequency. In practical application, the real parts of complex eigenvalues are used to estimate the possibility of squeal noise, if any real parts are positive or greater than zero, then brake system is considered as potentially instability and the possibility of happening squeal noise is more greater. The equation of calculating system complex modals is shown as follows

$$(M\lambda^2 + C\lambda + K\lambda)\phi = 0 \qquad (1)$$

where M 、 C 、 K are mass matrix, damping matrix and stiffness matrix, respectively, ϕ is the eigenvector, λ is complex eigenvalue. if λ_i is a complex eigenvalue as order i, then λ_i may be expressed as follows

$$\lambda_i = \beta_i + j\omega_i \qquad (2)$$

where β_i, ω_i are the real parts and imaginary parts, respectively, and the imaginary parts ω_i is just right the frequency of complex modal as order i.

The response of system bay be expressed as follows

$$u(t) = \sum_i \phi_i \exp(\beta_i + j\omega_i)t \qquad (3)$$

The equation (3) show that if any real pars of complex modals is greater than zero, then the response $u(t)$ is added persistently along with the time, that means the system is unstable and the squeal noise may be induced.

3. Finite element model of tread brake system

Finite element model of brake system is shown in Figure 1, it is comprised of wheel, brake shoe, brake shoe salver and brake beam. The each components are partitioned by use of the hexahedron and pentahedron element, the numbers of total elements are 47082. A contact of surface to surface is defined between the wheel surface and brake shoe surface, the coulomb friction model is used; the tied constrains are imposed between the brake shoe and brake shoe salve and between the brake shoe salve and the beam, respectively; the fixed constraints are imposed at the axel hole of wheel. When the braking is required, a pressure is produced between the brake shoe and the wheel because of a braking force acting on the brake beam, and therefore the braking will be achieved. The material property are listed in Table 1.

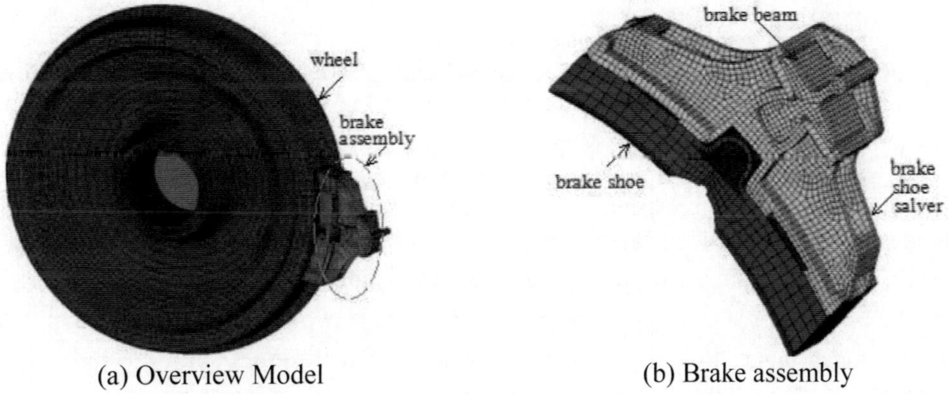

(a) Overview Model (b) Brake assembly

Figure 1. Finite element model of brake system.

Table 1. Material property

component	material	elasticity modulus (MPa)	Poisson ratio	density Kg/m3
wheel	steel	2.1e11	0.3	7800
Brake shoe	complex	8.0e8	0.4	1800
Brake shoe salve	steel	2.0e11	0.3	7800
Brake beam	steel	2.0e11	0.3	7600

4. Computing results and analysis

The complex modals of the above finite element model are computed by using of the ABAQUS software, the effect of both the pressure angle of brake assemble and the friction factor on the complex modal are analysed.

4.1. The initial analysis

The initial state of the braking is follows: the brake force is 25.3kN, the direction of brake force is point to the wheel center, it means $\alpha=0$,it is shown Figure 2; the running speed of railway freight is 8.28km/H, it means that the angle speed of wheel is 5rad/s, the friction factor of between the wheel surface and brake shoe is 0.3. The front 200 complex modals are computed, the vibration model of complex modal foe 135 order is shown Figure 3. The real parts of frequencies of complex modals computed are all zero, this means that the brake system is stabile, the possibility of bringing brake squeal is lesser.

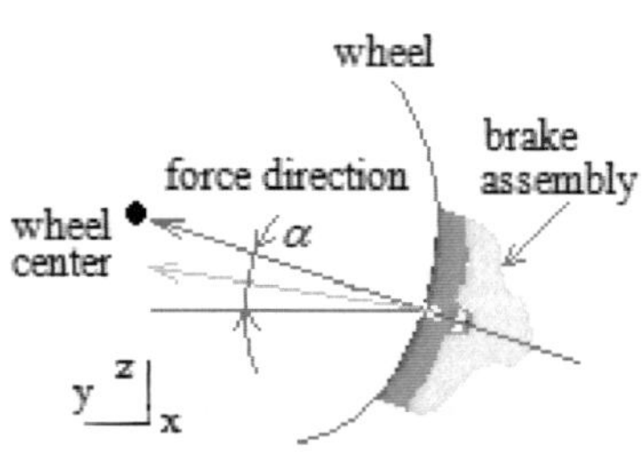

Figure 2. Pressure angle of brake system.

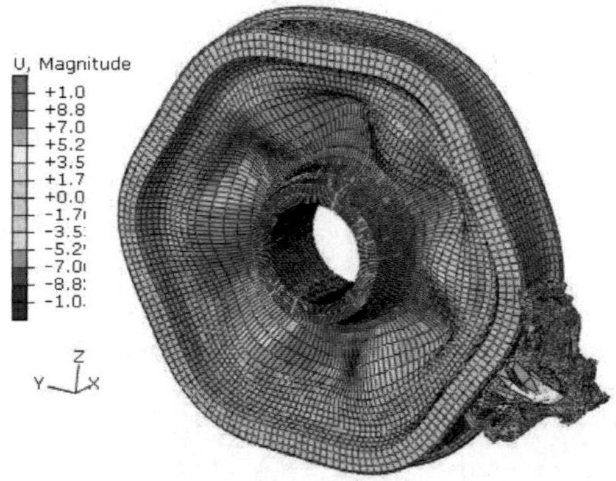

Figure 3. The 125th complex modal with a Frequency of 2635.1Hz.

4.2. The effect of pressure angle on the Squeal Noise

In running process of railway freight, the direction of braking force may be changed because the poor design, poor operation condition and various vibration, this can results that the direction of braking force is not point to the wheel center. When the pressure angles are both $\alpha=7^0$ and $\alpha=14^0$, the complex modals of brake system are computed ,respectively, the real parts and imaginary parts are shown in Figure 4 and Figure 5,respectively. The Figure 4 and Figure 5 show that the possibility of bringing brake squeal are added when the brake force is not point to the wheel center , the real parts of three and six complex modals are positive values for $\alpha=7^0$ and $\alpha=14^0$, respectively, and in case of $\alpha=14^0$ the number of the real parts with the positive values are two times as $\alpha=7^0$, the most value of real part is also much bigger than that of $\alpha=7^0$. This shows that the instability of brake system are deteriorated with the increasing of pressure angle.

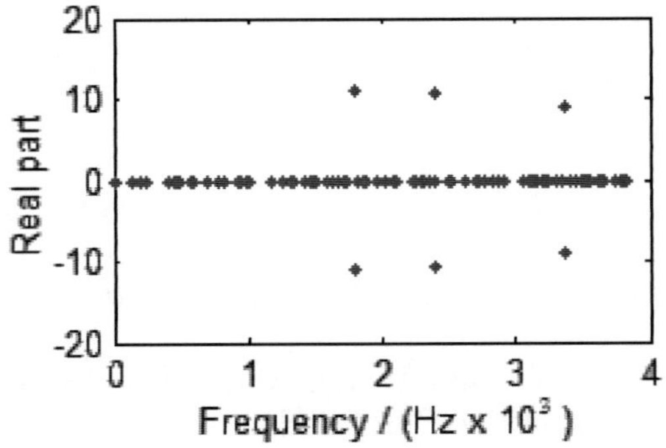

Figure 4. The complex modal with $\alpha=7^0$.

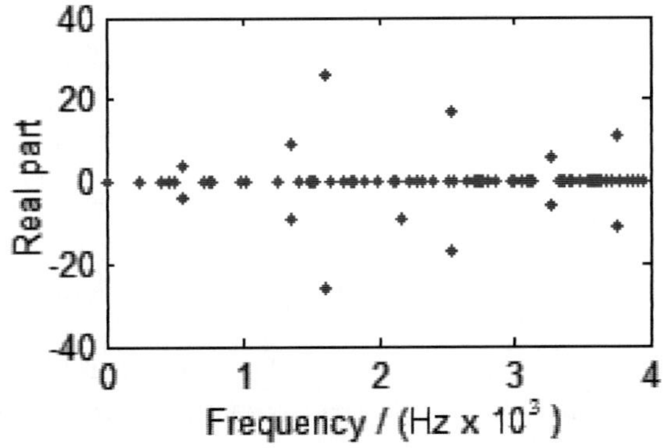

Figure 5. The complex modal with $\alpha=14^0$.

4.3. The effect of friction factor on the Squeal Noise

In case of $\alpha=7^0$, the complex modals of brake system with the 0.35 and 0.4 for friction factor are computed, the results show that the number of complex modals with a positive values for real parts are added than original one , and most values of real parts with a positive values are added, this means that the possibility of bringing brake squeal are added with along the addition of friction factor. Therefore, in order to increase the stability and reduce the possibility of bringing brake squeal, the friction factor should be properly chosen.

5. Conclusions

The pressure angle or the direction of brake force have the important influence on the stability and squeal noise of brake system, the direction of brake force deviating from the wheel center may be induced because of poor running condition. The more greater the pressure angle is deviated from the wheel center is, the instability of brake system is higher, the more greater the possibility of bringing brake squeal is. The possibility of bringing brake squeal are added with along the addition of friction factor.

Acknowledgement

This work was financially supported by The State Key Laboratory Open Project of Traction Power in China (TPL1402).

References

[1] Cao H Z 2004 Design and analysis of 160 km/ h freight car braking system *China Railway Vehicle* **1(42)** 16-20.
[2] Zeng X K, Zhao J J 2006 Development of braking system on freight cars and the present condition of operation in China *RAILWAY LOCOMOTIVE &CAR* **26(3)** 10-15.
[3] Guan D H, Su X D 2004 An overview on brake vibrations and noise *Engineering Mechanics* **21(4)** 150 -155.
[4] AbuBakar A R, Ouyang H J 2006 Complex eigenvalue analysis and dynamic transient analysis in predicting disc brake noise *International Journal of Vehicle Noise and Vibrations* **2(2)** 143- 155.
[5] Junior M T 2008 Analysis of brake squeal noise using the finite element method: A parametric study *Applied Acoustics* **69** 147 -162.
[6] Dai H Y 2006 Analysis on chatter in brake system of railway vehicle *Journal of Engineering Design* **13(1)** 41-45.
[7] Li B, Wang Q B, Qu B 2012 The noise analysis of disc brake based on ansys soft *Journal of Guizhou University* **29(3)** 112-115.
[8] Sheng Y S, Ma L, Sun G H 2007 Brake noise oriented complex modal analysis is of disc brake

with FEM *Machinery Design & Manufacture* **11** 87-90.

[9] Huang Z H, Wan X, Lei W 2016 Disc brake nonlinear model analysis and improvement base on complex modal *Modern Manufacturing Engineering* **12** 89-93.

[10] Liu B W, Yang Y, Xiong X 2009 The research on brake noise of automobile *Tribology* **4(29)** 385-392.

[11] Su X D, Di G H 2003 Reduction of brake squeal using substructure dynamic optimization *Automotive Engineering* **25 (2)** 167-170.

[12] Yang G J 2010 Review of structural vibration squeal of drum brake *Noise and vibration control* **1** 01-06.

[13] Hou W J, Chen G X 2013 Research on brake squeal of a railway freight tread brake system using the finite element method *LUB RICATION ENGINEE RING* **38(9)** 23-26.

IOP Conf. Series: Materials Science and Engineering **220** (2017) 012041 doi:10.1088/1757-899X/220/1/012041

Servo-controlling structure of five-axis CNC system for real-time NURBS interpolating

Chen Liangji[1], Guo Guangsong[2] and Li Huiying[1*]

[1]School of Mechanical Engineering, Tianjin Polytechnic University, Tianjin, 300387, China

[2]School of Mechatronics Engineering, Zhengzhou University of Aeronautics, Zhengzhou, 450015, China

*E-mail: lihuiying@tjpu.edu.cn

Abstract. NURBS (Non-Uniform Rational B-Spline) is widely used in CAD/CAM (Computer-Aided Design / Computer-Aided Manufacturing) to represent sculptured curves or surfaces. In this paper, we develop a 5-axis NURBS real-time interpolator and realize it in our developing CNC(Computer Numerical Control) system. At first, we use two NURBS curves to represent tool-tip and tool-axis path respectively. According to feedrate and Taylor series extension, servo-controlling signals of 5 axes are obtained for each interpolating cycle. Then, generation procedure of NC(Numerical Control) code with the presented method is introduced and the method how to integrate the interpolator into our developing CNC system is given. And also, the servo-controlling structure of the CNC system is introduced. Through the illustration, it has been indicated that the proposed method can enhance the machining accuracy and the spline interpolator is feasible for 5-axis CNC system.

1. Introduction

Many commercial CAD systems currently provide their correspondents the capability of defining sculptured surfaces. Among the mathematical techniques used for designing specific surfaces, the NURBS is one that attracts a lot of attention[1~3]. NURBS method offers a common mathematical form for describing both standard analytical shapes and free-form surfaces[3].

In order to get a part with sculptured surfaces, a CAD system is firstly chosen to design the part with NURBS method, and then a CAM system is used to calculate tool paths for cutting the part[4]. The tool paths are typically approximated with piecewise line or circular segments. As a result, the corresponding NC codes are generated and the motion controller can be applied to guide the CNC machines to execute accurately cutting[5].

However, the above approximation approach may result in several problems such as the large size of the CNC program, discontinuity of velocity and acceleration and so on[6,7].

In order to overcome the disadvantages of conventional CNC machining methods and also keep the advantages of CAD systems (use NURBS to represent curves), the need to develop general real-time NURBS command generators for CNC machines cannot be overlooked[8]. In addition, the high-speed and high-accuracy controlled machining of parts with NURBS surfaces is generally finished in 5-axis coordinated CNC machine tools[9]. Consequently, the CNC system should be provided with the ability to directly real-time generate 5-axis servo-controlling signals to drive the cutting tool along the NURBS curve relative the part.

The rest of the paper is organized in the following way. Section 2 describes how a NURBS curve is represented in a CAD system. Section 3 introduces the generating process of NC code. Section 4 shows the data structures of NC translators. Section 5 presents the realization method of 5-axis spline interpolator. Section 6 demonstrates a real example to validate the proposed method. Section 7 concludes the paper.

2. NURBS curve representation

Most CAD/CAM systems use parametric forms to represent curves or surfaces. The general form of a parametric curve in three-dimensional (3D) space can be expressed, in terms of a free parameter $0 \le u \le 1$, as

$$\boldsymbol{C}(u) = x(u)\boldsymbol{i} + y(u)\boldsymbol{j} + z(u)\boldsymbol{k} \tag{1}$$

where u is an arbitrary parameter, $\boldsymbol{i}, \boldsymbol{j}, \boldsymbol{k}$ is the unit vector of axis orientation of x, y, z respectively.
A p-degree NURBS curve is defined as follows

$$\boldsymbol{C}(u) = [x(u)\ y(u)\ z(u)]^{\mathrm{T}} = \frac{\sum_{i=0}^{n} N_{i,p}(u) W_i \boldsymbol{P}_i}{\sum_{i=0}^{n} N_{i,p}(u) W_i} \qquad 0 \le u \le 1 \tag{2}$$

Among which $\boldsymbol{P}_i$ is the ith 3-D control point; W_i is the corresponding weight factor of $\boldsymbol{P}_i$; $(n+1)$ is the number of control points. $N_{i,p}(u)$ is the so-called normalized B-spline basis function, where p is its degree, and can be calculated by the following recursive formula

$$\begin{cases} N_{i,0} = \begin{cases} 1 & (u_i \le u < u_{i+1}) \\ 0 & (\text{ otherwise }) \end{cases} \\ N_{i,p}(u) = \dfrac{u - u_i}{u_{i+p} - u_i} N_{i,p-1}(u) + \dfrac{u_{i+p+1} - u}{u_{i+p+1} - u_{i+1}} N_{i+1,p-1}(u) \\ \dfrac{0}{0} = 0 \quad (\text{prescribed}) \end{cases}$$

3. Tool paths and NC code generation

General CAD/CAM software such as Unigraphics cannot generate NC code including the introduced 5-axis coordinated double NURBS spline interpolation format, thus the CAD/CAM module needs to be developed by ourselves. The flow chart of generating NC code by the CAD/CAM module is shown as Figure 1.

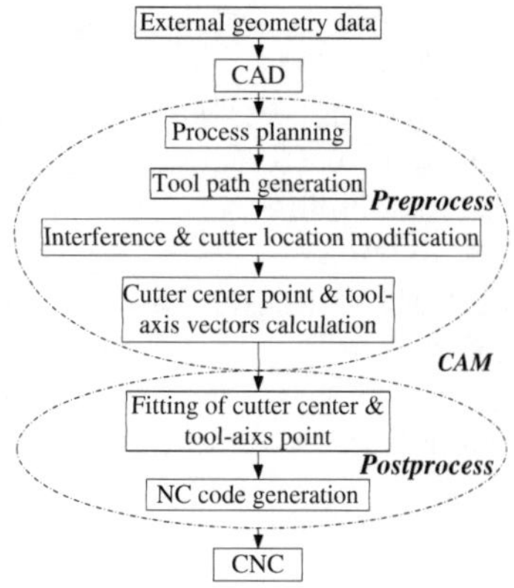

Figure 1. Generation of NC code.

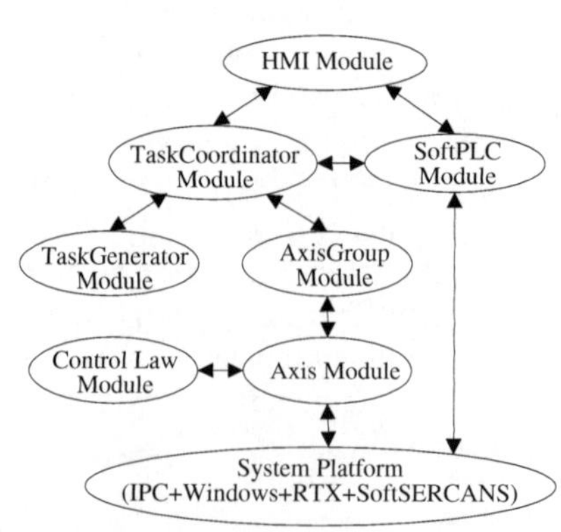

Figure 2. Modular structure of the CNC system.

The CAD part in the module supplies interfaces connecting other CAD/CAM software to make full use of the geometry modeling functions of excellent CAD/CAM softwares. After pre-process procedure such as tool path generation, interference checking, the CAM part in the module calculates the cutter center point coordinates and the corresponding tool-axis unit vectors, which are then fitted into cutter center point spline and orientation spline in the post-process procedure and the NC code including the double NURBS spline curve interpolation format is generated and output.

4. Realization of the translator

The modular structure of the developing CNC system is shown as Figure 2, among which, TaskGenerator module is realized in Microsoft's COM technology and is responsible for the translation of NC programs, and the AxisGroup module is realized in VenturCom's RTDLL (real-time dynamical link library) technology and is responsible for the real-time interpolation and other real-time tasks.

It is to be integrated a 5-axis NURBS translator to the CNC system. Modules need to be changed are only TaskGenerator module and AxisGroup module. Data structures (described in C++) used in TaskGenerator module is given as follows:

```
typedef enum {LINE, DNURBS} SingleStep_TYPE;
typedef struct dnurbs_struct{
  double start_rate,traverse_rate, end_rate;
  int p;
  double r[20],kv[26];
  double clx[20],cly[20],clz[20];
  double tclx[20],tcly[20],tclz[20];
} dnurbs;
typedef struct singleStep_struct{
  SingleStep_TYPE singleStep_type;
  line line_struct;
  dnurbs dnurbs_struct;
} singleStep;
typedef std::deque<singleStep> singleStep_deque;
```

The dnurbs struct is added in the module for the translation of 5-axis NURBS code. The information extracted from each motion segment after translation is put into singleStep_deque and is transferred into the AxisGroup module for interpolation.

5. Real-time 5-axis spline interpolation

Footnotes should be avoided whenever possible. If required they should be used only for brief notes that do not fit conveniently into the text.

The task of real-time NURBS spline interpolation is to calculate next interpolation period's CL data that including tool-tip location coordination and tool-axis orientation and then transfer the CL data into machine tool's motion commands such as X,Y,Z,A,C through post-procession.

5.1. Real-time calculation of CL data

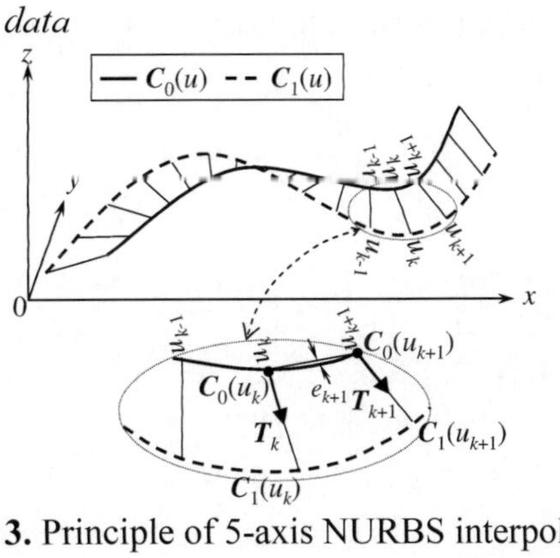

Figure 3. Principle of 5-axis NURBS interpolation.

IOP Conf. Series: Materials Science and Engineering **220** (2017) 012041 doi:10.1088/1757-899X/220/1/012041

As shown in Figure 3, $\boldsymbol{C}_0(u)$ and $\boldsymbol{C}_1(u)$ is the cutter center point spline and the orientation spline, respectively. The two splines are constructed in the same knot vector. Assumed that parameter u is function of time t, that is, $u=u(t)$. By using Taylor's expansion of the parameter u with respect to time t to obtain the first order approximation interpolation algorithm, the first order approximation up to the first derivatives is

$$u_{k+1}=u_k+T_s\frac{\mathrm{d}u}{\mathrm{d}t}\Big|_{t=t_k} \tag{3}$$

where T_s is interpolation period, u_k and u_{k+1} are corresponding parameters of current and next time t_k and t_{k+1}. The feedrate of tool-tip point along $\boldsymbol{C}_0(u)$ is defined by

$$\boldsymbol{V}(u)=\frac{\mathrm{d}\boldsymbol{C}_0(u)}{\mathrm{d}t}=\left(\frac{\mathrm{d}\boldsymbol{C}_0(u)}{\mathrm{d}u}\right)\left(\frac{\mathrm{d}u}{\mathrm{d}t}\right) \tag{4}$$

and the curve speed at the parameter u_k is

$$V(u_k)=\left\|\frac{\mathrm{d}\boldsymbol{C}_0(u)}{\mathrm{d}t}\right\|_{u=u_k} \tag{5}$$

According to equations (4) and (5), the first derivative of u with t is obtained as

$$\frac{\mathrm{d}u}{\mathrm{d}t}\Big|_{t=t_k}=\frac{V(u_k)}{\left\|\frac{\mathrm{d}\boldsymbol{C}_0(u)}{\mathrm{d}u}\right\|_{u=u_k}} \tag{6}$$

Therefore, the first-order interpolation algorithm is obtained by substituting equation (6) into equation (3), and the latter can be expressed as follows

$$u_{k+1}=u_k+\frac{T_sV(u_k)}{\left\|\frac{\mathrm{d}\boldsymbol{C}_0(u)}{\mathrm{d}u}\right\|_{u=u_k}} \tag{7}$$

The first derivative of $\boldsymbol{C}_0(u)$ with u is obtained as

$$\frac{\mathrm{d}\boldsymbol{C}_0(u)}{\mathrm{d}u}=\frac{\sum_{i=0}^{n}N'_{i,p}(u)W_i\boldsymbol{P}_i}{\sum_{i=0}^{n}N_{i,p}(u)W_i}-\frac{\sum_{i=0}^{n}N'_{i,p}(u)W_i\sum_{i=0}^{n}N_{i,p}(u)W_i\boldsymbol{P}_i}{\left(\sum_{i=0}^{n}N_{i,p}(u)W_i\right)^2} \tag{8}$$

where the general algorithm for 1st order derivative of $N_{i,p}(u)$ is

$$N'_{i,p}(u)=p\left[\frac{N_{i,p-1}(u)}{u_{i+p-1}-u_i}-\frac{N_{i+1,p-1}(u)}{u_{i+p}-u_{i+1}}\right] \tag{9}$$

$\boldsymbol{C}_0(u_{k+1})$ and $\boldsymbol{C}_1(u_{k+1})$ are obtained by substituting the calculated u_{k+1} into $\boldsymbol{C}_0(u)$ and $\boldsymbol{C}_1(u)$. Suppose the tool-axis unit vector is $\boldsymbol{T}_{k+1}$, so

$$\boldsymbol{T}_{k+1}=\frac{C_1(u_{k+1})-C_0(u_{k+1})}{\left\|C_1(u_{k+1})-C_0(u_{k+1})\right\|} \tag{10}$$

5.2. Real-time post-procession of CL data

To here, we have the following CL data:

$$\boldsymbol{C}_0(u_{k+1})=(o_x, o_y, o_z) \quad \boldsymbol{T}_{k+1}=(t_x, t_y, t_z)$$

For 5-axis machine of table tilting/rotating type (Figure 4), the following post-procession method is presented to transfer the CL data into the machine's motion commands in real-time.

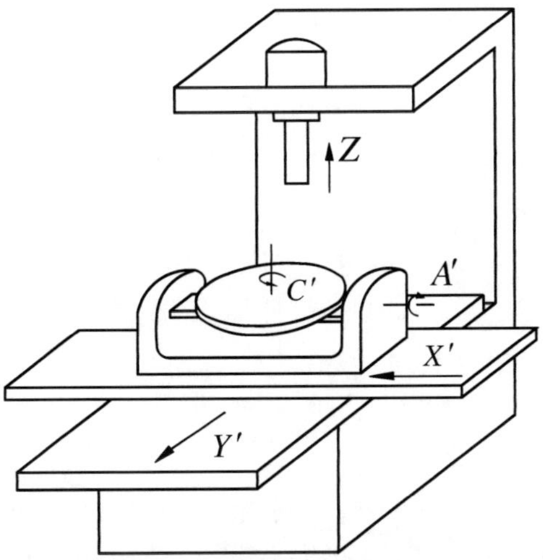

Figure 4. A type of 5-axis machine tool.

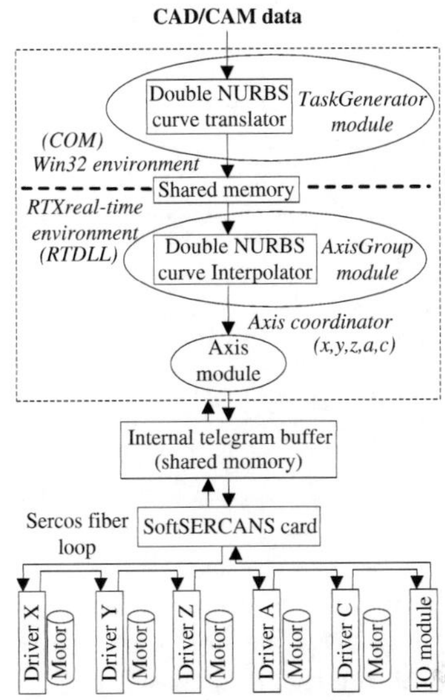

Figure 5. Servo-controlling structure of CNC system.

(a)if $t_x = t_y = 0, t_z = 1$, then $A'=0, C'=0$

(b)else if $t_x = 0, t_y \neq 0$,then $A' = \arcsin(-t_y), C' = 0$

(c)else if $t_y = 0, t_x \neq 0$,then $A' = \arcsin(t_x), C' = \pi/2$

(d)else

$$A' = \arccos(t_z)$$

$$C' = \begin{cases} \arccos(-t_y/\sin A') & t_x > 0 \\ 2\pi - \arccos(-t_y/\sin A') & t_x < 0 \end{cases}$$

5.3. Servo-controlling structure

The proposed format's double NURBS NC code is generated in CAD/CAM and disposed by the translator. Then the obtained data is sent to the interpolator in which it is disposed according to the given interpolation period and thereafter, the calculated target position is treated with control law. For the CNC system adopts SoftSERCANS technology (IEC61491-SERCOS technology's software realization, as shown in Figure 2), the final data is put into an internal telegram buffer (a block of shared memory), and the SoftSERCANS card exchanges data with it. Then the data telegram containing moving instructions is sent to drivers by SoftSERCANS through fiber optic ring and thus motors are driven. The controlling structure is shown as Figure 5.

6. Example and realization

The 5-axis NURBS NC code is shown as follows:

```
N30 G06.5 P3
X47.7773 Y71.0202 Z58.8006 U47.8066 V71.9791 W59.0828 R1 K0 F300
X44.3229 Y75.6066 Z54.4778 U44.3651 V76.5494 W54.8138
X37.2164 Y84.6443 Z44.5755 U37.2554 V85.5421 W45.0342
X28.7567 Y101.5653 Z29.7151 U28.7081 V102.3535W30.3361
X-1.6963 Y163.6796 Z5.8369 U-1.8572 V163.9906 W6.7762 K0.5046 K1
```

It's obvious the proposed method can greatly reduce the size of NC files; especially when in high precision machining, the size change will be more obvious. Meanwhile, the CNC system's pre-reading one part program segment equals pre-reading tens or even hundreds of linear interpolation segments; thus, the "prospective property" of CNC system can be solved easily.

In Figure 6, the curve inflection point is near the parameter 0.42, where its curvature is the smallest. Compared with traditional linear approximation machining mode, the proposed method can greatly ameliorate the machining performance, improve machining accuracy and the machining accuracy is far less influenced by the curve curvature change.

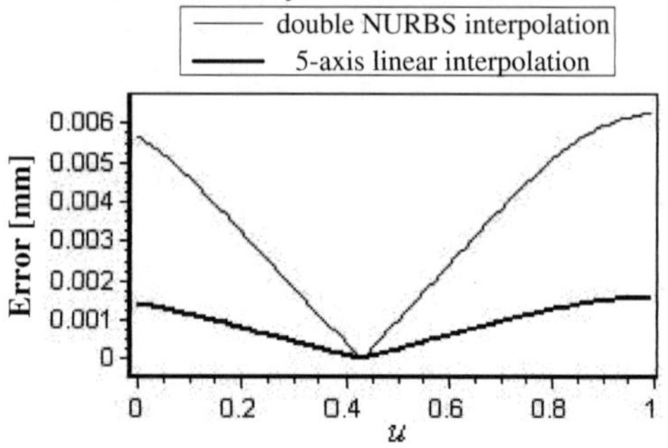

Figure 6. Error comparison of two methods.

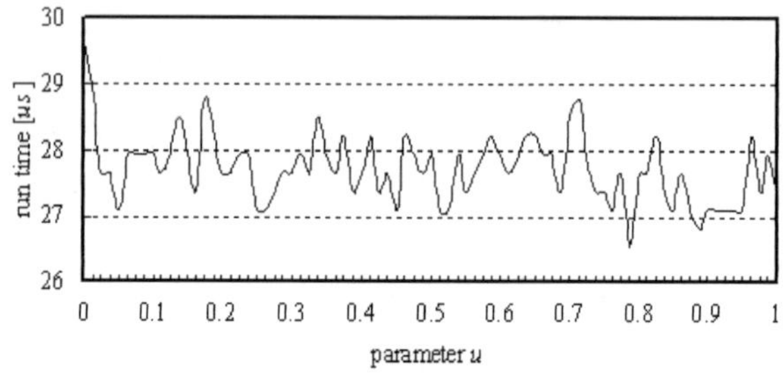

Figure 7. Run time of the proposed method.

The proposed method's real-time property is tested in a computer (Pentium IV 2.4GHz, RAM 512M) and the result is shown in Figure 7, the calculation is based on the second order approximation algorithm and the beginning of a time segment is the calculation of the basis function and the end is the finish of each axis' position calculation. It can be found the average run time in a double NURBS interpolation period is near 28μs and the time expense occupies a small proportion in a CNC system of which the interpolation period is 1 ms or even 500μs.

7. Conclusions

On the basis of analyzing the defects of the existing 5-axis linear interpolation method used in the machining of sculptured surface, 3-axis coordinated NURBS interpolation format is extended and a spline expressing the tool-axis orientation is added in the paper, and double NURBS curve interpolation method is put forward, the NC code generating method is introduced and the method to realize the interpolation function in an open CNC system developed by ourselves is given. Therefore, the proposed double NURBS interpolation method can not only be used in the real CNC system, but also have prospective future.

Acknowledgments

This work was funded by the National Natural Science Foundation of China (No. 51641509).

References

[1] Tsehaw Y and Ranga N 2003 *Computer-Aided Design* **35** 1249.

[2] Marchenko M, Tae J K and Lee S H 2004 *International Journal of Machine Tools & Manufacture* **44** 237.

[3] Piegl L 1991 *IEEE Computer Graphics & Application* **11** 55.

[4] Zhang Q G and Greenway R B 1998 *Robotics and Computer- Integrated Manufacturing* **14** 27.

[5] Yeh S S and Hsu P L 2002 *Computer-Aided Design* **34** 229.

[6] Mahadevan B 2003 *Computer-Aided Design* **35** 359.

[7] J Zhang, Zhang L, Zhang K and Mao J 2016 *International Journal of Advanced Manufacturing Technology* **83** 2015.

[8] Lo C C 1999 *International Journal of Machine Tools & Manufacture* **39** 471.

[9] Farouki R T and Tsai Y F 2001 *Computer-Aided Design* **33** 155.

Detection of Drug Effects on Brain Activity using EEG-P300 with Similar Stimuli

Arjon Turnip[1]*, Dwi Esti K. [1], M Faizal Amri[1], Artha I. Simbolon[1], M. Agung Suhendra[1], Shelly IsKandar[2], and Firman F. Wirakusumah[2]

[1]Technical Implementation Unit for Instrumentation Development, Indonesian Institute of Sciences, Indonesia
[2]Faculty of Medicine, Universitas Padjadjaran, Indonesia

*arjon.turnip@lipi.go.id, jujhin@gmail.com

Abstract. Drug addiction poses a serious problem to our species. The worry that our significant family might be involved in drug use and are concerned to know how to detect drug use. Examinations of thirty taped EEG recordings were performed. The subjects consist of three group: addictive, methadone treatment (rehabilitation), and control (normal) which 10 subjects for each group. Statistical analysis was performed for the relative frequency of wave bands. The higher average amplitude is obtained from the addiction subjects. In the comparison with the signals source, channels P3 provide slightly higher average amplitude than other channels for all of subjects.

1. Introduction

Drug addiction poses a serious problem to our species. Consider the disastrous effects caused by the abuse of one of our oldest drugs, alcohol: automobile accidents, fetal alcohol syndrome, cirrhosis of the liver, Korsakoff's syndrome, increased rate of heart disease, and increased rate of intracerebral haemorrhage [1]. Moreover, drug addiction has been associated with a broad range of cognitive deficits in domains, including emotional regulation and motivation, attention and flexibility, working memory, learning, and decision making [2]. For example, cocaine addiction can cause psychotic behavior, brain damage, and death from overdose; and competition for lucrative and illegal markets terrorizes neighborhoods, subverts political and judicial systems, and causes many violent deaths. The drugs stimulate brain mechanisms responsible for positive reinforcement. In addition, most of them also reduce or eliminate unpleasant feelings, some of which are produced by the drugs themselves. What actually seems to happen is that the occurrence of an appetitive stimulus activates a reinforcement mechanism in the brain that increases the likelihood of the most recent response in the present situation. All natural reinforcers that have been studied so far (such as food for a hungry animal, water for a thirsty one, or sexual contact) have one physiological effect in common: They cause the release of dopamine in the nucleus accumbens [3].

Perhaps we worry that our significant family might be involved in drug use and are concerned to know how to detect drug use. Regardless of whom the person is and what his or her relationship is to us, it is important to understand how to detect drug use so we can get help for our child, loved one, or business associate. A drug test is a technical analysis of a biological specimen, for example urine, hair, blood, breath, sweat, or oral fluid/saliva to determine the presence or absence of specified parent drugs or their metabolites. However, those technical analysis are not veasible applied without medical personel assistance. Previouss studies have shown that drug abuse is associated with altered brain function [3-10]. Electroencephalogram (EEG) power analysis are one of veasible tool for examining the effects of drugs on brain function. Due to its wide availability, relatively low-cost, superior temporal resolution, easy implementation, and non-invasiveness, intensive research has been performed on EEG as tool for cognitive processes in research and clinical diagnose [11-18].

The EEG pattern is constantly changing depending on mental activity, relaxation, drowsiness and sleep and this dynamic process is therefore an index of cortical activation, cognitive function and consciousness. The EEG may thus be considered an intermediate phenotype for complex behaviors and psychopathology in which arousal is implicated, such as anxiety, depression and addiction. Using the EEG signals to gain access to a neural correlate of brain activity could possibly be useful in the investigation of its relationship to the clinical syndrome of drug dependence, and ultimately in the development or selection of medications. The brain of drug abusers presents various structural and neurophysiological abnormalities, some of which may predate drug consumption onset. However, how these changes translate into modifications in functional brain connectivity is still poorly understood. To characterize functional connectivity patterns, the EEG activity from 20 drug abusers and 10 age-matched control subjects performing a simple counting task are recorded.

2. Data aquisition

Thirty male subjects were recruited from Hasan Sadikin General Hospital (HSGH) of Bandung-Indonesia. The subjects consist of three group: addictive, methadone treatment (rehabilitation), and control (normal) which 10 subjects for each group. At the time of testing none of the subjects was taking prescribed or non-prescribed medications. The healthy control subjects who never used any drug were recruited from treatment staff. Most drug dependent subjects had a history of additional drug use and ten subjects had ever been in methadone maintenance treatment. The study was approved by the Ethical Committee of the institution in which the the experiment was performed. All subjects provided written informed consent.

Subjects were asked to participate in a study concerning the measurement of brain function. They were informed that participation involved EEG measurements. Both the patients and control subjects received a remuneration of IDR 150.000. All subjects provided written informed consent. The experiment started with a short explanation of the procedure and informed consent was obtained. Personal data and history of drug use were recorded by the experimenter. Then the subject completed the questionnaires. After completion the subject was seated in the EEG chair and electrodes were attached. Instructions were to sit relaxed and concentrated on counting the total appear number in the screen. The appearance sequence of the stimuli (shown in Fig. 1) were randomly flashed. After EEG measurements, electrodes were removed and subjects received their financial compensation.

The EEG signals were measured with a digital Mitsar 202 amplifier using Electrocap Ag/AgCI electrodes at the 9 scalp sites according to the International 10/20 System (Fp1, Fp2, F7, F3, Fz, F4, F8, P3, P4) with green color as shown in Fig. 2. Ground was placed between A1 and A2. All signals were digitized without filtering on a PC with WinEEG software with a sample rate of 500 Hz and 16-bit A/D conversion. Off-line, an average reference channel was computed (all channels except A1/A2), and used for all analysis. For all electrodes, impedance was kept under 5 KΩ. With modern high input-impedance amplifiers such as used in the present study, and accurate digital filters for power line noise, high-quality EEG can be recorded. The experiment setup is given in Fig. 3.

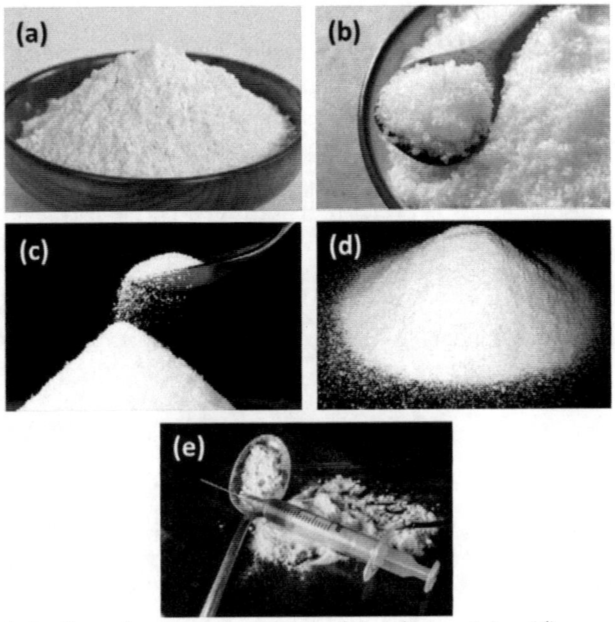

Figure 1. The flashed similar stimuli represented of the drugs: (a) – (d) as non target and (e) as a target.

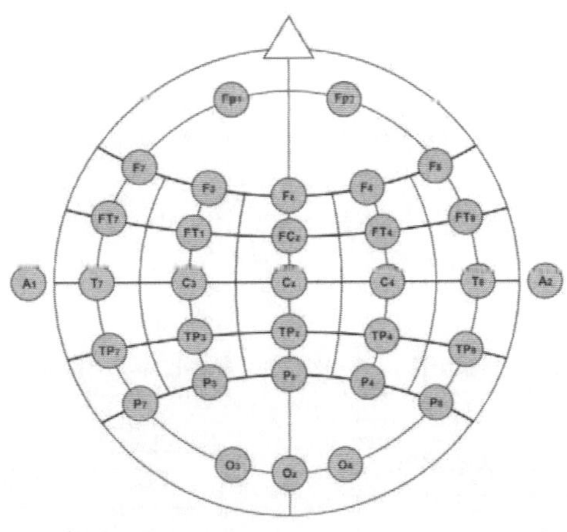

Figure 2. The electrodes position according to the International 10/20 System.

Figure 3. Experiment setup.

3. Signal processing

The flowchart of the signal processing is given in Fig. 4. One session for each subject consist of 20 seconds (i.e., each ten blank and number). One trial or epoch length is processed for 2 seconds. Prior to feature extraction, several preprocessing operations including filtering were applied to the raw data. For filtering, a 6th-order band pass filter (BPF) with cutoff frequencies of 3 Hz and 30 Hz was used.

Wavelet theory is a relatively new concept developed. The properties of the wavelet are given as follow: The time complexity is linear. Wavelet transform can be done perfectly with the time that is linear; Wavelet coefficients that are rarely elected. Practically, most of the wavelet coefficients of low value or zero. This condition is very beneficial, especially in areas of compression or data compression. Wavelet can be adapted to different types of functions, such as functions that are not continuous, and the function defined on bounded domains [19]. In general, a common function of wavelet is defined as [20 - 23].

$$\psi_{s,t}(t) = | s | \frac{1}{2} \psi \left[\left(\frac{t - \tau}{s} \right) \right] \quad (1)$$

where s and τ , $s \neq 0$, denotes the scale and translation parameters, and t indicates time. In the continuous wavelet transform, the signal is analyzed using a set of basic functions that are related to scaling and simple transition. The development of the CWT is presented in the following equation.

$$X_{WT}(\tau, s) = \frac{1}{\sqrt{| s |}} \int x(t) \psi \left(\frac{t - \tau}{s} \right) dt \quad (2)$$

Wavelet decomposition is embodied in the input signal and the filtered, lowpass fileter generates a waveforms called approximation and highpass filter generates a random waves called detail. The correlation of both filter with wavelet functions arranged in a hierarchy scheme called multiresolusi decomposition, where decomposition separates the signal into details at different scales and approximation.

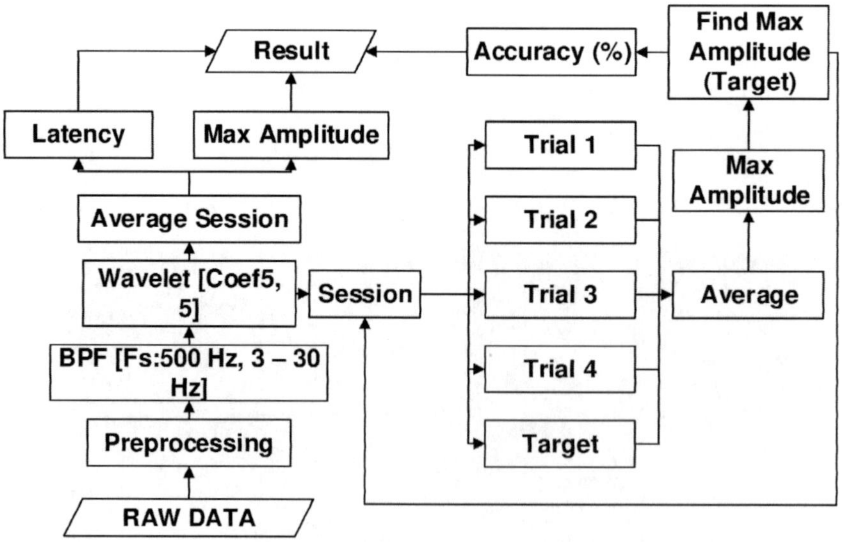

Figure 4. Signal processing processes

4. Results and discussions

IOP Conf. Series: Materials Science and Engineering **220** (2017) 012042 doi:10.1088/1757-899X/220/1/012042

The recorded EEG signals of addiction subject 2 from 10 channels (Fp1, Fp2, F7, F3, Fz, F4, F8, P3, PzP4) is shown in Fig. 5. For the raw data of all subject are generally similar which are mostly corrupted by noise. Fig. 6 indicate the filtered signals using band pass filter which the noise signals have been higly remove. The input signal is the raw EEG data then decomposed by the decomposition of the signal with noise on wavelet bases where there is information on signal wavelet coefficient. Denoising can be obtained by thresholding wavelet coefficients. Denoising is done by separating the wavelet coefficients with the download threshold. Wavelet group is used to pass the signal through wavelet transform blocks for the decomposition process, then the signal will be decomposed into wavelet coefficients which will be in thresholding and denoising where the process occurs. Thresholding function to eliminate noise and preserve the information that is important to the maximum signal.

Fig. 7 is the extracted signal using wavelet denoising. By applying the fast Fourier transform (FFT) algorithm, the peak amplitude of P300 component for each considered channels is obtained as Fig. 8. The higher amplitude is obtained from channels P3 and P4. The drug abuse has the lower amplitude in prefrontal cortex. People with a long history of drug abuse not only show the same deficits on tasks that involve the prefrontal cortex as do people with lesions of this region, they also show structural abnormalities of this region.

Fig. 9 and Fig. 10 are the comparison of the average latency and amplitude of the P300 component (addiction, control, and rehabilitation) for each channels. The control subjects has the shorter latency compare with the others group onthe target. It is indicates that, the control subject is faster to make a decision and others especially the rehabilitaion group are need more time to make a decision. Compare with the control one, the addiction subject need more power to get focus or concentrate. These EEG abnormalities may reflect underlying changes in brain function due to long-term drug use. The high amplitude is provided by the addiction subjects and the lower amplitude is provided by the rehabilitation subjects. It is indicate that the brain activities especially in the prefrontal cortex of the adicitive group is higher. It is suspected that the adictive subjects were still affected by the drug.

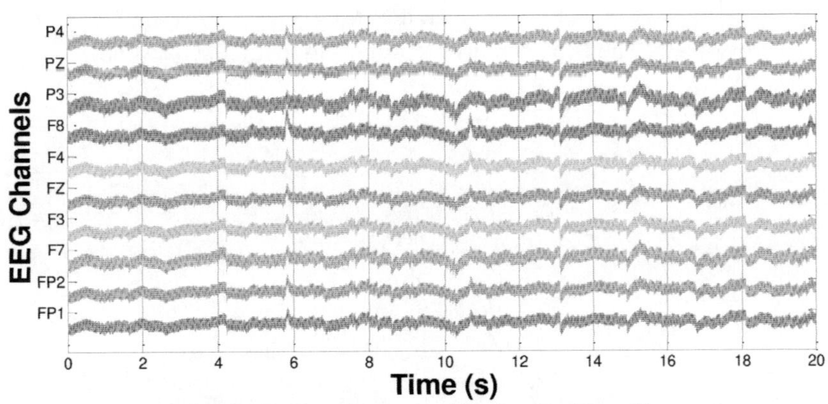

Figure 5. Raw data of addiction (subject 2).

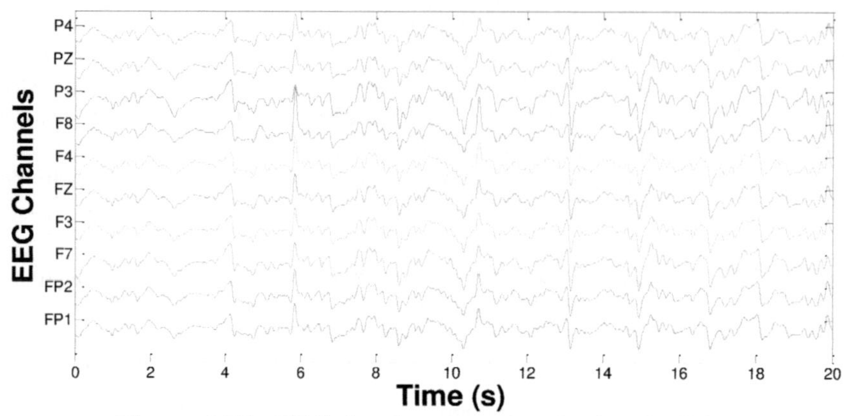

Figure 6. The EEG signals preprocessed using the BPF.

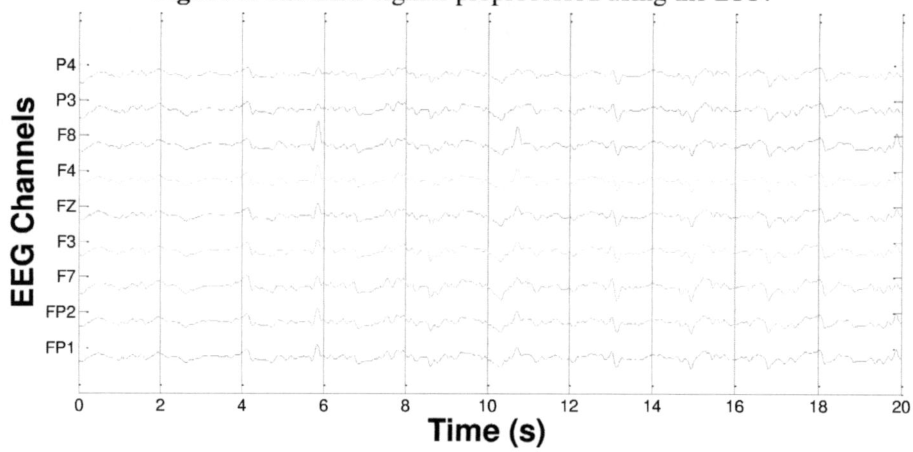

Figure 7. The EEG signals preprocessed using the Wavelet.

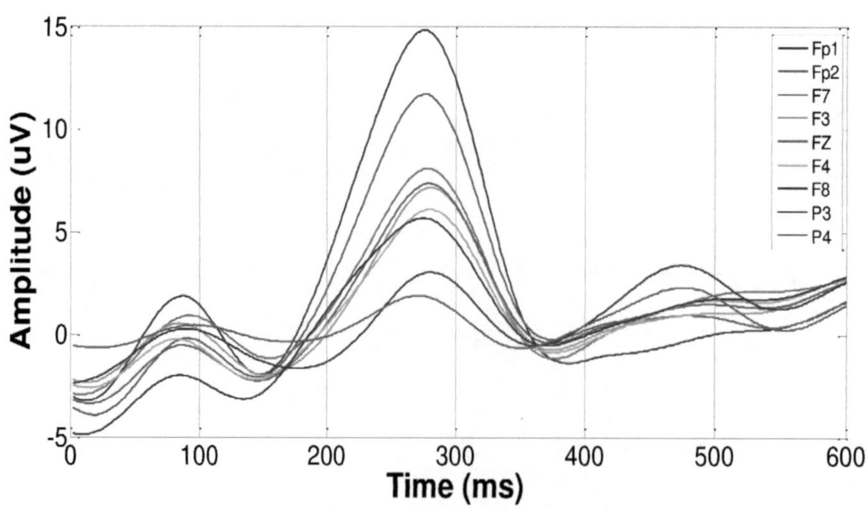

Figure 8. Extracted EEG-P300 signals from each considered channels.

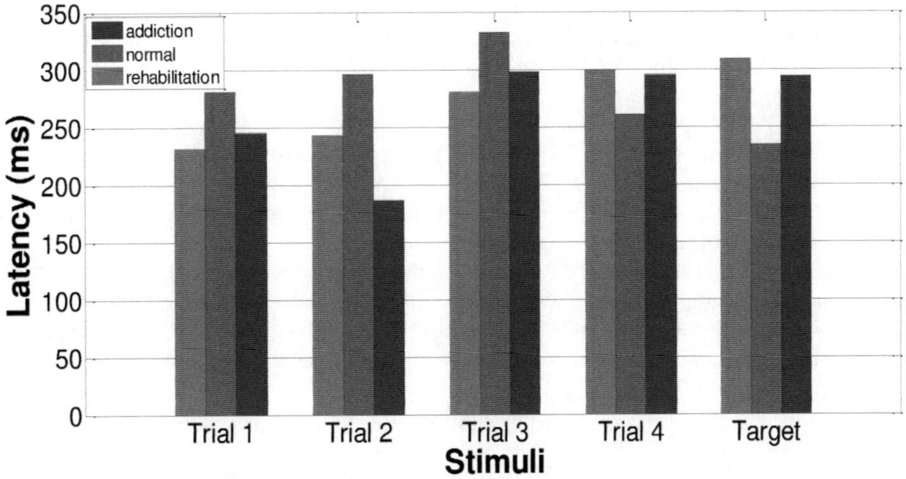

Figure 9. The comparison of the average latency (addiction, control, and rehabilitation) for each channels

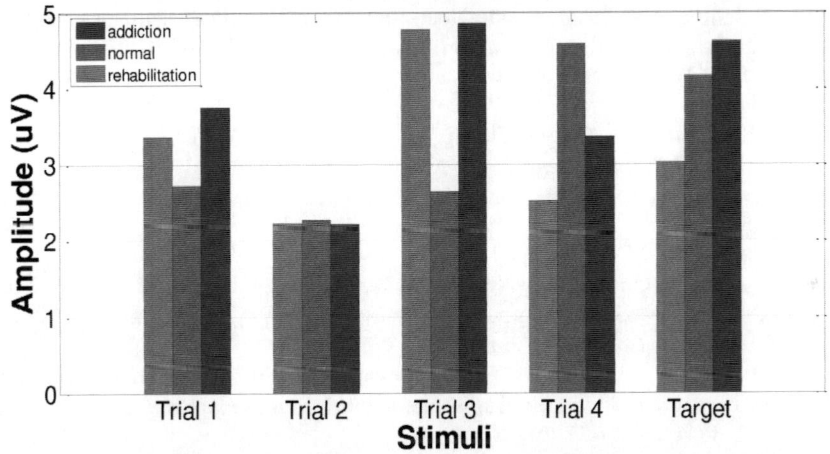

Figure 10. The comparison of the average amplitude (addiction, control, and rehabilitation) for each channels

5. Conclusions

The current study shows that excessive drug use subjects have an enhanced relative average amplitude compared to healthy control subjects. It has been suggested that a physiological relevance of the brain signal as a reflection of neural activity and the evidence of abnormal neural activity in drug dependence suggest the possibility that the brain signals might reflect abnormal neurobiology in drug dependence. These results indicate that chronic drug use may lead to distinct patterns of cognitive impairment that may be associated with dysfunction of different components of cortico-striatal circuitry.

Acknowledgments

This research was supported by the thematic program through the Bandung Technical Management Unit for Instrumentation Development (Deputy for Scientific Services), and the flagship program through the Research Center for Physics (Deputy for Engineering Sciences) funded by Indonesian Institute of Sciences and Faculty of Medicine (Universitas Padjadjaran), Indonesia.

References

[1] N. R. Carlson, Physiology of behaviour Person, Pearson Education, Inc., 2013.
[2] R. Z. Goldstein and N. D. Volkow, "Dysfunction of the Prefrontal Cortex in Addiction: Neuroimaging Findings and Clinical Implications", *Nat. Rev. Neurosci.* Vol. 12, pp. 652–669, 2011.
[3] F. J. White, "Synaptic regulation of mesocorticolimbic dopamine neurons", *Annual Review of Neuroscience*, Vol. 19, pp. 405–436, 1996.
[4] R. Coullaut-Valera, I. Arbaiza, R. Bajo, R. ARRUE, M. E. L´Opez, J. Coullaut-Valera, A. Correas, D. L´Opez-Sanz, F. Maestu, and D. Papo, "Drug Polyconsumption is Associated with Increased Synchronization of Brain Electrical-Activity at Rest and in a Counting Task", *International Journal of* Neural Systems, Vol. 24, No. 1, 2014.
[5] R. Z. Goldstein and N. D. Volkow, "Dysfunction of the prefrontal cortex in addiction: Neuroimaging findings and clinical implications", *Nat. Rev. Neurosci.* Vol. 12 , pp. 652–669, 2011.
[6] V. Bassareo, M. A. De Luca, and G. Di Chiara, "Differential expression of motivational stimulus properties by dopamine in nucleus accumbens shell versus core and prefrontal cortex", *J. Neurosci.* vol. 22, pp. 4709–4719, 2002.
[7] K. D. Ersche et al., "Abnormal brain structure implicated in stimulant drug addiction", *Science*, vol. 335, pp. 601–604, 2012.
[8] K. R. Alper, L. S. Prichep, S. Kowalik, M. S. Rosenthal, and E. R. John, "Persistent QEEG abnormality in crack cocaine users at 6 months of drug abstinence", *Neuropsychopharmacology*, vol. 19 , pp. 1–9, 1998.
[9] M. Vejmelka, M. Palus and K. Susm´akov´a, "Identification of nonlinear oscillatory activity embedded in broadband neural signals", *Int. J. Neural Syst.*, vol. 20, pp. 117–128, 2010.
[10] M. Ahmadlou and H. Adeli, "Functional community analysis of brain: A new approach for EEG-based investigation of the brain pathology", *NeuroImage*, volo. 58, pp. 401–408, 2011.
[11] U. R. Acharya, S. V. Sree, S. Chattopadhyay, and J. S. Suri, "Automated diagnosis of normal and alcoholic EEG signals", *Int. J. Neural Syst.*, vol. 22, pp. 1250011, 2012.
[12] D. M. Cole et al., "Nicotine replacement in abstinent smokers improves cognitive withdrawal symptoms with modulation of resting brain network dynamics", *NeuroImage*, vol. 52, pp. 590–599, 2010.
[13] A. Turnip, K.-S. Hong, and M.-Y. Jeong, "Real-time feature extraction of P300 component using adaptive nonlinear principal component analysis", *BioMedical Engineering OnLine*, vol. 10, 2011.
[14] A. Turnip and K. S. Hong, "Classifying mental activities from EEG-P300 signals using adaptive neural network", *Int. J. Innov. Comp. Inf. Control*, vol. 8(7), 2012.
[15] A. Turnip, S. S. Hutagalung, J. Pardede, and, D. Soetraprawata, "P300 detection using multilayer neural networks based adaptive feature extraction method", *International Journal of Brain and Cognitive Sciences*, vol. 2, no. 5, pp. 63-75, 2013.
[16] A. Turnip and M. Siahaan, "Adaptive Principal Component Analysis based Recursive Least Squares for Artifact Removal of EEG Signals", *Advanced Science Letters*, vol. 20, no.10-12, pp. 2034-2037(4), October 2014.
[17] A. Turnip and D. E. Kusumandari, "Improvement of BCI performance through nonlinear independent component analisis extraction", *Journal of Computer*, vol. 9, no. 3, pp. 688-695, March 2014.
[18] A. Turnip, Haryadi, D. Soetraprawata, and D. E. Kusumandari, A "Comparison of Extraction Techniques for the rapid EEG-P300 Signals", *Advanced Science Letters*, vol. 20, no. 1, pp. 80-85(6), January, 2014.
[19] Mohd Zul-Hilmi Bin mohamad, "Denoising of an experimental acoustic emission (AE) data using one dimensional (1-D) Wavelet Packet Analysis", 2009.
[20] O.A. Rosso, M.T. Martin, A. Figliola., K. Keller, A. Plastino, "EEG analysis using wavelet-based information tools", *Journal of neuroscience methods*, vol. 153, pp. 163-182, 2006.
[21] H. Leung, K. Schindler, A.Y.Y Chan, A.Y.L. Lau, K. L. Leng, "Wavelet-denoising of electroencephalogram and the absolut slope method: A new tool to improve electroencephalographic localization and lateralization", *clinical neurophysiology*, vol. 120, pp.1273-1281, 2009.
[22] R. Q. Quiroga, H.Garcia, "Single-trial event-related potentials with wavelet denoising", *clinical* neurophysiology, vol. 114, pp. 376-390, 2003.
[23] L.Pasti, B. Walczak, D.L. Massart, P. Reschiglian, "Optimization of signal denoising in discrete wavelet transform", *Chemometrics and Intelligent Laboratory Systems*, vol. 48, pp. 21-34, 1999.

AUTHOR INDEX

AUTHOR INDEX

Institute of Physics
Dirac House, Temple Back
Bristol BS1 6BE UK

ISSN: 1757-8981
ISBN 978-1-5108-4510-7

CURRAN ASSOCIATES INC.
proceedings
.com